Introduction to NFL Analytics with R

It has become difficult to ignore the analytics movement within the NFL. An increasing number of coaches openly integrate advanced numbers into their game plans, and commentators, throughout broadcasts, regularly use terms such as air yards, CPOE, and EPA on a casual basis. This rapid growth, combined with an increasing accessibility to NFL data, has helped create a burgeoning amateur analytics movement, highlighted by the NFL's annual Big Data Bowl. Because learning a coding language can be a difficult enough endeavor, **Introduction to NFL Analytics with R** is purposefully written in a more informal format than readers of similar books may be accustomed to, opting to provide step-by-step instructions in a structured, jargon-free manner.

Key Coverage:

- Installing R, RStudio, and necessary packages
- Working and becoming fluent in the tidyverse
- Finding meaning in NFL data with examples from all the functions in the nflverse family of packages
- Using NFL data to create eye-catching data visualizations
- Building statistical models starting with simple regressions and progressing to advanced machine learning models using tidymodels and eXtreme Gradient Boosting

The book is written for novices of R programming all the way to more experienced coders, as well as audiences with differing expected outcomes. Professors can use **Introduction to NFL Analytics with R** to provide data science lessons through the lens of the NFL, while students can use it as an educational tool to create robust visualizations and machine learning models for assignments. Journalists, bloggers, and arm-chair quarterbacks alike will find the book helpful to underpin their arguments by providing hard data and visualizations to back up their claims.

Bradley J. Congelio is an Assistant Professor in the College of Business at Kutztown University of Pennsylvania, where he teaches the popular Sport Analytics course.

CHAPMAN & HALL/CRC DATA SCIENCE SERIES

Reflecting the interdisciplinary nature of the field, this book series brings together researchers, practitioners, and instructors from statistics, computer science, machine learning, and analytics. The series will publish cutting-edge research, industry applications, and textbooks in data science.

The inclusion of concrete examples, applications, and methods is highly encouraged. The scope of the series includes titles in the areas of machine learning, pattern recognition, predictive analytics, business analytics, Big Data, visualization, programming, software, learning analytics, data wrangling, interactive graphics, and reproducible research.

Recently Published Titles

Geographic Data Science with R
Visualizing and Analyzing Environmental Change
Michael C. Wimberly

Hands-On Data Science for Librarians
Sarah Lin and Dorris Scott

Data Science for Water Utilities
Data as a Source of Value
Peter Prevos

Practitioner's Guide to Data Science
Hui Lin and Ming Li

Natural Language Processing in the Real World
Text Processing, Analytics, and Classification
Jyotika Singh

Telling Stories with Data
With Applications in R
Rohan Alexander

Big Data Analytics
A Guide to Data Science Practitioners Making the Transition to Big Data
Ulrich Matter

Data Science for Sensory and Consumer Scientists
Thierry Worch, Julien Delarue, Vanessa Rios De Souza and John Ennis

Data Science in Practice
Tom Alby

Introduction to NFL Analytics with R
Bradley J. Congelio

For more information about this series, please visit: https://www.routledge.com/Chapman--HallCRC-Data-Science-Series/book-series/CHDSS

Introduction to NFL Analytics with R

Bradley J. Congelio

CRC Press is an imprint of the
Taylor & Francis Group, an **informa** business
A CHAPMAN & HALL BOOK

First edition published 2024
by CRC Press
62385 Executive Center Drive, Suite 320, Boca Raton, FL 33431, U.S.A.

and by CRC Press
4 Park Square, Milton Park, Abingdon, Oxon, OX14 4RN

CRC Press is an imprint of Taylor & Francis Group, LLC

ISBN: 978-1-032-42795-9 (hbk)
ISBN: 978-1-032-42775-1 (pbk)
ISBN: 978-1-003-36432-0 (ebk)

DOI: 10.1201/9781003364320

Typeset in LM Roman
by KnowledgeWorks Global Ltd.

Publisher's note: This book has been prepared from camera-ready copy provided by the authors.

Book cover illustration by Corinne Deichmeister.

For Joey and Quinn.

... as proof that anything is possible.

Contents

Preface

On April 27, 2020, Ben Baldwin hit send on a tweet that announced the birth of `nflfastR`, an R package designed to scrape NFL play-by-play data, allowing the end-user to access it at speeds quicker than similar predecessors (hence the name).

Thanks to the work of multiple people (@mrcaseB, @benbbaldwin, @TanHo, @LeeSharpeNFL, and @thomas_mock ... to name just a few), the process of getting started with analytics using NFL data is now easier than ever.

That said, and without getting too far into the weeds of the history, the above-mentioned people are responsible in some shape or form for the current status of the `nflverse`, which is a superb collection of data and R-based packages that allows anybody the ability to access deeply robust NFL data as far back as the 1999 season.

The `nflverse` as we know it today was initially birthed from the `nflscrapR` project, which was started by the Carnegie Mellon University student and professor duo of Maksim Horowitz and Sam Ventura. After Horowitz graduated – and got hired by the Atlanta Hawks – the `nflscrapR` package was taken over by fellow CMU student Ron Yurko (who would go on to receive his Ph.D. from the Statistics and Data Science program and, at the time of this book's writing, is an Assistant Teaching Professor in the Department of Statistics and Data Science at CMU). The trio's work on `nflscrapR` led to a peer-reviewed paper titled "nflWAR: A Reproducible Method for Offensive Player Evaluation in Football." Ultimately, the `nflscrapR` project came to an end when the specific .json feed used to gather NFL data changed. At this point, Ben Baldwin and Sebastian Carl had already built upon the `nflscrapR` project's foundations to create `nflfastR`. Yurko officially marked the end of the `nflscrapR` era and the beginning of the `nflfastR` era with a tweet on September 14, 2020:[1]

As a reply to his first tweet about the `nflfastR` project, Baldwin explained that he created the original function to scrape NFL data for the creation of his NFL analytics website. Thankfully, he and Carl did not keep the creation to themselves and released `nflfastR` to the public. Because of the "open source" nature of R and R packages, a laundry list of companion packages quickly

[1] Thanks to Ben Baldwin for chatting with me on Discord and providing this brief understanding of the backstory.

developed alongside `nflfastR`. The original `nflfastR` package is now part of the larger `nflverse` of packages that drive the NFL analytics community on Twitter and beyond.

The creation of the `nflverse` allowed for anybody interested in NFL analytics to easily access data, manipulate it to their liking, and release their visualizations and/or metrics to the wider public. In fact, it is now a regular occurrence for somebody to advance their R programming ability because of the `nflverse` and then go on to win the Big Data Bowl. As of the 2022 version of the Big Data Bowl, over "30 participants have been hired to work in data and analytics roles in sports, including 22 that were hired in football" (*Big Data Bowl*, n.d.). Most recently, the Chargers hired 2020 participate Alex Stern and the Chiefs hired Marc Richards, a member of the winning 2021 team, as a Football Research Analyst.

Kevin Clark, in a 2018 article for *The Ringer*, explained that despite not being as obvious as the sabermetrics movement in baseball, the analytics movement in the NFL is "happening in front of you all the time." The use of analytics in the NFL did, however, predate Clark's article. In 2014, Eagles head coach Doug Pederson explained that all decisions made by the organization – from game planning to draft strategy – are informed by hard data and analytics. Leading this early adoption of analytics, and reporting directly to team Vice President Howie Roseman, were Joe Douglas and Alec Halaby, "a 31-year-old Harvard grad with a job description" that had an emphasis on "integrating traditional and analytical methods in football decision-making." The result? A "blending of old-school scouting and newer approaches" that were often only seen in other leagues, such as the NBA and MLB (Rosenthal, 2018). Pederson believed in and trusted the team's approach to analytics so much that a direct line of communication was created between the two during games, with the analytics department providing the head coach with math-based recommendations for any scenario Pederson requested (Awbrey, 2020).[2]

In just under five years time since the publishing of that article, it has become hard to ignore the analytic movement within the NFL. Yet, there is still so much growth to happen in the marriage between the NFL and advanced metrics. For example, there is no denying that the sabermetrics movement drastically "altered baseball's DNA" Heifetz (2019). Moreover, as explained in Seth Partnow's outstanding *The Midrange Theory: Basketball's Evolution in the Age of Analytics*, the analytics movement in the NBA essentially killed the midrange shot (briefly: it is more beneficial to try to work the ball in directly under the basket (for a high-percentage shot) or to take the 3-pointer, as the possible additional point is statistically worth more despite the lower success probability as opposed to a 2-point midrange shot).

[2]Thanks, again, to Ben Baldwin for providing his personal knowledge about the "early days" of the Eagles' analytics department.

Compared to both the NBA and MLB, the NFL is playing catch up in analytics driving changes equivalent to the death of the midrange shot or the plethora of additional tactics and changes to baseball because of sabermetrics. Joe Banner, who served as the President of the Eagles from 2001 to 2012 and then the Chief Executive Officer of the Browns from 2012 to 2013, explained that some of the hesitation to incorporate analytics into NFL game planning was a result of the game being "very much driven by conventional wisdom to an extreme degree" (Fortier, 2020). Perhaps nobody encapsulates this better than Pittsburgh Steelers Head Coach Mike Tomlin. When asked about his position on analytics during the 2015 season, Tomlin explained:

> I think that's why you play the game. You can take analytics to baseball and things like that but football is always going to be football. I got a lot of respect for analytics and numbers, but I'm not going to make judgements based on those numbers. The game is the game. It's an emotional one played by emotional and driven men. That's an element of the game you can't measure. Often times decisions such as that weigh heavily into the equation (Kozora, 2015).

Given that Tomlin's quote is from 2015, perhaps the Steelers pivoted since and are now more analytically inclined. That does not seem to be the case. In a poll of NFL analytics staffers conducted by ESPN, the Steelers were voted as one of the least analytically advanced teams in the league.

There is a large gap between the least analytically inclined teams (Washington, Tennessee, Cincinnati, New York Giants, and Pittsburgh) and those voted as the most analytically inclined (Baltimore, Cleveland, Philadelphia, and Houston). In the ESPN poll, the Browns were voted as the analytics department producing the highest level of work. One of those polled spoke to the fact that much of this outstanding work is a result of General Manager Andrew Berry being a "true believer," explaining that he is one of the "rare guys you'll come across in life where you think to yourself, 'Man, this guy thinks at a different level. Just pure genius.' He's one of them."

In his article for the *Washington Post*, Sam Fortier argues that many teams became inspired to more intimately introduce analytics into game planning and on-field decisions after the 2017 season. On their run to becoming Super Bowl Champions, the Philadelphia Eagles were aggressive on 4th down, going for it 26 times during the season and converting on 17 of those for a conversion percentage of 65.4%. An examination and visualization of the data highlights the absolutely staggering increase in 4th down aggressiveness among NFL head coaches from 2000 to 2021:

There has been a 96.3% increase in the number of 4th down attempts from just 2000 to 2021 (see Figure 1). In fact, the numbers may actually be higher as I was quite conservative in building the above plot by only considering those 4th down attempts that took place when the offensive team had between a

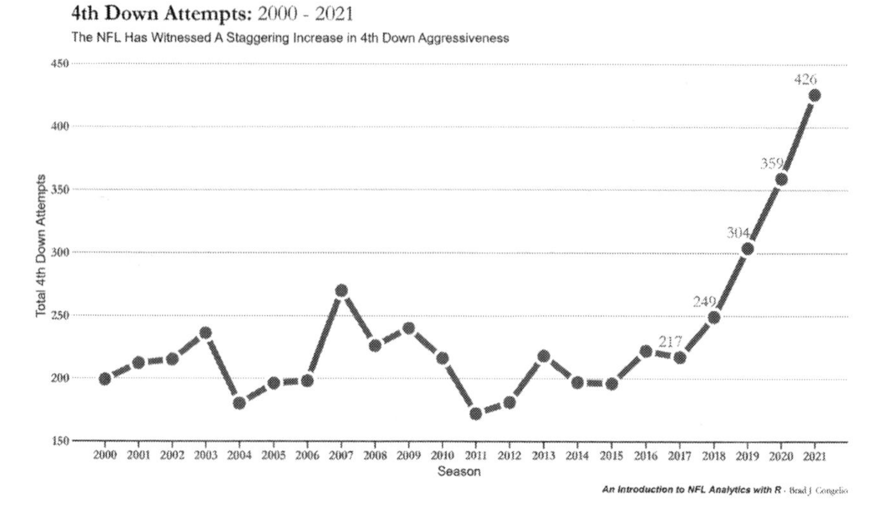

Figure 1: Number of 4th down attempts: 2000–2021

5-to-95% winning probability and those prior to the two-minute warning of either half. Even with those conservative limitations, the increase is staggering. The numbers, however, support this aggression. During week one of both the 2020 and 2021 seasons, *not* going for it on 4th down "cost teams a cumulative 170 percentage points of win probability" (Bushnell, 2021).

Ben Baldwin, using the `nfl4th` package that is part of the `nflverse`, tracked this shift in NFL coaching mentality regarding 4th down decisions by comparing 2014's "go for it percentage" against the same for 2020 (see Figure 2). When compared to the 2014 season, NFL coaches are now much more in agreement with analytics on when to "go for it" on 4th down in relation to the expected gain in win probability.

It should not be surprising then, considering Mike Tomlin's quote from above and other NFL analytics staffers voting the Steelers as one of the least analytically driven teams in the league, that Pittsburgh lost the most win probability by either kicking or punting in "go for it" situations during the 2020 NFL season. On the other end, the Ravens and Browns – two teams voted as the most analytically inclined – are the two best organizations at knowing when to "go for it" on 4th down based on win probability added (see Figure 3). There seems to be a defined relationship between teams buying into analytics and those who do not.

The NFL's turn toward more aggressive 4th-down decisions is just one of the many analytics-driven changes occurring in the league. Another significant example is Defense-Adjusted Value Over Average (or DVOA), a formula created

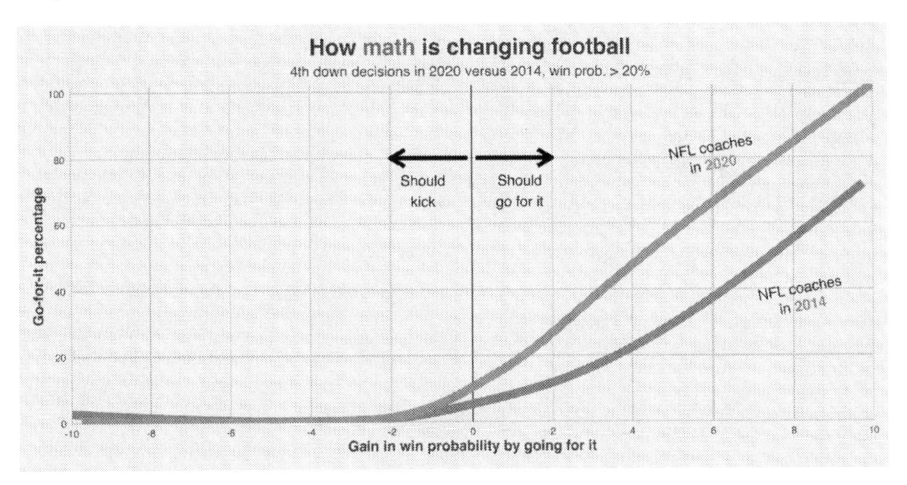

Figure 2: Credit: Ben Baldwin

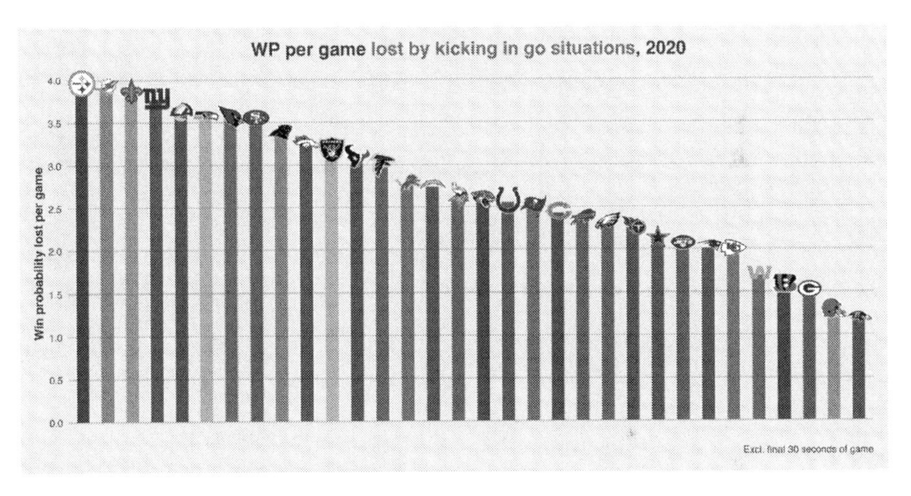

Figure 3: Credit: Ben Baldwin

by Aaron Schatz, now the editor in chief of Football Outsiders, that sought to challenge the notion that teams should, first, establish the running game in order to open up the passing game. Some of these changes are apparent on television screens on Sunday afternoons in the fall, while others are occurring behind the scenes (analytics departments working on scouting and draft preparation, for example). Indeed, the use of analytics in the NFL is not as tightly ingrained as we see in other prominent leagues. And we must remember that there are certainly continued holdouts among some NFL coaches (like Mike Tomlin).

Despite some coaching holdouts on fully embracing analytics, the "thirst for knowledge in football is as excessive as any other sport and the desire to get the most wins per dollar is just as high." As the pipeline of data continues to grow, both internally in the league and data that becomes publicly accessible, "smart teams will continue to leave no rock unturned as they push the limits on how far data can take them." Joe Banner explained that while the NFL has long been a league of coaches saying "well, that is the way we've always done it," the league is ripe for a major shift (Bechtold, 2021).

Banner's belief that those teams searching for every competitive advantage will "leave no rock unturned" is the driving force behind this book. For all intents and purposes, the age of analytics in the NFL is still in its infancy. Turning back, again, to the 2017 season, the Eagles' management praised and credited the team's analytics department as part of the reason they were able to win Super Bowl LII. Doing so, Danny Heifetz argues, "changed the language of football." The NFL, he explains, is a "copycat league" and, as witnessed with the increase in 4th down aggressiveness since 2017, teams immediately began to carbon copy Philadelphia's approach to folding traditional football strategy with a new age analytics approach. Because of the modernity of this relationship between long-held football dogmas and analytics, nobody can be quite sure what other impacts it will create on the gamesmanship of football.

However, as Heifetz opines, both the NBA and MLB can serve as a roadmap to where analytics will take the NFL. Importantly, the NFL's relationship with analytics is still in its "first frontier of what will likely be a sweeping change over the next two decades." Because of this, we cannot be sure what the next major impact analytics will make, nor when it may occur. But, with the ever-growing amount of publicly accessible data, it is only a matter of time until it is discovered. For example, in an interview with Heifetz, Brian Burke – one of the forefather's of NFL analytics and now a writer for ESPN – expressed his belief that the next major impact will be "quantifying how often quarterbacks make the correct decision on the field."

It seems that every new NFL season results in an amateur analyst bringing a groundbreaking model and/or approach to the table. Unlike, for example, MLB where there is little left to discover in terms of sabermetrics and new approaches to understanding the game and its associated strategy, the NFL is – for lack of a better phrase – an open playing field. With more and more data becoming available to the public, it is now easier than ever to investigate your own ideas and suspicions and to create your own models to confirm your intuition.

For example, I am originally from the greater Pittsburgh area and am a big Steelers fan (which certainly explains some of the Steelers-centric examples I use in the writing of this book). I was adamant in my belief that Pittsburgh's TJ Watt should win the 2021 Defensive Player of the Year award, despite many others calling for Los Angeles' Aaron Donald to claim the title. In an effort to prove my point, I sought out to design what I coined Adjusted Defensive

Impact. To begin, I wanted to explore the idea that not all defensive sacks are created equal, as a player's true impact is not always perfectly represented by top-level statistics.

To account for that, I opted to adjust and add statistical weight to sack statistics. This was done over multiple areas. For instance, not all players competed in all 17 regular-season games in 2021. To adjust for this, I took the total of games played in the data (2,936) and divided by 17 (a full season) to achieve a weighted adjustment of 0.0058. TJ Watt played in just 15 games in 2021. His adjusted equation, therefore, is (17-'games') * 0.0058. The result? He gets a bit more credit for this adjustment than, say, Myles Garrett who played all 17 regular-season games.

Going further with the model, I created a weighted adjustment for solo sacks (0.90), a negative weighted adjustment (−0.14) for any sack charted as "unblocked," and a weighted adjustment to account for how many times a player actually rushed the QB compared to how many defensive snaps they played. Using data from the SIS Data Hub, the full code is below:

```
options(digits = 2)

pass.data <- pass_rush_data %>%
  select(Player, Team, Games, `Pass Snaps`, `Pass Rushes`,
         `Solo Sacks`, `Ast. Sacks`, `Comb. Sacks`,
         `Unblocked Sacks`, Hurries, Hits) %>%
  rename(total.snaps = `Pass Snaps`,
         total.rushes = `Pass Rushes`,
         solo.sacks = `Solo Sacks`,
         asst.sacks = `Ast. Sacks`,
         comb.sacks = `Comb. Sacks`,
         unblocked.sacks = `Unblocked Sacks`,
         player = Player,
         team = Team,
         games = Games,
         hurries = Hurries,
         hits = Hits)

pass.data$rush.percent <- pass.data$total.rushes / pass.data
     $total.snaps

calculated.impact <- pass.data %>%
  group_by(player) %>%
  summarize(
    adjusted.games = (17 - games) * 0.0058,
```

```
adjusted.solo = solo.sacks * 0.9,
adjusted.unblocked = unblocked.sacks / -0.14,
adjusted.rush.percent = 0.81 - rush.percent,
combined.impact = sum(adjusted.games +
                        (solo.sacks * 0.9) +
                        (unblocked.sacks * -0.14) +
                        adjusted.rush.percent))
```

The end result? Taking into account the above adjusted defensive impact, TJ Watt was absolutely dominant during the 2021 season (see Figure 4).

All of these examples – from Ben Baldwin's 4th-down model, to Football Outsiders' DVOA, to my attempt to further quantify defensive player impact – are just the leading edge of the burgeoning analytics movement in the NFL. Moreover, the beauty of analytics is that you do not have to be a mathematician or statistics buff in order to enter the fray. All it takes is a genuine curiosity to explore what Bob Carroll, Pete Palmer, and John Thorn coined as the "Hidden Game of Football" and the desire to learn, if you have not already, the R programming language.

Who This Book Is For

Writing a book that is wholly dependent on the R programming language to achieve the end goals is not an easy task, as there are likely two types of people reading this: those with familiarity in R and those without. If you are part of the former group, you can likely skip Chapter 2 as it is a primer on installing R and learning the language of the `tidyverse`. On the other hand, if you are part of the latter group, you should skip ahead to Chapter 2 before even looking at Chapter 1, which serves as an introduction to NFL analytics with examples. That said, this book can serve multiple audiences:

1. Those interested in learning how to produce NFL analytics regardless of their current knowledge of the R programming language.

2. Professors who instruct data science courses can provide lessons through the lens of sport or, perhaps better, create their own Sport Analytics courses designed around the book. Moreover, future plans for this book include instruction reference guides to include PowerPoint templates, assignments/project instructions and grading rubrics, and quiz/exam content for online management systems (D2L, Canvas, Moodle, etc.).

ADJUSTED & WEIGHTED DEFENSIVE IMPACT

PLAYER	WEIGHTED GAMES	WEIGHTED SOLO SACKS	WEIGHTED UNBLOCKED SACKS	WEIGHTED RUSH PERCENT	ADJUSTED COMBINED IMPACT
T.J. Watt	0.0116	23.3	-0.14	-0.074	23.1
Robert Quinn	0.0058	17.8	-0.14	-0.103	17.5
Nick Bosa	0.0000	16.7	0.00	-0.182	16.5
Myles Garrett	0.0000	15.6	0.00	-0.165	15.4
Trey Hendrickson	0.0058	14.4	-0.14	-0.120	14.2
Micah Parsons	0.0058	13.3	-0.28	0.294	13.4
Cameron Jordan	0.0058	13.3	-0.14	-0.161	13.0
Matt Judon	0.0000	12.2	-0.42	0.024	11.8
Markus Golden	0.0058	12.2	-0.42	-0.051	11.8
Harold Landry	0.0000	12.2	-0.56	0.033	11.7
Joey Bosa	0.0058	11.1	0.00	-0.111	11.0
Aaron Donald	0.0000	11.1	0.00	-0.179	10.9
Yannick Ngakoue	0.0000	11.1	-0.14	-0.184	10.8
Chandler Jones	0.0116	11.1	-0.42	-0.090	10.6
Shaquil Barrett	0.0116	10.0	0.00	-0.045	10.0
Preston Smith	0.0058	10.0	0.00	-0.098	9.9
Cameron Heyward	0.0000	10.0	0.00	-0.190	9.8
Von Miller	0.0116	10.0	-0.28	-0.065	9.7
Carlos Dunlap	0.0000	8.9	0.00	-0.080	8.8
Rashan Gary	0.0058	8.9	0.00	-0.165	8.7

Data: SIS Data Hub
Table: An Introduction to NFL Analytics with R | @BradCongelio

Figure 4: Adjusted defensive impact results

3. Students are able to use this book in multiple ways. For example, in my Sport Analytics course, students are first introduced to R and the `tidyverse` using the built-in R data sets (`mtcars`, `iris`, and `nycflights13`). While those data sets certainly serve a purpose, I have found that students grasp the concepts and language of the `tidyverse` more quickly when the class turns to using data from the `nflverse`. Because of this, students can use this book to start learning the R programming language using football terms (passing yards, first downs, air yards, completion percentage over expected) as opposed to the variables they may find much less interesting housed in the built-in data. Moreover, students already fluid in R can use this book to construct machine learning models using football data, for example, as part of assignments in data science/analytics courses.

4. Journalists, bloggers, and arm-chair quarterbacks alike can use the book to underpin their arguments and to provide hard data to backup their claims.

It is also important to note that it is not expected for you to be an expert in statistics, coding, or data modeling in order to find this book useful. In fact, I self-taught myself the R programming language after discovering its potential usefulness in my personal academic research. I only "became more serious" about advancing my understanding of the language's nuances, and more advanced methods, after discovering the `nflscrapR` package and using it as my learning tool for the deep dive into R. My decision to pursue an academic certificate in the language was driven by the creation of my Sport Analytics course, as the certificate provided the proof of "academic training" that was necessary to move the course through the university's curriculum committee. Nonetheless, because of this self-taught journey – and despite being an academic myself – I find that I largely avoid the use of "complicated jargon" that is so evident in most formal writing. A core goal of this book was to write it in an accessible format, providing the information and step-by-step instructions in a more informal format than readers of similar books may be accustomed to. Learning a coding language can be difficult enough, so I do not see the need for readers to also decipher overly-complicated prose. This book is not written from atop the ivory tower, and for that I will not apologize.

Because of this, and regardless of which chapter you begin with, I believe that this book achieves its main goal: to provide a gentle introduction to doing NFL analytics with R.

As this book is published online, it allows me to continue real-time development of it, so please make note of the following:

- *Introduction to NFL Analytics with R* is published through CRC Press.

- Please feel free to contribute to the book by filing an issue or making a pull request at the book's GitHub repository: *Introduction to NFL Analytics with R* Github Repository

- The majority of the chapters conclude with exercises. In some cases, prepared data will be provided with a link to download the file. In other cases, you are expected to use the `nflverse` to collect the data yourself. In either case, the answers to the exercises are include in the book's Git repository: Answers to Chapter Exercises.

- Soon there will be a YouTube series to go along with the written version of this book. In brief, the videos will include my going over each chapter, step by step, with additional instruction and/or approaches.

- Are you an instructor hoping to create or grow your Sport Analytics course? Future plans for this book include the creation of Instructor Materials to include an example syllabus plus structured lesson plans, exercises, assignments, quizzes, and exams. As well, templates for lectures will be included in PowerPoint form so you may edit to fit your specific needs.

Overview of Chapters

- **Chapter 1** introduces the process of investigative inquiry, and how it can be used to formulate a question to be answered through the use of data and R. Specifically, the chapter provides a broad overview of what is possible with analytics and NFL data by exploring the impact that having a high number of unique offensive line combinations had on the 2022 Los Angeles Rams. The process of data collection, manipulation, preparation, and visualization are introduced to seek the answer regarding how changes in an offensive line can impact various attributes of an offense. The chapter concludes with another working example, allowing readers to explore which teams from the 2022 season most required an explosive play per drive to score a touchdown.

- **Chapter 2** covers the process of downloading both R and RStudio, as well as the necessary packages to do NFL analytics. As one of the most important chapters in the book (especially for those new to the R programming language), readers take a deep dive into wrangling NFL data with the `tidyverse` package. To begin, readers will learn about the `dplyr` pipe (`%>%`) and use, in exercises, the six most important verbs in the `dplyr` language: `filter()`, `select()`, `arrange()`, `summarize()`, `mutate()`, and `group_by()`. At the conclusion of the chapter, multiple exercises are provided to allow readers to practice using the `dplyr` verbs, relational operators within the `filter()` function and create "new stats" by using the `summarize()` function. Moreover, readers will determine the relationship between the `dplyr` language and important variables within the `nflverse` data such as `player_name` and `player_id`, which is important for correct manipulation and cleaning of data.

- **Chapter 3** examines the numerous and, often, bewildering amount of functions "underneath the hood" of the packages that make up the `nflverse`. For example, `load_pbp()` and `load_player_stats()` are included in both `nflfastR` and `nflreadr`. However, `load_nextgen_stats()`, `load_pfr_stats()`, and `load_contracts()` are all part of just `nflreadr`. Because of this complexity, readers will learn how to efficiently operate within the `nflverse`. Moreover, Chapter 3 provides a plethora of examples and exercises related to all of the various functions included.

- **Chapter 4** moves readers from data cleaning and manipulation to an introduction to data visualization using `ggplot2`. As well, Chapter 4 provides further instruction on `nflverse` functions such as `clean_player_names()`, `clean_team_abbrs()`, and `clean_homeaway()`. As well, to prep for data visualization, readers will be introduced to the `teams_colors_logos` and `load_rosters` functions as well as the `nflplotR` package, which provides "functions and geoms to help visualization of NFL related analysis" (Carl, 2022). Readers will produce multiple types of visualizations, including `geom_bar`, `geom_point`, `geom_density`, and more. As well, readers will learn to use `facet_wrap` and `facet_grid` to display data over multiple seasons. For visualizations that include team logos or player headshots, instruction will cover both how to do the coding manually using `teams_colors_logos` or `load_rosters` and to use the `nflplotr` package to avoid the need to use `left_join` to merge `teams_colors_logos` to your data frame. At the conclusion of the chapter, readers will be introduced to the `gt` and `gtExtras` packages for creating sleek tables as well as be walked through the creation of their first Shiny App.

- **Chapter 5** introduces advanced methods in R using `nflverse` data, with a specific focus on modeling and machine learning. To streamline the process of learning, readers will be introduced to `tidymodels`, a "collection of packages for modeling and machine learning using `tidyverse` principles" (Silge, n.d.). Readers will first be introduced to the modeling process by creating and running a simple linear regression model. After, regressions are built upon with multiple linear regressions, binary regressions, and binomial regression. Readers will then begin working with more advanced methods of machine learning, such as k-means clustering and building an XGBoost model.

About the Author

I (Bradley Congelio) am currently an Assistant Professor in the College of Business at Kutztown University of Pennsylvania. Aside from my core area of instruction, I also teach the very popular Sport Analytics (SPT 313) course.

I earned my Ph.D. from the University of Western Ontario and received a specialized certificate in R for Data Analytics from the University of California, San Diego in 2021. I am a proud undergraduate alumni of West Liberty University and am a strong advocate of a broad-based liberal arts education.

My research focuses on using big data, the R programming language, and analytics to explore the impact of professional stadiums on neighboring communities. I use the proprietary Zillow ZTRAX database as well as U.S. Census and other forms of data to create robust, applied, and useful insight into how best to protect those living in areas where stadiums are proposed for construction.

As well, my work in sport analytics, specifically the NFL, has been featured on numerous media outlets, including *USA Today* and *Sports Illustrated*.

Finally, my most recent academic, peer-reviewed publications include:

1. Congelio, B. (2022). "Examining the Impact of New Stadium Construction on Local Property Prices Using Data Analytics and the Zillow ZTRAX Database." *Journal of Business, Economics, and Technology* Spring 2022, 39–55.

2. Congelio, B. (2021). "Monitoring the Policing of Olympic Host Cities: A Novel Approach Using Data Analytics and the LA2028 Olympic Summer Games." *Journal of Olympic Studies* 2(2), 129–145.

3. Congelio, B. "Predicting the Impact of a New Stadium on Surrounding Neighborhoods Through the Use of a k-means Unsupervised Clustering Algorithm." Currently under peer review.

4. Congelio, B. "Examining Megaevent's Impact on Foot Traffic to Local Businesses Using Mobility and Demographic Aggregation Data." Currently writing and funded by a $15,000 grant.

Why a Book Instead of Working in Analytics?

I am sometimes asked why I spend time in the classroom teaching this material rather than taking my domain knowledge to the "industry side" and working in the NFL or an otherwise NFL-connected outlet.

The honest and, likely boring, answer is this: I love teaching. My favorite experience in the classroom yet is always in my Sport Analytics course. The frustration and sense of helplessness is palpable in the first weeks of the semester as students attempt to wrap their heads around, what a former student called, "this [censored] foreign language." I insist that they keep pushing through the exercises and assignments. Often, there is a line out my door and extending down the hallway during office hours comprised of just students from the Sport Analytics class.

And then something amazing happens.

Typically about halfway through the semester, I start seeing the light bulbs go off. Instead of cursing in anger at the "foreign language," students begin randomly cursing in excitement as the flow of the `tidyverse` language "clicks." Once that happens, it is off to the races because, once they understand speaking in `tidyverse`, learning more difficult packages (like `tidymodels`) seems doable.

And that is why I teach. That moment where I realize my lecturing, assisting, explaining, and gentle nudging are all finally paying dividends – not for me, though. For the students.

This book serves as an extension of that classroom experience. As a reader of this book, you are now a "student" and I hope you do not hesitate to reach out to me if you ever have any questions or, more importantly, *when (not if)* you have that "light bulb moment" and everything begins to click for you.

Technical Details

This book was written using RStudio's Visual Editor for R Markdown. It was published using the `Quarto` publishing system built on `Pandoc`. As well, the packages used in this book are listed in Table 0.1.

Finally, please note that this book uses the dplyr pipe operator (%>%) as opposed to the new, built-in pipe operator released with version 4.1 of R (|>). It is likely that you can work through the exercises and examples in this book by using either operator. I maintain my use of the dplyr pipe operator for no other reason than a personal (and problematic) dislike of change.

License

Table 0.1: Packages Used in This Book

Package	Version	Source
arrow	10.0.1	CRAN (R 4.1.3)
bonsai	0.2.1	CRAN (R 4.1.3)
caret	6.0-94	CRAN (R 4.1.3)
cowplot	1.1.1	CRAN (R 4.1.1)
cropcircles	0.2.1	CRAN (R 4.1.3)
doParallel	1.0.17	CRAN (R 4.1.3)
dplyr	1.1.1	CRAN (R 4.1.3)
extrafont	0.19	CRAN (R 4.1.3)
factoextra	1.0.7	CRAN (R 4.1.1)
geomtextpath	0.1.1	CRAN (R 4.1.3)
ggcorrplot	0.1.4	CRAN (R 4.1.3)
ggfx	1.0.1	CRAN (R 4.1.3)
ggimage	0.3.1	CRAN (R 4.1.3)
ggplot2	3.4.2	CRAN (R 4.1.3)
ggpmisc	0.5.2	CRAN (R 4.1.3)
ggrepel	0.9.2	CRAN (R 4.1.3)
ggridges	0.5.4	CRAN (R 4.1.3)
ggtext	0.1.2	CRAN (R 4.1.3)
glue	1.6.2	CRAN (R 4.1.3)
gt	0.8.0	CRAN (R 4.1.3)
gtExtras	0.4.5	CRAN (R 4.1.3)
lightgbm	3.3.5	CRAN (R 4.1.3)
magick	2.7.3	CRAN (R 4.1.1)
nflfastR	4.5.1	CRAN (R 4.1.3)
nflreadr	1.3.2	CRAN (R 4.1.3)
nflverse	1.0.2	https://nflverse.r-universe.dev (R 4.1.3)
nnet	7.3-16	CRAN (R 4.1.1)
RColorBrewer	1.1-3	CRAN (R 4.1.3)
reshape2	1.4.4	CRAN (R 4.1.1)
scales	1.2.1	CRAN (R 4.1.3)
tidymodels	1.0.0	CRAN (R 4.1.3)
tidyverse	2.0.0	CRAN (R 4.1.3)
vroom	1.6.1	CRAN (R 4.1.3)
webshot	0.5.4	CRAN (R 4.1.3)

1

An Introduction to NFL Analytics and the R Programming Language

1.1 Introduction

It might seem odd to begin an introductory book with coding and visualization in Chapter 1, while placing information about learning the basics of the `tidyverse` in a later chapter. But there is a good reason for this pedagogical approach being utilized in this book. As explained by Hadley Wickham and Garrett Grolemund in their outstanding book *R for Data Science*, the process of reading in and then cleaning data is not exactly the most exciting part of doing analytics. As evidence suggests, early excitement about and integration into a topic increases the likelihood of following up and learning the "boring" material.

Because of this, I follow the approach of Wickham and Grolemund and provide data that is already, for the most part, "tidied" and ready to be used. We will, however, in later chapters, pull raw data directly from its source (such as `nflreadr`, Pro Football Reference, and Sports Info Solutions) that requires manipulation and cleaning before any significant analysis can begin.

> **! Important**
>
> I am assuming, while you may not have a full grasp of the `tidyverse` yet, **that you do currently have base R and RStudio installed**. If you do not, more detailed instructions are provided in Chapter 2. If you would rather jump right into this material, you can download base R and RStudio at the following links. Once both are installed, you can return to this point in the chapter to follow along.
>
> To download/install base R: cran.rstudio.com
> To download/install RStudio: RStudio Desktop
> (scroll to bottom of page for Mac options)

DOI: 10.1201/9781003364320-1

Moreover, as briefly outlined in the Preface, we move through the process of learning NFL analytics via a close relationship with the investigative inquiry. In this instance, we will define the process of investigative inquiry as one that seeks both knowledge and information about a problem/question through data-based research. To that end, we will routinely use the process throughout this book to uncover insights, patterns, and trends relating to both players and teams that serve to help us answer the problem/question we are examining.

While it can – and should – be entertaining to develop visualization and models around arbitrarily picked statistics and metrics, it is important to remember that the end goal of the process is to glean useful insights that, ultimately, can be shared with the public. Much like the work done by a data analyst for a Fortune 500 company, the work you produce as a result of this book should do two things: (1) provide deeper insight and knowledge about NFL teams and players and (2) effectively communicate a story.

This is why the process of investigative inquiry is ingrained, as much as possible, into every example provided in the coming chapters. In doing so, the standard outline for completing an investigate inquiry is modified to fit the needs of this book – specifically, the addition of communicating your findings to the public at the end.

1.2 The Investigate Inquiry Outline

1. **Identify the problem or question**. The first step in any investigative inquiry is to clearly define the problem or question that you are trying to answer. Many times, fans have questions about their individuals favorite team and/or players. For example, the 2022 Los Angeles Rams – the defending Super Bowl Champions – were eliminated from playoff contention with three weeks remaining in the season. With the early exit, the Rams tied the 1999 Denver Broncos for the earliest elimination from playoff contention for any prior Super Bowl Champion. The Rams' early elimination can be explained by the high number of injuries during the season, including Matthew Stafford, Cooper Kupp, and Aaron Donald. However, another factor that was routinely discussed during the season was the team's inability to keep offensive linemen healthy. In this specific example, in terms of *identifying the problem or question*, a potential problem or question to explore is: how many unique combinations of offensive linemen did the 2022 LA Rams use, and what sort of impact did this have on the team's playmaking ability? Have other teams

in recent history faced the same amount of offensive line turnover yet still made the playoffs? As you can see, there are a number of different avenues in which the *problem or question* surrounding the Rams' offensive line injury issues can be explored.

2. **Gather data**. With a question or problem determined, we now turn to the process of finding and gathering the necessary data to find answers. Unfortunately, data is not always going to be available to answer your investigate inquiry. For example, the NFL's tracking data is only released in partial form during the annual Big Data Bowl. In the event that your question or problem requires data that is not available, you must loop back to Step 1 and reconfigure your question to match available data. In the case of the 2022 LA Rams' offensive line, access to data that can answer the question is available through two cores avenues: the `load_participation()` and `load_snap_counts()` functions within the **`nflverse`** family of packages.

3. **Clean and prepare the data**. It is not often that the data you obtain to answer your question will be perfectly prepared for immediate analysis. As will be explored below, the data collected to explore the Rams' offensive line combinations required both (1) a critical thought process on how to best solve oddities in the data while still producing reliable information and (2) cleaning and preparation to make the changes as a result of that critical thinking process. As you progress through the many examples and exercises in this book, you will often be presented with prepared data sets that require you to determine the best approach to data manipulation through this critical thinking and cleaning/preparation process.

4. **Analyze the data**. After problem-solving to ensure the data is as reliable and consistent as possible, we can turn to analyzing the data. In this case, since we are concerned with unique combinations of offensive linemen, we can quickly get results by using the `n_distinct()` function within `dplyr`.

5. **Visualize the data**. There are generally two options for visualizing data: plotting with `ggplot()` or creating a table with `gt` and the outstanding companion package `gtExtras`. Considering the following can help determine whether to present your findings in chart or table format.

    - **The type of data you are working with**. If you have a large amount of numerical data that needs to be compared or analyzed, a table may be the most effective way to present this information. On the other hand, if you want to highlight trends or patterns in your data, a chart can help illustrate the information in a more clear manner.

- **The purpose of your visualization**. You must consider what you ultimately want to communicate with your visualization. If you want to provide a detailed breakdown of your data, a table is usually more appropriate. However, if you want to show the overall trend or pattern in your data, a chart is going to be more effective.

- **The audience for your visualization**. As you determine whether to use a chart or a table, think about who will be viewing your visualization and what level of detail they need. If your audience is familiar with the data and needs to see precise values, a table may be a better choice. If your audience is not as familiar with the data and you want to highlight the main trends or patterns, a chart may be more effective.

6. **Interpret and communicate the results**. Lastly, it is time to communicate your results to the public. Whether this be through Twitter, a team blog, or a message board, there are numerous items to consider when preparing to build your story/narrative for sharing. This will be covered further in Chapter 4 as well.

With a clear direction via the investigative inquiry process, we can turn to taking a deeper dive into the LA Rams' 2022 offensive linemen issue.

1.3 Investigating the Rams' 2022 Offensive Line

The "Super Bowl hangover" is real.

At least for the loser of the big game.

Since the AFL and NFL merged in 1970, a total of 15 of the 51 losers of the Super Bowl went on to miss the playoffs the following season, while 13 failed to even achieve a winning record. Teams coming off a Super Bowl victory have generally fared better, with the winners putting together a .500 record or better 45 out of 51 times.

Of those six teams to not achieve a .500 record after winning the Super Bowl, only a few have been as downright terrible as the 2022 Los Angeles Rams.

As explained by Mike Ehrmann, the Rams' poor Super Bowl defense is "what happens when a laundry list of things go wildly wrong at the same time" (Kirschner, 2022). As outlined above in our investigative inquiry outline, one of the core items on the Rams' laundry list of bad luck was the absurd amount of offensive linemen ending up on the injured list. This, combined with losing Andrew Whitworth to retirement after the Super Bowl, led to quarterback Matthew Stafford being sacked on 8.6% of his dropback attempts (a rate that nearly doubled from the previous season).

Given that context, just *how* historically bad was the Rams' 2022 offensive line turnover? We can being diving into the data to find our results and build our story.

1.3.1 Unique Offensive Line Combinations: How to Collect the Data?

To begin obtaining and preparing the data to determine the number of unique offensive line combinations the Rams had in the 2022 season, we turn to two possible options: the `load_participation()` and `load_snap_counts()` functions within the `nflreadr` package. The `load_participation()` function will return, if `include_pbp = TRUE`, a list of every player ID number that was on the field for each play, whereas `load_snap_counts()` returns – on a per game basis – the percentage of snaps each player was on the field for.

In the end, using `load_snap_counts()` creates the most accurate, reliable, and straightforward way in each to collect unique offensive line combinations. The `load_participation()` function results in several oddities in the data (not with the collection of it by the `nflverse` maintainers, but with individual NFL team strategies and formations). To highlight this, the following code selects the first offensive play for each team, in each game, of the 2022 season.

```
participation <- nflreadr::load_participation(2022,
   include_pbp = TRUE)
rosters <- nflreadr::load_rosters(2022) %>%
  select(full_name, gsis_id, depth_chart_position)

oline_participation <- participation %>%
  filter(play_type %in% c("pass", "run")) %>%
  group_by(nflverse_game_id, possession_team, fixed_drive)
   %>%
  filter(fixed_drive == 1 | fixed_drive == 2) %>%
  filter(row_number() == 1) %>%
  select(nflverse_game_id, play_id, possession_team,
```

```
        offense_personnel, offense_players) %>%
  dplyr::mutate(gsis_id = stringr::str_split(offense_players,
    ";")) %>%
  tidyr::unnest(c(gsis_id)) %>%
  left_join(rosters, by = c("gsis_id" = "gsis_id"))

oline_participation <- oline_participation %>%
  filter(depth_chart_position %in% c("T", "G", "C")) %>%
  group_by(nflverse_game_id, possession_team) %>%
  mutate(starting_line = toString(full_name)) %>%
  select(nflverse_game_id, possession_team,
         offense_personnel, starting_line) %>%
  distinct()
```

While the output using the `load_participation()` function is correct, a quick examination of the `offense_personnel` column causes concern about the viability of this approach to calculate the total number of unique offensive line combinations. A grouping and summing of the `offense_personnel` column highlights the issue.

```
oline_participation %>%
  group_by(offense_personnel) %>%
  summarize(total = n())
```

```
# A tibble: 14 x 2
   offense_personnel                                total
   <chr>                                            <int>
 1 1 RB, 0 TE, 4 WR                                     4
 2 1 RB, 1 TE, 3 WR                                   240
 3 1 RB, 2 TE, 2 WR                                   171
 4 1 RB, 3 TE, 1 WR                                    19
 5 2 QB, 1 RB, 1 TE, 2 WR                               4
 6 2 RB, 0 TE, 3 WR                                     1
 7 2 RB, 1 TE, 2 WR                                    89
 8 2 RB, 2 TE, 1 WR                                    14
 9 3 RB, 1 TE, 1 WR                                     1
10 6 OL, 1 RB, 0 TE, 3 WR                               2
11 6 OL, 1 RB, 1 TE, 2 WR                              12
12 6 OL, 1 RB, 2 TE, 1 WR                               1
13 6 OL, 2 RB, 0 TE, 2 WR                               1
14 7 OL, 0 RB, 0 TE, 0 WR,1 P,1 LS,1 DL,1 K             1
```

```
oline_participation
```

```
# A tibble: 560 x 4
# Groups:   nflverse_game_id, possession_team [557]
   nflverse_game_id possession_team offense_personnel starting_line
   <chr>            <chr>           <chr>             <chr>
 1 2022_01_BAL_NYJ  NYJ             1 RB, 2 TE, 2 WR  George Fant, Al~
 2 2022_01_BAL_NYJ  BAL             2 RB, 1 TE, 2 WR  Ben Powers, Mor~
 3 2022_01_BUF_LA   BUF             1 RB, 1 TE, 3 WR  Ryan Bates, Dio~
 4 2022_01_BUF_LA   LA              1 RB, 1 TE, 3 WR  Rob Havenstein,~
 5 2022_01_CLE_CAR  CAR             1 RB, 1 TE, 3 WR  Ikem Ekwonu, Ta~
 6 2022_01_CLE_CAR  CLE             1 RB, 1 TE, 3 WR  Jedrick Wills, ~
 7 2022_01_DEN_SEA  SEA             1 RB, 2 TE, 2 WR  Phil Haynes, Ab~
 8 2022_01_DEN_SEA  DEN             1 RB, 2 TE, 2 WR  Garett Bolles, ~
 9 2022_01_GB_MIN   MIN             1 RB, 1 TE, 3 WR  Christian Darri~
10 2022_01_GB_MIN   GB              1 RB, 2 TE, 2 WR  Royce Newman, J~
# i 550 more rows
```

Of concern are rows 10 through 14. In 15 different cases, a team ran its first play of the game with six offensive linemen. And, in one case, the resulting data indicates that the Dallas Cowboys ran their first play in week 5 against the LA Rams with seven offensive linemen, one punter, one long snapper, and a kicker.

In the first case, the data is correct that the teams ran their first offensive play with six offensive linemen. For example, in its week 3 game against the Steelers, the data list the Cleveland Browns as having started Jack Conklin (tackle), Jedrick Wills Jr. (tackle), Joel Bitonio (guard), Michael Dunn (guard), Wyatt Teller (guard), and Ethan Pocic (center). Viewing the NFL's All-22 film of this specific play confirms that, indeed, all six offensive linemen were on the field for the Browns' first snap of the game (see Figure 1.1).

In the second case, Dallas' offense personnel on its "first play" from scrimmage is the result of the Cowboys returning a fumble for a touchdown on the Rams' first offensive possession with a botched snap on the ensuing extra point attempt. Because of that, the extra point attempt is no longer scored as an `extra_point` in the `play_type` variable within the play-by-play data, but a rushing attempt. As a result of this oddity, the data is correct in listing Dallas' first offensive play as coming out of an extra-point personnel grouping.

Both of these examples are problematic as a team's "starting offensive line" is considered to be just five players: the left tackle, the left guard, the center, the right guard, and the right tackle. In order to correctly determine the number of combinations used, we need to first determine the five most-commonly used

Figure 1.1: Steelers vs. Browns – six offensive linemen

offensive linemen for each team. Because of the off-the-wall situations that can occur in football, building offensive line combinations through personnel groupings in the play-by-play data is tricky, at best.

To avoid these situations, we can turn to the `load_snap_counts()` function with the `nflreadr` package to determine the number of unique offensive line combinations. The process to do so occurs over several steps and involves decision-making on our end on how best to accurately represent the five core offensive linemen for each team.

```
oline_snap_counts <- nflreadr::load_snap_counts(seasons = 2022)

oline_snap_counts <- oline_snap_counts %>%
  select(game_id, week, player, position, team, offense_pct) %>%
  filter(position %in% c("T", "G", "C")) %>%
  group_by(game_id, team) %>%
  arrange(-offense_pct) %>%
  dplyr::slice(1:5) %>%
  ungroup()

oline_snap_counts
```

```
# A tibble: 2,840 x 6
   game_id          week player               position team  offense_pct
   <chr>           <int> <chr>                <chr>    <chr>       <dbl>
 1 2022_01_BAL_NYJ     1 Ben Powers           G        BAL          1
 2 2022_01_BAL_NYJ     1 Morgan Moses         T        BAL          1
 3 2022_01_BAL_NYJ     1 Kevin Zeitler        G        BAL          1
 4 2022_01_BAL_NYJ     1 Tyler Linderbaum     C        BAL          1
 5 2022_01_BAL_NYJ     1 Patrick Mekari       G        BAL          0.57
 6 2022_01_BAL_NYJ     1 Max Mitchell         T        NYJ          1
 7 2022_01_BAL_NYJ     1 Laken Tomlinson      G        NYJ          1
 8 2022_01_BAL_NYJ     1 Alijah Vera-Tucker   G        NYJ          1
 9 2022_01_BAL_NYJ     1 George Fant          T        NYJ          1
10 2022_01_BAL_NYJ     1 Connor McGovern      C        NYJ          1
# i 2,830 more rows
```

First, we obtain snap count data from the 2022 season and write it into a data frame titled `oline_snap_counts`. After, we select just the necessary columns and then filter out the `position` information to include only tackles, guards, and centers. After grouping each individual offensive line by `game_id` and his respective `team`, we arrange each player's snap count in descending order using `offense_pct`.

And this is where a decision needs to be made on how to best construct the five starting offensive linemen for each team. By including `slice(1:5)`, we are essentially selecting *just* the five offensive linemen with the most snap counts in that singular game.

Are these five players necessarily the same five that started the game as the two tackles, two guards, and one center? Perhaps not. But, hypothetically, a team's starting center could have been injured a series or two into the game, and the second-string center played the bulk of the snaps in that game.

Because of such situations, we can make the argument that the five offensive line players with the highest percentage of snap counts for each unique `game_id` are to be considered the combination of players used most often in that game.

Next, let's make the decision to arrange each team's offensive line, by game, in alphabetical order. Since we do not have a reliable way to include *specific* offensive line positions (that is, we have *just* tackle instead of *left* tackle or *right* tackle), we can build our combination numbers strictly based on the five downed linemen, regardless of specific position on the line of scrimmage.

After, we use the `toString()` function to place all five names into a single column (`starting_line`) and then filter out the data to include just one `game_id` for the linemen.

```
oline_snap_counts <- oline_snap_counts %>%
  group_by(game_id, team) %>%
  arrange(player, .by_group = TRUE)

oline_final_data <- oline_snap_counts %>%
  group_by(game_id, week, team) %>%
  mutate(starting_line = toString(player)) %>%
  select(game_id, week, team, starting_line) %>%
  distinct(game_id, .keep_all = TRUE)
```

The end result includes the `game_id`, the `week`, the `team` abbreviation, and the `starting_line` column that includes the names of the five offensive line players with the highest snap count percentage for that specific game.

```
# A tibble: 568 x 4
# Groups:   game_id, week, team [568]
   game_id          week team  starting_line
   <chr>           <int> <chr> <chr>
 1 2022_01_BAL_NYJ     1 BAL   Ben Powers, Kevin Zeitler, Morgan Mose~
 2 2022_01_BAL_NYJ     1 NYJ   Alijah Vera-Tucker, Connor McGovern, G~
 3 2022_01_BUF_LA      1 BUF   Dion Dawkins, Mitch Morse, Rodger Saff~
 4 2022_01_BUF_LA      1 LA    Brian Allen, Coleman Shelton, David Ed~
 5 2022_01_CLE_CAR     1 CAR   Austin Corbett, Brady Christensen, Ike~
 6 2022_01_CLE_CAR     1 CLE   Ethan Pocic, James Hudson, Jedrick Wil~
 7 2022_01_DEN_SEA     1 DEN   Cameron Fleming, Dalton Risner, Garett~
 8 2022_01_DEN_SEA     1 SEA   Abraham Lucas, Austin Blythe, Charles ~
 9 2022_01_GB_MIN      1 GB    Jake Hanson, Jon Runyan, Josh Myers, R~
10 2022_01_GB_MIN      1 MIN   Brian O'Neill, Christian Darrisaw, Ed ~
# i 558 more rows
```

With the data cleaned and prepared, we are now able to take our first look at the results. In the code below, we are grouping by all 32 NFL teams and then summarizing the total number of unique offensive line combinations used during the 2022 regular season.

```
total_combos <- oline_final_data %>%
  group_by(team) %>%
  summarize(combos = n_distinct(starting_line)) %>%
  arrange(-combos)
```

Despite much of the media focus being on the Rams' poor performance, given their title of defending Super Bowl Champions, the Arizona Cardinals had nearly as many unique offensive line combinations at the conclusion of the season.

With the data cleaned and prepared, let's use it to create a `ggplot` graph and compare the relationship between a team's number of unique offensive lines against its winning percentage. To complete this, we first need to join the winning percentages to our existing `total_combos` data frame.

1.3.2 Unique Offensive Line Combinations vs. Win Percentage

To bring in the individual winning percentages, we will use the `get_nfl_standings()` function from `espnscrapeR` and then combine the two sets of data on team abbreviations via a `left_join()`. Unfortunately, the team abbreviation returned from `espnscrapeR` does not match up with those used in the `nflverse` for both the Los Angeles Rams and the Washington Commanders (LAR vs. LA and WSH vs. WAS). As evidenced in the below code, correcting this issue is simple with the `clean_team_abbrs()` function in the `nflreadr` package.

```
records <- espnscrapeR::get_nfl_standings(season = 2022) %>%
  select(team_abb, win_pct)

records$team_abb <- nflreadr::clean_team_abbrs
  (records$team_abb)

total_combos <- total_combos %>%
  left_join(records, by = c("team" = "team_abb"))

total_combos
```

```
# A tibble: 32 x 3
   team  combos win_pct
   <chr>  <int>   <dbl>
 1 LA        13   0.294
 2 ARI       12   0.235
 3 CHI       10   0.176
 4 WAS       10   0.5
 5 DEN        9   0.294
 6 NO         9   0.412
 7 NYJ        9   0.412
 8 GB         8   0.471
 9 TB         8   0.471
10 BUF        7   0.812
# i 22 more rows
```

After collecting team records and merging them into the offensive line combination data, we can use `ggplot` to visualize the data. Individual team logos are used in place of the typical `geom_point` by using the `nflplotR` package.

> **! Important**
>
> Please note in the below `ggplot()` coding that we are using a custom theme, `nfl_analytics_theme()`. If you wish to replicate the below visualization using the theme, run the below code to place the theme into your RStudio environment allowing you to call it within `ggplot`. Be sure to install and run the `ggtext` package as the theme uses the package's `element_markdown()` function.

```
nfl_analytics_theme <- function(..., base_size = 12) {

  theme(
    text = element_text(family = "Roboto",
                        size = base_size,
                        color = "black"),
    axis.ticks = element_blank(),
    axis.title = element_text(face = "bold"),
    axis.text = element_text(face = "bold"),
    plot.title.position = "plot",
    plot.title = element_markdown(size = 16,
                                  vjust = .02,
                                  hjust = 0.5),
    plot.subtitle = element_markdown(hjust = 0.5),
    plot.caption = element_markdown(size = 8),
    panel.grid.minor = element_blank(),
    panel.grid.major =  element_line(color = "#d0d0d0"),
    panel.background = element_rect(fill = "#f7f7f7"),
    plot.background = element_rect(fill = "#f7f7f7"),
    panel.border = element_blank(),
    legend.background = element_rect(color = "#F7F7F7"),
    legend.key = element_rect(color = "#F7F7F7"),
    legend.title = element_text(face = "bold"),
    legend.title.align = 0.5,
    strip.text = element_text(face = "bold"))
}
```

```
ggplot(data = total_combos, aes(x = combos, y = win_pct)) +
  geom_line(stat = "smooth", method = "lm",
            linewidth = .7, color = "blue",
            alpha = 0.25) +
  nflplotR::geom_mean_lines(aes(v_var = combos, h_var = win_pct),
                            color = "black", size = .8) +
  nflplotR::geom_nfl_logos(aes(team_abbr = team), width = 0.065) +
  nfl_analytics_theme() +
  scale_x_reverse(breaks = scales::pretty_breaks(n = 12)) +
  scale_y_continuous(breaks = scales::pretty_breaks(n = 6),
             labels = scales::label_number(accuracy = 0.001)) +
  xlab("# of Unique Offensive Line Combinations") +
  ylab("Win Percentage") +
  labs(title = "**Unique Offensive Line Combinations vs.
    Win Percentage**",
       subtitle = "*Through Week #15*",
       caption = "*An Introduction to NFL Analytics with R*<br>
       **Brad J. Congelio**")
```

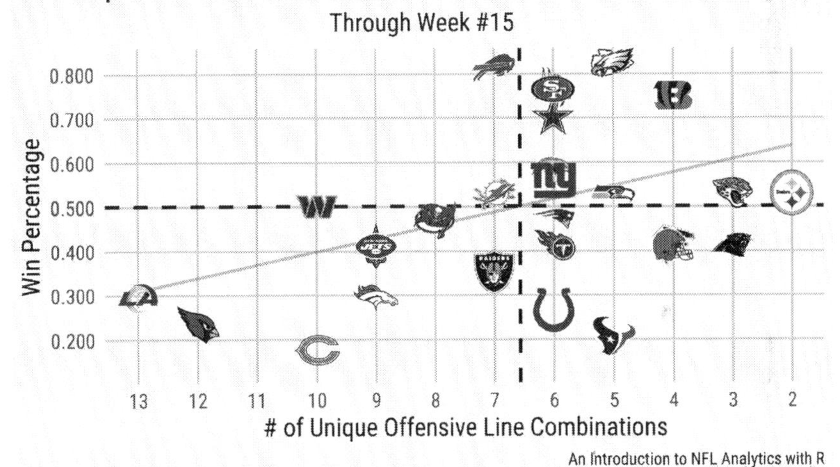

Figure 1.2: Number of unique combos vs. win percentage

As can be seen in the resulting graph – which shows little, if any, statistical correlation – there are still teams with worse records than the Rams and Cardinals that have fewer unique offensive line combinations (the Houston Texans and the Chicago Bears, for example). Perhaps there is a metric that correlates more strongly with a team's number of offensive line combinations.

To begin exploring that, we can hypothesize that more offensive line combinations leads to more quarterback pressures as the various member of the offensive line never have time to properly "gel."

1.3.3 Unique Offensive Line Combinations vs. Pressure Rate

Rather than calculate the data ourselves (which is the total number of dropbacks divided by the total number of pressures), we can turn away from `nflverse` data and retrieve the information from the SIS Data Hub. After downloading the spreadsheet, we can read it into the RStudio environment using `vroom` and then merge the information into the existing `total_combos` data frame by matching on the individual team abbreviations.

```
pressure_rate <- vroom("http://nfl-book.bradcongelio.com/
                        pressure-rate")

teams <- nflreadr::load_teams() %>%
  select(team_abbr, team_nick)

pressure_rate <- pressure_rate %>%
  left_join(teams, by = c("team" = "team_nick"))

total_combos <- total_combos %>%
  left_join(pressure_rate, by = c("team" = "team_abbr"))
```

```
# A tibble: 32 x 6
   team  combos win_pct season team.y      pressure_percent
   <chr>  <int>   <dbl>  <dbl> <chr>                  <dbl>
 1 LA        13   0.294   2022 Rams                    34.7
 2 ARI       12   0.235   2022 Cardinals               26.5
 3 CHI       10   0.176   2022 Bears                   42.3
 4 WAS       10   0.5     2022 Commanders              38.4
 5 DEN        9   0.294   2022 Broncos                 33.8
 6 NO         9   0.412   2022 Saints                  28.2
 7 NYJ        9   0.412   2022 Jets                    33.9
 8 GB         8   0.471   2022 Packers                 25.4
 9 TB         8   0.471   2022 Buccaneers              19.6
10 BUF        7   0.812   2022 Bills                   31.8
# i 22 more rows
```

With the pressure rate now merged with our unique offensive line combination data, we can make slight adjustments to our prior `ggplot` code to examine any potential relationship.

```
ggplot(data = total_combos, aes(x = combos, y =
       pressure_percent)) +
  geom_line(stat = "smooth", method = "lm",
            size = .7, color = "blue",
            alpha = 0.25) +
  nflplotR::geom_mean_lines(aes(v_var = combos,
      h_var = pressure_percent),
                            color = "black", size = .8) +
  nflplotR::geom_nfl_logos(aes(team_abbr = team), width =
      0.065) +
  scale_x_reverse(breaks = scales::pretty_breaks(n = 12)) +
  scale_y_reverse(breaks = scales::pretty_breaks(n = 6),
            labels = scales::percent_format(scale = 1,
                                         accuracy = 0.1)) +
  nfl_analytics_theme() +
  xlab("# of Unique Offensive Line Combinations") +
  ylab("Pressure Rate (per Dropback)") +
  labs(title = "**Unique Offensive Line Combinations vs.
                 Pressure Rate**",
       subtitle = "*2022 Season*",
       caption = "*An Introduction to NFL Analytics with
             R*< br>  **Brad J. Congelio**")
```

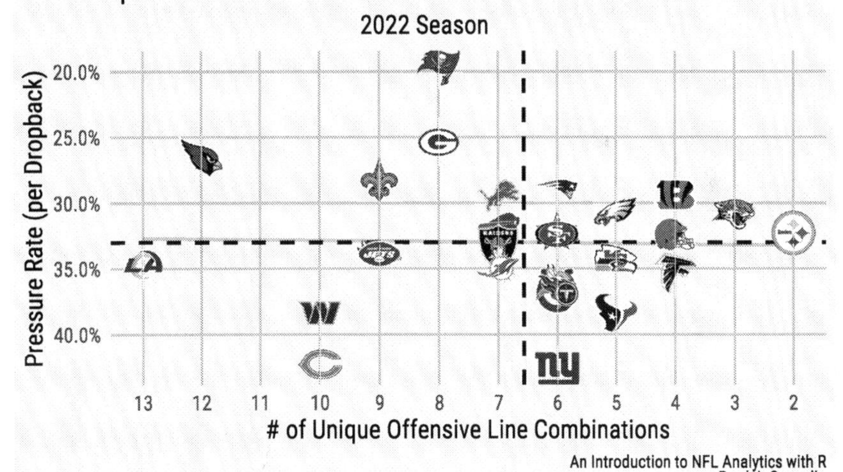

Figure 1.3: Number of unique combos vs. pressure percentage

Again, there does not seem to be any statistical correlation between a team's number of unique offensive line combinations and the pressure rate per dropback allowed. Rather than examining the pressure per dropback rate, perhaps there is a correlation between the number of unique offensive line combinations and the total number of quarterback hits allowed through the season? To determine the validity of that hypothesis, we can use data from `nflreadr` to collect total QB hits.

1.3.4 Unique Offensive Line Combinations vs. QB Hits Allowed

To begin, we can collect the QB hit data from `nflreadr` being sure to `group_by()` the `posteam` variable in order to calculate the number of QB hits allowed by each team's offensive line. After, we can merge the data into the `total_combos` data frame and produce the plot.

```
pbp <- nflreadr::load_pbp(2022)

qb_hits <- pbp %>%
  filter(!is.na(posteam)) %>%
  group_by(posteam) %>%
  summarize(total_qb_hits = sum(qb_hit == 1, na.rm = TRUE))

qb_hits_combined <- left_join(
  total_combos, qb_hits, by = c("team" = "posteam"))
```

After first filtering out any data that does not include `posteam` information, we can `group_by()` each individual offensive unit and then calculate the total sum of QB hits for each.

```
ggplot(data = qb_hits_combined, aes(x = combos,
           y = total_qb_hits)) +
    geom_line(stat = "smooth", method = "lm",
              size = .7, color = "blue",
              alpha = 0.25) +
  nflplotR::geom_mean_lines(aes(v_var = combos, h_var =
                total_qb_hits),
                            color = "black", size = .8) +
  nflplotR::geom_nfl_logos(aes(team_abbr = team), width =
              0.065) +
  scale_x_reverse(breaks = scales::pretty_breaks()) +
```

```
scale_y_reverse(breaks = scales::pretty_breaks()) +
nfl_analytics_theme() +
xlab("# of Unique Offensive Line Combinations") +
ylab("Total QB Hits Allowed") +
labs(title = "Unique Offensive Line Combinations vs.
        QB Hits Allowed",
     subtitle = "2022 Season",
     caption = "*An Introduction to NFL Analytics with
          R*<br> **Brad J. Congelio**")
```

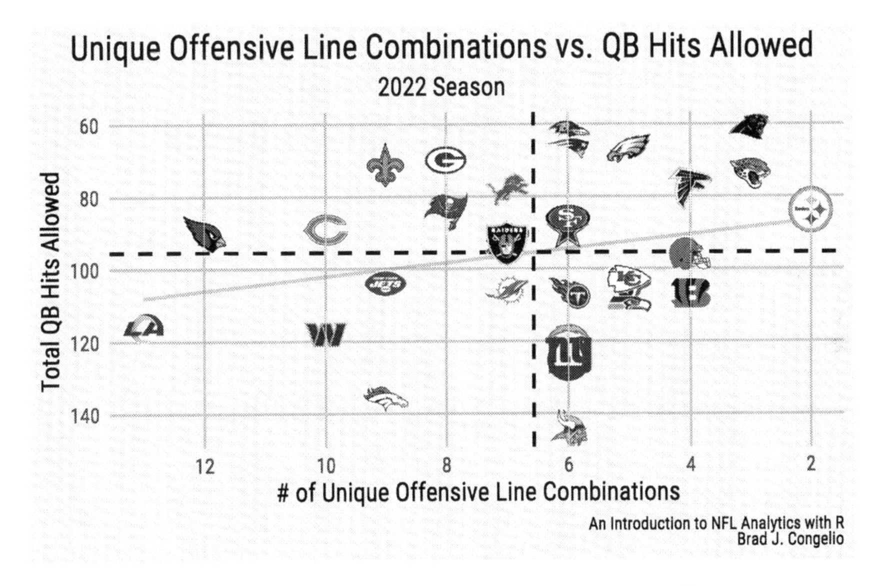

Figure 1.4: Number of unique combos vs. QB hits

There is again very little, if any, correlation between a team's unique number of offensive line combinations and the number of times its quarterback is hit on passing attempts.

1.3.5 Unique Offensive Line Combinations vs. Adjusted Sack Rate

In the previous examples, both pressure rate and QB hits are unadjusted metrics, meaning the results are not manipulated to account for context within the data. To add an example of a metric that *is* adjusted to our list of examples, we will use Adjusted Sack Rate from Football Outsiders. Aaron Schatz, in an article introducing the measurements in December of 2003, explained his findings that the league-wide sack rate varied based on both the down and yards to go (see Table 1.1).

Table 1.1: 2003 NFL Sack Rate – Aaron Schatz (FootballOutsiders.com)

Yards to Go	1–4	5–8	9–12	13–16	>17
1st Down	*1.5%*	*5.4%*	*4.8%*	*3.3%*	*5.6%*
2nd Down	*5.6%*	*4.7%*	*5.0%*	*6.1%*	*4.5%*
3rd/4th Down	*5.9%*	*8.2%*	*10.5%*	*7.6%*	*11.1%*

As a result of this, Schatz designed Adjusted Sack Rate, which accounts for the number of pass attempts, down, distance, and – importantly – the team's opponent. We can plot the Adjusted Sack Rate for each team from the 2022 season against the unique number of offensive line combinations.

```
adjusted_sack_rate <- vroom("http://nfl-book.bradcongelio.
   com/adj-sack-rate")
```

```
adjusted_sack_rate %>%
  filter(season == 2022) %>%
  ggplot(aes(x = combos, y = adj_sack_rate)) +
  geom_line(stat = "smooth", method = "lm",
            size = .7,
            color = "blue",
            alpha = 0.25) +
  nflplotR::geom_mean_lines(aes(v_var = combos, h_var =
                                 adj_sack_rate),
                                color = "black", size = .8) +
  nflplotR::geom_nfl_logos(aes(team_abbr = team), width =
                               0.065) +
  scale_x_reverse(breaks = scales::pretty_breaks()) +
  scale_y_reverse(breaks = scales::pretty_breaks(),
                  labels = scales::percent_format(scale =
                                                   1)) +
  xlab("# of Unique Offensive Line Combinations") +
  ylab("Adjusted Sack Rate") +
  labs(title = "**Unique Offensive Line Combinations vs.
       Adjusted Sack Rate**",
       subtitle = "*2022 Season*",
       caption = "*An Introduction to NFL Analytics with R*<br>
       **Brad J. Congelio**") +
  nfl_analytics_theme()
```

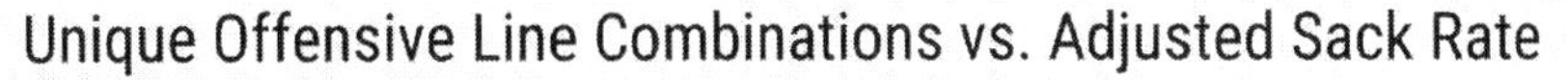

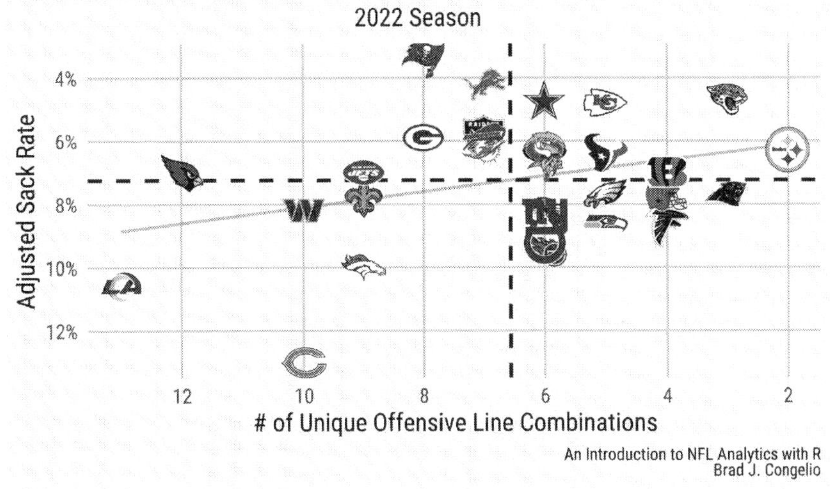

Figure 1.5: Number of unique combos vs. adjusted sack rate

The graph shows that, in general, those teams with higher numbers of unique offensive line combinations trend toward the lower-left quadrant and some – the Rams and Bears – are significantly below the league average, indicating a worse adjusted sack rate (with the Arizona Cardinals being an outlier). Compared to the prior metrics (pressure rate and QB hits), it seems that the adjusted sack rate may be impacted by the consistency of a team's offensive linemen.

Using the `adjusted_sack_rate` data, we can easily visualize the same results over the last three NFL seasons to see if there is any consistency between the offensive line combinations and a team's adjusted sack rate.

```
adjusted_sack_rate %>%
  ggplot(aes(x = combos, y = adj_sack_rate)) +
  geom_line(stat = "smooth", method = "lm",
            size = .7,
            color = "blue",
            alpha = 0.25) +
  nflplotR::geom_mean_lines(aes(v_var = combos, h_var =
                adj_sack_rate),
                           color = "black", size = .8) +
  nflplotR::geom_nfl_logos(aes(team_abbr = team), width =
                      0.065) +
  scale_x_reverse(breaks = scales::pretty_breaks()) +
```

```
scale_y_reverse(breaks = scales::pretty_breaks(),
          labels = scales::percent_format(scale = 1)) +
xlab("# of Unique Offensive Line Combinations") +
ylab("Adjusted Sack Rate") +
labs(title = "**Unique Offensive Line Combinations vs.
     Adjusted Sack Rate**",
     subtitle = "*2020 - 2022 Seasons*",
     caption = "*An Introduction to NFL Analytics with
             R*<br>
     **Brad J. Congelio**") +
nfl_analytics_theme() +
facet_wrap(~season, nrow = 2)
```

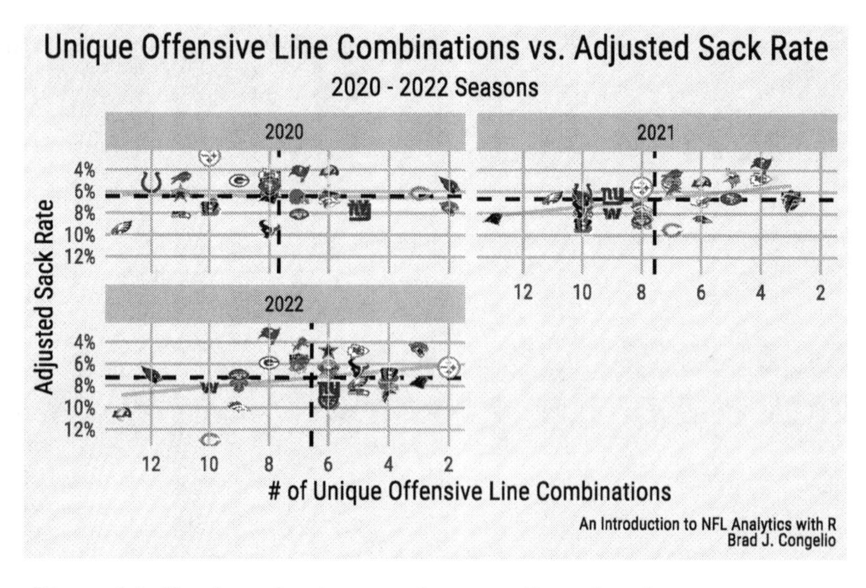

Figure 1.6: Number of unique combos vs. adjusted sack rate per season

Given the result of just the 2022 season, the output for the 2021 season remains consistent, perhaps even more so with less spread from the line of best fit. In 2021, the Carolina Panthers had the most unique offensive line combinations as well as one of the worst adjusted sack rates in the league, while those teams with fewer combinations generally trend toward the upper-right quadrant of the plot.

However, the consistency between the 2022 and 2021 seasons is not found in the 2020 season. The Philadelphia Eagles had the most unique combinations in 2020 (13 in total), but were all but tied with the Houston Texans for worst adjusted sack rate (9.4% to 9.5%). Moreover, the Pittsburgh Steelers, Buffalo

Bills, and Indianapolis Colts – despite all having more than ten unique offensive line combinations – are among the best in adjusted sack rate.

Why is there this sudden departure from the consistency in 2022 and 2021?

The answer is within the context of data, in that Ben Roethlisberger – the long-time quarterback for the Steelers – had one of the fastest Time to Throw scores in modern NFL history, getting rid of the ball, on average, in just 2.19 seconds after the snap. Philip Rivers, of the Colts, was nearly as quick to release the ball with an average of 2.39 seconds. It proved to be quite difficult for opposing defenses to get to either Roethlisberger or Rivers given these incredibly quick snap-to-release times.

It should be noted that the same contextual explanation does not hold true for Buffalo. In the 2020 season, Josh Allen released the ball, on average, 3.04 seconds after the snap, which is plenty of time for the opposing defense to generate pressure and record sacks. Because of this, if we were to add statistical weight using the average time to throw, the Buffalo Bills would remain an outlier in this specific season, while both the Steelers and Colts would likely regress closer to the mean for those teams with higher numbers of offensive line combinations.

While pressure rate and QB hits did not prove to impacted by a team's number of line combinations, there is a broad relationship (though not strong enough to begin using the term "correlation") between the number of combinations and adjusted sack rate.

To continue this exploration, let's pivot away from exploring the impact of offensive line combinations on the quarterback and examine any potential correlation with the running game and run blocking, while using metrics that are adjusted, as the additional context and weighting seemed to be helpful in finding relationships. We can gather statistics provided by Football Outsiders concerning the performance of offensive lines.

1.3.6 Unique Offensive Line Combinations vs. Adjusted Line Yards

1. **Adjusted Line Yards** – derived by Football Outsiders with a regression analysis, this metric takes the total of a team's total rushing attempts and assigns a quantitative value to the offensive line's impact. Any run that ends with a loss of yards results in 120% of the value being placed on the offensive line, while a 0–4 yard rush is 100%, a 5–10 yard rush is 50%, and anything over 11 yards is a 0% value. As explained by Football Outsiders, if a running back breaks free for a 50-yard gain, just how much of that is the offensive

line responsible for? Adjusted Line Yards produce a numeric value as the answer to that question.

2. **Power Rank** – a ranking between 1 (best) and 32 (worst), power rank is the result of the power success metric, which is the percentage of rushing attempts on 3rd or 4th down, with two or less yards to go, that resulted in either a 1st down or a touchdown.
3. **Stuffed Rank** – again provided in the data frame as a ranking between 1 and 32, stuffed rank is the percentage of rushing attempts where the running back was tackled at, or behind, the line of scrimmage.

To begin, we can gather the necessary data into a data frame called `fb_outsiders_oline`.

```
fb_outsiders_oline <- vroom("http://nfl-book.bradcongelio.
        com/fbo_oline")

fb_outsiders_oline
```

```
# A tibble: 32 x 5
   team  combos adj_line_yards power_rank stuffed_rank
   <chr>  <dbl>          <dbl>      <dbl>        <dbl>
 1 LV         7           4.93         23            6
 2 GB         8           4.85         18            2
 3 KC         5           4.82         31            9
 4 SF         6           4.7          26           22
 5 ATL        4           4.68         17            5
 6 PHI        5           4.66          7            7
 7 DET        7           4.66         20           18
 8 MIA        7           4.61         30           14
 9 CAR        3           4.56         14           11
10 PIT        2           4.54          1            4
# i 22 more rows
```

You can see in the output of the data that `power_rank` and `stuffed_rank` are both in a "ranked format," meaning the team with the best power success score (the Pittsburgh Steelers) is ranked 1st, while the team with the worst (the Minnesota Vikings) is ranked 32nd. The `adj_line_yards` variable is provided in its unranked, raw format. Because of this, let's first use both `power_rank` and `stuffed_rank` and use the `cowplot` package to view them together, and then construct the plot for `adj_line_yards` separately since it operates on a differing y-axis scale.

```
power_rank_plot <- ggplot(fb_outsiders_oline,
                          aes(x = combos, y = power_rank)) +
  nflplotR::geom_mean_lines(aes(v_var = combos,
                                h_var = power_rank),
                            color = "black",
                            size = .8) +
  nflplotR::geom_nfl_logos(aes(team_abbr = team), width = 0.065) +
  scale_x_reverse(breaks = scales::pretty_breaks()) +
  scale_y_reverse(breaks = seq(1, 32, 2)) +
  labs(title = "**Unique Offensive Line Combinations vs.
       Power Ranking**",
       subtitle = "*2022 Season  |  FootballOutsiders.com*") +
  xlab("# of Unique Offensive Line Combinations") +
  ylab("Power Ranking (1 = best, 32 = worst)") +
  nfl_analytics_theme()

stuffed_rank_plot <- ggplot(fb_outsiders_oline,
                            aes(x = combos, y = stuffed_rank)) +
  geom_smooth(method = "lm", se = FALSE, size = 0.8, alpha = .08) +
  nflplotR::geom_mean_lines(aes(v_var = combos,
                                h_var = stuffed_rank),
                            color = "black",
                            size = .8) +
  nflplotR::geom_nfl_logos(aes(team_abbr = team), width = 0.065) +
  scale_x_reverse(breaks = scales::pretty_breaks()) +
  scale_y_reverse(breaks = seq(1, 32, 2)) +
  labs(title = "**Unique Offensive Line Combinations vs.
       Stuffed Rank**",
       subtitle = "*2022 Season  |  FootballOutsiders.com*")+
  xlab("# of Unique Offensive Line Combinations") +
  ylab("Stuffed Ranking (1 = best, 32 = worst)") +
  nfl_analytics_theme()

plot_grid(power_rank_plot, stuffed_rank_plot, ncol = 1)
```

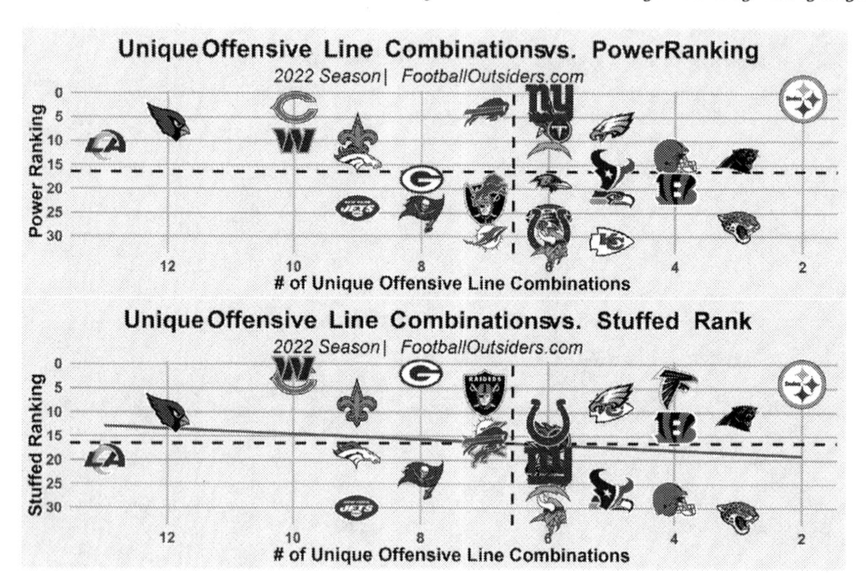

Figure 1.7: Number of unique combos vs. adjusted sack rate per season

Based on the results, it does not seem that the number of unique offensive line combinations has an impact on either power rank or stuffed rank. The Los Angeles Rams and Arizona Cardinals, despite have the two highest unique line combinations, are among the best in the league in the metric. In fact, the opposite seems to be true in that the Minnesota Vikings, with just six unique line combinations during the 2022 season, are last in the league in power ranking. The Rams performed in bit worse when comparing line combinations to stuffed rank, but the Cardinals are still among the best in the league. The Vikings again are at the bottom, only under performed by the Jacksonville Jaguars (who had just three unique combinations throughout the season).

```
ggplot(fb_outsiders_oline, aes(x = combos, y =
                               adj_line_yards)) +
  geom_smooth(method = "lm", se = FALSE, size = 0.8, alpha =
                                .08) +
  nflplotR::geom_mean_lines(aes(v_var = combos, h_var =
                                adj_line_yards),
                               color = "black", size = .8) +
  nflplotR::geom_nfl_logos(aes(team_abbr = team), width =
                                0.065) +
  scale_x_reverse(breaks = scales::pretty_breaks()) +
  scale_y_continuous(breaks = scales::pretty_breaks()) +
  labs(title = "**Unique Offensive Line Combinations vs.
```

```
        Adjusted Line Yards**",
        subtitle = "*2022 Season  |  FootballOutsiders.com*",
        caption = "*An Introduction to NFL Analytics with
         R*<br> **Brad J. Congelio**") +
    xlab("# of Unique Offensive Line Combinations") +
    ylab("Adjusted Line Yards") +
    nfl_analytics_theme()
```

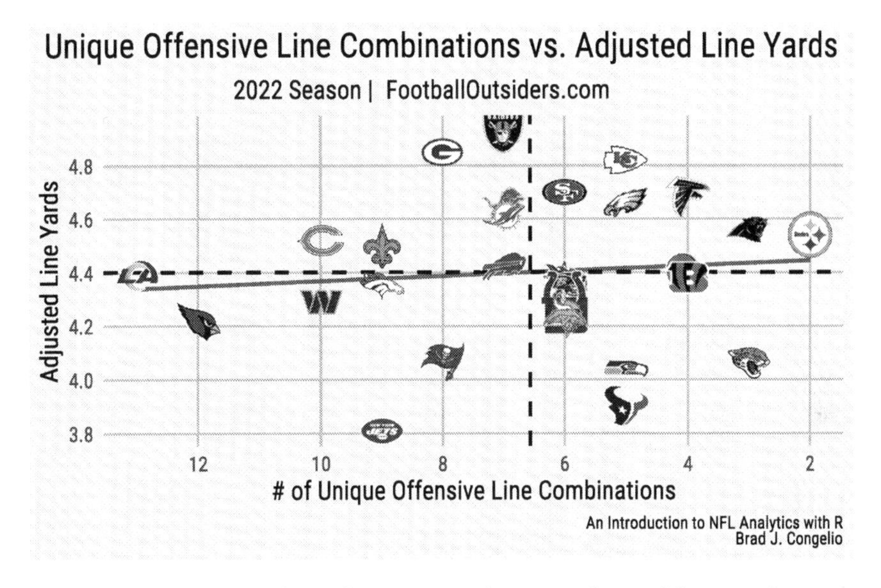

Figure 1.8: Number of unique combos vs. adjusted line yards

Comparing a team's unique offensive line combinations to adjusted line yards is more promising than power and stuffed rank, as the team's with the highest amount of combinations begins to fall under the league-average line for ALY. There are still several teams that, despite having a low number of combinations through the season, performed poorly such as the Houston Texans, Seattle Seahawks, and the Jaguars.

So what do we make of this?

Given that adjusted line yards is constructed to measure the responsibility of the offensive line, it is logical to also want to consider the "performance" of the running back. A running back that is able to avoid early hits, or break tackles and earn yards after contact, can overcome an under performing offensive line (whether that be from lack of talent or lack of cohesiveness from a high number of combinations). In the description of the adjusted line yards metrics, Football Outsiders confirms this, writing that a "team with a very good running back will appear higher no matter how bad their line, and a team with a great line will appear lower if the running back is terrible."

For the last exploration in this topic, we can bring in `Elusive Rating`, which is a signature statistic from Pro Football Focus that quantifies the "success and impact of a runner with the ball independently of the blocking. Let's bring in the data that contains all of the information from `fb_outsiders_oline`, but with the addition of an averaged `elusive` rating for each team from PFF.

```
oline_combo_elusive <- vroom("http://nfl-book.bradcongelio.
      com/ol-elusive")
```

Because we want to determine if the elusiveness of a team's stable of running backs directly impacts the offensive line's Adjusted Line Yards, we can create a new variable that uses the team's `elusive_rating` to, in an elementary fashion, add weight to the `adj_line_yards`. While there are numerous approaches to doing so (such as ranking, categorical weighting, logarithmic weighting, using coefficients from a regression model, etc.), we will conduct a more simple type of feature engineering by first using the `scale()` package to standardize the values, resulting in each team receiving what is called a "z-score" (or the total standard deviations away from the mean). After, we will create the `weighted_aly` metric by subtracting each team's z-score from the existing `adj_line_yards`, and then plot the results.

```
oline_combo_elusive <- oline_combo_elusive %>%
  mutate(adjusted_elu = scale(avg_elu),
         weighted_aly = adj_line_yards - adjusted_elu)

online_combo_test <- oline_combo_elusive %>%
  mutate(weighted_aly = adj_line_yards - adjusted_elu)

ggplot(oline_combo_elusive, aes(x = combos, y = weighted_aly)) +
  geom_smooth(method = "lm", se = FALSE, size = 0.8, alpha = .08) +
  nflplotR::geom_mean_lines(aes(v_var = combos, h_var =
          weighted_aly),
                            color = "black", size = .8) +
  nflplotR::geom_nfl_logos(aes(team_abbr = team), width = 0.065) +
  scale_x_reverse(breaks = scales::pretty_breaks()) +
  scale_y_continuous(breaks = scales::pretty_breaks()) +
  labs(title = "**Unique Offensive Line Combinations vs.
       Weighted Adjusted Line Yards**",
       subtitle = "*2022 Season  |  Adjusted with PFF's 'Elusive'*",
       caption = "*An Introduction to NFL Analytics with R*<br>
       **Brad J. Congelio**")+
  xlab("# of Unique Offensive Line Combinations") +
  ylab("Weighted Adjusted Line Yards") +
  nfl_analytics_theme()
```

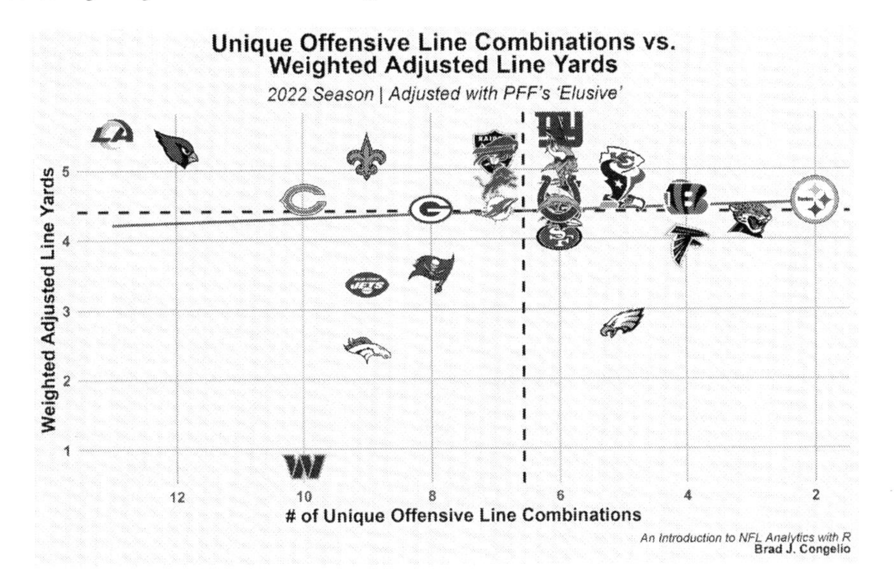

Figure 1.9: Number of unique combos vs. weighted elusiveness

Despite having two of the bottom four scores in `elusive_rating` (21.4 and 21.8), the Rams and the Cardinals are in the league's top five for `Weighted Adjusted Line Yard`. This is an indication that each team's offensive line is proficient at run blocking and that any perceived shortcoming in the ground game was the result of a lack of elusive running backs. Conversely, the Washington Commanders unweighted `Adjusted Line Yards` was 4.29, but when account for the high elusiveness score among the team's running backs, the metric dipped to a league-worst 2.14 weighted adjusted line yards.

While there is certainly more examination that could be conducted, these preliminary results seem to indicate that a higher number of unique offensive line combinations has little impact on both the passing and rushing games, with the exception of a general relationship with the number of combinations and a team's adjusted sack rate. Even then, there are caveats within the data that force us to consider further ways to contextualize the data such as a quarterback's average time to release.

1.4 Exploring Explosive Plays per Touchdown Drive

In November of 2022, Warren Sharp – of Sharp Football Analysis – tweeted perhaps the most ridiculous statistic of the 2022 NFL season.

Heading into their week 9 bye week, the Steelers' longest play resulting in a touchdown was just eight yards, or less than what is required for a 1st down. Perhaps making it worse, the touchdown came early in the season, on a pass from Mitch Trubisky to Pat Freiermuth in a week 2 contest against the New England Patriots.

However, the statistic as presented by Sharp does not tell the complete story as it isolates the entirety of a touchdown drive to a single play. While the team's longest touchdown scoring play was just eight yards, this does not mean that there was a lack of explosive, high-yardage plays earlier in the drive that helped the Steelers get closer to the endzone.

To determine if this is the case, we can construct a metric that explore explosive plays per touchdown drive during the 2022 season.

```
pbp <- nflreadr::load_pbp(2022)

explosive <- pbp %>%
  filter(!is.na(posteam) &
           !is.na(yards_gained)
         & fixed_drive_result == "Touchdown") %>%
  filter(special == 0 & fumble == 0 & interception == 0) %>%
  group_by(posteam, game_id, drive) %>%
  summarize(max_yards = max(yards_gained)) %>%
  mutate(explosive_play = if_else(max_yards >= 20, 1, 0)) %>%
  ungroup() %>%
  group_by(posteam) %>%
  summarize(tds_no_explosive = sum(explosive_play == 0),
            tds_explosive = sum(explosive_play == 1),
            total_drives = sum(tds_no_explosive +
              tds_explosive),
            percent_no_exp = tds_no_explosive / total_drives,
            percent_w_exp = tds_explosive / total_drives) %>%
  select(posteam, percent_w_exp, percent_no_exp)
```

After collecting the play-by-play information for the 2022 NFL season, we use `filter()` to gather only those `fixed_drive_result` rows that include `Touchdown` and then remove any play that is a special teams play and includes a fumble or interception. This is done as we only want those touchdowns scored by the offense, not a fumble or interception return for a score. Because of this, the results of total touchdowns are lower than the "official" number from Pro Football Reference, as the site's "Touchdown Log" includes this data.

We then use `group_by()` on `posteam`, `game_id`, and `drive` in order to create the column `max_yards` with `summarize()`. The resulting numeric value in `max_yards` represents the largest amount of yards gained on a single play, per drive. The important `explosive_play` metric is returned, in a binary `1` or `0` format, based on whether or not the `max_yards` in each drive was greater than or equal to 20 yards. With this calculation complete, we `ungroup()` the data and then group it only by `posteam` to determine the results. After finding the percent of drives with an explosive play and those without, we can plot the results.

```
ggplot(explosive, aes(y = reorder(posteam, percent_w_exp),
                      x = percent_w_exp)) +
  geom_col(aes(color = posteam, fill = posteam), width = 0.5) +
  nflplotR::scale_color_nfl(type = "secondary") +
  nflplotR::scale_fill_nfl(alpha = 0.5) +
  nfl_analytics_theme() +
  theme(axis.text.y = element_nfl_logo(size = .65)) +
  scale_x_continuous(expand = c(0,0),
                     breaks = scales::pretty_breaks(n = 5),
                     label = scales::percent_format()) +
  labs(title = "**Which Team Has The Highest % of
       Explosive Plays per TD Drive?**",
       subtitle = "*2022 Season*",
       caption = "*An Introduction to NFL Analytics with R*<br>
       **Brad J. Congelio**") +
  xlab("Percent of TD Drives with an Explosive Play (20+ Yards)") +
  ylab("")
```

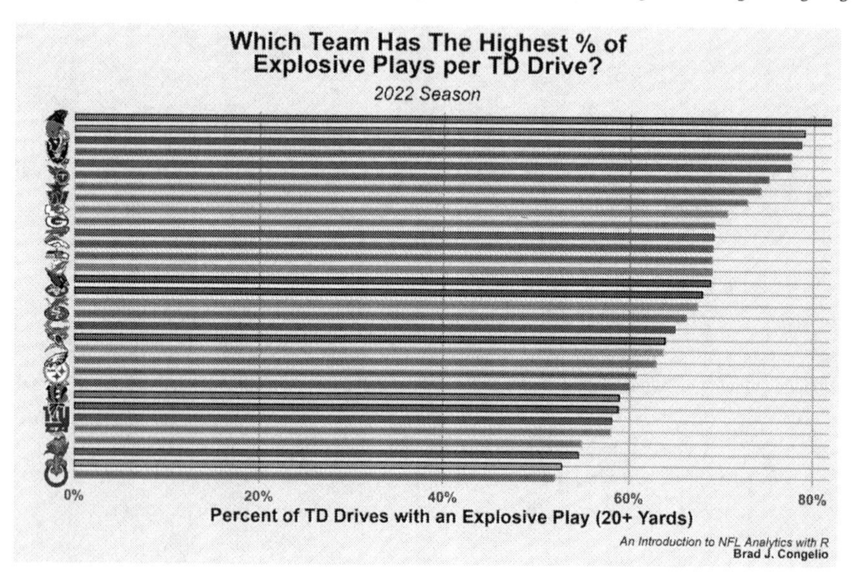

Figure 1.10: Explosive plays per touchdown drive

The Carolina Panthers had a play that gained 20 or more yards on nearly 82% of their touchdown drives, while the Indianapolis Colts had a league-worst 52% of touchdown drives including explosive plays. Notably, the Steelers are the near bottom as well with result of just under 61%. Given the statistic in Sharp's tweet, we can make an assumption that Pittsburgh (as well as other the other teams with lower percentages) methodically "marched down" the field on the way to scoring touchdowns while teams such as the Panthers and Cleveland Browns heavily relied on at least one big "homerun" play per touchdown drive in order to move the ball.

2

Wrangling NFL Data in the `tidyverse`

2.1 Downloading R and RStudio

Prior to downloading R and RStudio, it is important to explain the difference between the two, as they are separate pieces of our analytics puzzle that are used for differing purposes. R is the core programming language used for statistical computing and graphics. R provides a wide range of statistical and graphical techniques and has a large community of users who develop and maintain a multitude of packages (essentially libraries of pre-written functions) that extend the capabilities and ease of coding. While R can be run from your computer's command line, it is also has an integrated development environment (IDE) in RStudio that provides a graphical user interface for working with R scripts, data files, and packages.

RStudio is free to download and use and provides a user-friendly interface for writing R code, organizing projects and files, and working with both big and small data. Regularly updated by the Posit team, RStudio includes many features that are specifically designed to making working with R easier, including syntax highlighting, code suggestions, robust debugging tools, and a built-in package manager.

> **! Important**
>
> It is important that you download and successfully install R before proceeding to install RStudio on your computer.

DOI: 10.1201/9781003364320-2

2.1.1 Downloading R

1. To download R, visit the official R website at https://www.r-project.org/
2. Click on the "CRAN" link at the top of the page, directly underneath the word "Download." This will bring you to the Comprehensive R Archive Network.
3. Each mirror that hosts a copy of R is sorted by country. Select a mirror that is geographically close to you.
4. Click on the version of R that is appropriate for your operating system (Linux, macOS, or Windows).
5. Select the "base" option to download an updated version of R to your computer.
6. Open the downloaded file and follow the provided installation instructions.

2.1.2 Downloading RStudio

1. To download RStudio, visit the official RStudio website at https://posit.co
2. Click on the "Download RStudio" button in the top-right of the page.
3. Scroll down and select "Download" within the "RStudio Desktop - Free" box.
4. Open the downloaded file and follow the provided installation instructions.

2.1.3 The Layout of RStudio

When you open RStudio for the first time, you will see the interface laid out as in Figure 2.1 (sourced from the official RStudio documentation from posit).

As you can see, RStudio provides a graphical interface for working with R and is sorted into four main panes, each of which serves a specific purpose.

1. **The source pane**: In the upper-left, the source pane is where you write and run your R code. It serves as the main working pane within RStudio.
2. **The console pane:** the console pane serves multiple functions, including allowing you to interact directly with R by typing commands and receiving the output. Additionally, any errors outputted by code

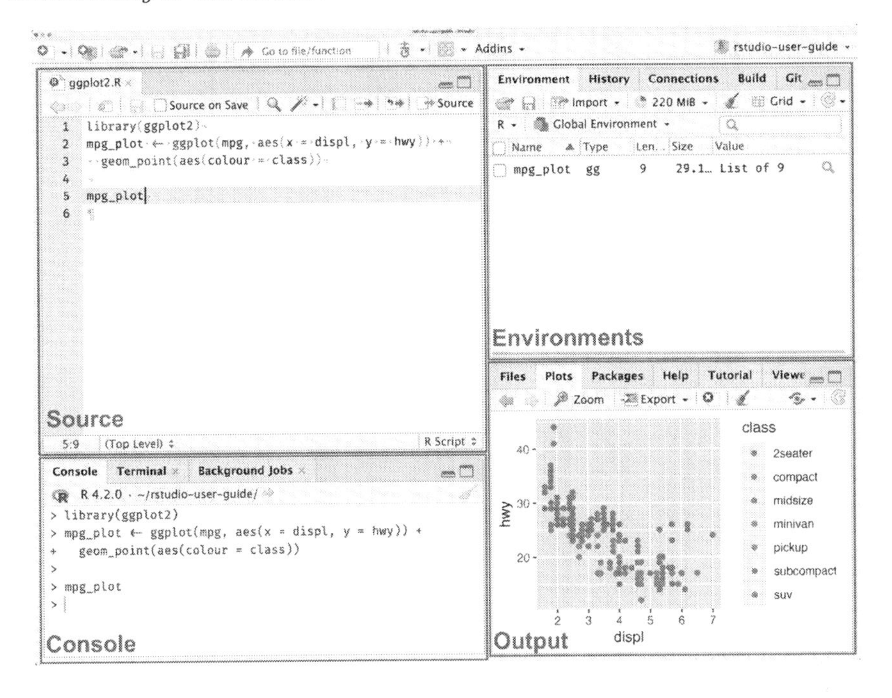

Figure 2.1: RStudio layout

ran in the source pane will be detailed in the console, allowing you to troubleshoot and debug.

3. **Environment and history pane**: this pane, in the upper-right, displays information about the current R environment and command history. Perhaps more important, it displays the information of all the R-created objects currently stored in your computer's memory including data sets, vectors, lists, etc.
4. **Files, plots, packages, and help pane**: in the bottom-right, this pane provides access to numerous RStudio tools and resources including the ability to browse and navigate through the files and folders on your computer and view the output of plots and graphics. As well, the "Packages" tab gives you the ability to manage any R packages that you have installed on your system and to view each packages' help documentation.

2.1.4 Running Code in RStudio

To begin writing code in RStudio you first need to open a new R script. To do so, select "File -> New File -> R Script." A new script tab, titled `Untitled 1` will open in your RStudio's source pane. To run a line of code – or multiple lines of code – you can do one of two options:

1. Place your cursor directly at the end of the last line of code, or highlight all the code you wish to run, and press "Ctrl + Enter" (Windows) or "Command + Enter" (Mac).
2. Place your cursor directly at the end of the last line of code, or highlight all the code you wish to run, and then use your mouse to click the "Run" button in the source pane's toolbar.

As a working example, let's do a simple addition problem within the source pane:

```
2 + 2
```

After following one of the two above options for running the addition problem, the output in your console should appear as seen in Figure 2.2.

Figure 2.2: Example RStudio console output

> **Tip**
>
> It is important to notice the > after the output in the console. That indicates that the coding process has completely run and RStudio is ready to run the next task that you submit. If you have an error in your code, you will notice that the > is replaced by what my students refer to as the "hanging plus sign," +.

> You receive the + sign when RStudio is expecting a continuation of the code. Several issues can cause this to happen, including forgetting to provide an equal number of opening and closing parenthesis or mistakenly including a pipe, %>%, after the last line of code.
>
> In any case, when you see the + in the console, simply use your mouse to click within the console and hit your keyboard's escape key. Doing so exits the prompt and resets the console to include the > symbol.

2.2 Installing and Loading Necessary Packages

Installing and loading packages is an important part of working with RStudio, as they provide additional functionality that allow you to more efficiently conduct data analysis and/or modeling. Several packages are widely used through this book, including the ever-important **tidyverse**, the **nflverse** family of packages to retrieve NFL statistics and data, and many others such as **tidymodels** when we tackle building and testing advanced models in Chapter 5.

To begin, let's install both **tidyverse** and **nflverse**. In your source pane, you can install a package by using the **install.packages()** function. To complete the code, you simply supply the package name within quotation marks.

```
install.packages("tidyverse")
install.packages("nflverse")
```

After running the code, your console pane will output what is going on "behind the scenes." When complete, you will again see the > symbol within the console. At this point, you are able to load the packages you just installed.

```
library(tidyverse)
library(nflverse)
```

2.3 The tidyverse and Its Verbs

The `tidyverse`, now installed and loaded within RStudio, is a collection of R packages designed for data manipulation, visualization, and analysis. It was developed by Hadley Wickham, the Chief Scientist at Posit, and a varied team of contributors. The goal of the `tidyverse` is to provide a consistent, easy-to-understand set of functions and syntax for working with data in R.

The core principle of the `tidyverse` is "tidy data," which is the development team's belief in creating a standard way of organizing data sets. To that end, a "tidy" data set is one that is comprised of observations (rows) and variables (columns) with each variable being a distinct piece of information and each observation being a unit of analysis.

Installing and loading the `tidyverse` results eight of the core packages automatically being loaded and ready to use:

1. **dplyr:** "dplyr provides a grammar of data manipulation, providing a consistent set of verbs that solve the most common data manipulation challenges."
2. **tidyr:** "tidyr provides a set of functions that help you get to tidy data. Tidy data is data with a consistent form: in brief, every variable goes in a column, and every column is a variable."
3. **readr:** "readr provides a fast and friendly way to read rectangular data (like csv, tsv, and fwf). It is deigned to flexibly parse many types of data found in the wild, while still cleanly failing when data unexpectedly changes."
4. **purrr:** "purrr enhances R's functional programming (FP) toolkit by providing a complete and consistent set of tools for working with functions and vectors. Once you master the basic concepts, purrr allows you to replace many for loops with code that is easier to write and more expressive."
5. **tibble**: "tibble is a modern re-imagining of the data frame, keeping what time has proven to be effective, and throwing out what it has not. Tibbles are data.frames that are lazy and surly; they do less and complain more forcing you to confront problems earlier, typically leading to cleaner, more expressive code."
6. **stringr:** "stringr provides a cohesive set of functions designed to make working with strings as easy as possible. It is built on top of stringi, which uses the ICU C library to provide fast, correct implementations of common string manipulations."

7. **forcats:** "forcats provides a suite of useful tools that solve common problems with factors. R uses factors to handle categorical variables, variables that have a fixed and known set of possible values."
8. **ggplot2:** "ggplot2 is a system for declaratively creating graphics, based on The Grammar of Graphics. You provide the data, tell ggplot2 how to map the variables to aesthetics, what graphical primitives to use, and it takes care of the details" (Wickham, 2022).

Aside from the core eight packages, the `tidyverse` will also install a multiple of other packages such as `rvest` (for web scraping), `readxl` (for reading Excel sheets in the RStudio environment), `lubridate` (a very powerful tool for working with times and dates), and `magrittr` (the package that provides the pipe `%>%`). As well, prior versions of the `tidyverse` utilized the `modelr` package. Modeling is now handled in the `tidyverse` by the `tidymodels` package.

2.4 The Flow of the `tidyverse`

The underlying design of coding in the `tidyverse`, aside from the `dplyr` verbs, are both the assignment statement (`<-`) and the pipe (`%>%`). Please note, as mentioned in the book's Preface, that I still use the pipe (`%>%`) that is part of the `magrittr` package and not the native pipe operator (`|>`) included in the 4.1 release of R. The choice of pipe operator you use is your decision to make, as either will work seamlessly within the examples and activities provided in this book.

As I explain to my Sport Analytics students, the language and flow of the `tidyverse` can seem like a foreign language at first. But, it is important that you stick with it because, sooner rather than later, the light bulb above your head will go off. Before detailing the in's and out's of the `tidyverse` in the below section, let's first dissect an example of the `tidyverse` workflow.

```
pbp <- nflreadr::load_pbp(2022) %>%
  filter(posteam == "PHI" & rush == 1) %>%
  group_by(rusher) %>%
  summarize(success_rate = mean(success))
```

The given example involves multiple iterations of the `tidyverse` paradigm. At the outset of my Sport Analytics course, when introducing the concepts of the `tidyverse`, I emphasize that it is possible to "talk your way through" the process from the beginning to your end goal (especially once you have you have a more comprehensive understanding of the `dplyr` verbs, which are expounded upon in the subsequent section). The following stepwise method illustrates this using the above example of code:

1. We first create a data set, denoted by `pbp`, by utilizing the `load_pbp` function from the `nflreadr` package. To talk through this, you can say "`pbp`is an copy of `nflreadr::load_pbp(2022)`." While R purists may laugh at teaching the `tidyverse` in such language, it does indeed work. Going forward, the assignment operator (`<-`) simply implies that "*something is*." In this case, our `pbp` data frame *is* the container for the play-by-play data we are collecting from `nflreadr`.
2. We then move into our first pipe operator (`%>%`). Again, R language purist will likely develop a eye twitch upon reading this, but I explain to my students that the pipe operator serves as a "*... and then*" command. In terms of the "talk it out" method above, the flow would be: "`pbp` is a copy of `nflreadr::load_pbp(2022)` and then ..."
3. After the pipe operator (or the first "and then ..." command), we move into our first `dplyr` verb. In this case, we are using the `filter()` verb to select just the Philadelphia Eagles as the offensive team and just offensive plays that are rush attempts. With another pipe operator, we are including a second "*... and then*" command.
4. To finish the example, we are grouping by each individual rusher on the Eagles "*and then*" summarize the average success rate for each rusher.

To put it together, "talking it out" from beginning to end results in:

"First, create a data set called `pbp` that is a copy of `nflreadr::load_pbp(2022)` *and then* `filter()` for all instances, where the `posteam` is PHI and `rush == 1` *and then* `group_by()` each individual rusher, *and then* summarize the average success rate for each rusher into a new column titled `success_rate`."

To showcase this visually, the "talking through" method is inputted into the example code below:

```
pbp <- "is" nflreadr::load_pbp(2022) %>% "... and then"
  filter(posteam == "PHI" & rush == 1) %>% "... and then"
  group_by(rusher) %>% "... and then"
  summarize(success_rate = mean(success))
```

2.5 Working with NFL Data and the dplyr Verbs

Of the packages nestled within the `tidyverse`, `dplyr` is perhaps the most important in terms of wrangling and cleaning data. As mentioned above, `dplyr` is a powerful tool for data manipulation in R as it provides a key set of functions, known as verbs, that are designed to be easy to use and understand. The verbs can be used to filter, group, summarize, rearrange, and transform all types of data sets. For those just starting their NFL analytics endeavors in the R programming language, the following four `dplyr` verbs are perhaps the most important. Specific examples of working with these verbs, as well as others, follow below.

1. `filter()`: the `filter()` verb allows you to subset data based on certain criteria. For example, you can use `filter()` to keep only those rows in a data set, where a certain variable meets a certain conditions (i.e., more than 100 completed passes). Moreover, the `filter()` verb can be used in conjunction with logical operators such as `&` and `|` to create more complex criteria.
2. `group_by()`: the `group_by()` verb allows you to group a data set by one or more variables. It is a useful tool when you want to perform an operation on each group, such as calculating a summary statistic (i.e., intended air yards per quarterback) or when creating a plot.
3. `summarize()`: the `summarize()` verb allows you to reduce a data set to a single summary value. The `summarize()` verb is often used in conjunction with the `group_by()` function, allowing you to group the data by one or more variables. The `summarize()` verb allows for a wide range of summary statistics, including means, medians, standard deviations, and more. You can also use it to calculate custom summary statistics.
4. `mutate()`: the `mutate()` verbs allows you to create new variables within your data while also preserving existing ones.

2.5.1 NFL Data and the `filter()` Verb

The `filter()` verb allows you to extract specific rows from your dataset based on one, or multiple, supplied conditions. The conditions are supplied to the `filter()` verb by using logical operators, listed in the below table, that ultimately evaluate to either `TRUE` or `FALSE` in each row of the dataset. The `filter()` process returns a data set that includes only those rows that meet the specified conditions.

Logical Operator	Meaning
==	equal to
!=	not equal to
<	less than
<=	less than or equal to
>	greater than
>=	greater than or equal to
!	not
&	and
|	or
%in%	includes
c()	used to combine arguments into a vector
is.na	checks for missing values
!is.na	is not missing specific values

In order to work through specific examples of the above logical operators, we will use 2022 play-by-play data from **nflreadr**. To begin, let's read in the data:

Note

Please note that a more detailed overview of reading in **nflreadr** data is provided in Chapter 3. For the purposes of learning about the `filter()` verb, please make sure you have both the **tidyverse** and **nflreadr** loaded by running the following:

```
library(tidyverse)
library(nflreadr)
```

If you have difficult with the above step, please see the Installing and Loading Necessary Packages section above.

```
pbp <- nflreadr::load_pbp(2022)
```

After running the above code, you will have a data set titled `pbp` placed into your RStudio environment consisting of 50,147 observations over 372 variables. With the data set loaded, let's create a new data set titled `ne_offense` that contains *only those plays where the New England Patriots are the offensive team.*

```
ne_offense <- pbp %>%
  filter(posteam == "NE")

ne_offense
```

```
# A tibble: 1,323 x 372
   play_id game_id    old_game_id home_team away_team season_type  week
     <dbl> <chr>      <chr>       <chr>     <chr>     <chr>       <int>
 1      44 2022_01_~ 2022091106  MIA       NE        REG             1
 2      59 2022_01_~ 2022091106  MIA       NE        REG             1
 3      83 2022_01_~ 2022091106  MIA       NE        REG             1
 4     109 2022_01_~ 2022091106  MIA       NE        REG             1
 5     130 2022_01_~ 2022091106  MIA       NE        REG             1
 6     154 2022_01_~ 2022091106  MIA       NE        REG             1
 7     175 2022_01_~ 2022091106  MIA       NE        REG             1
 8     196 2022_01_~ 2022091106  MIA       NE        REG             1
 9     236 2022_01_~ 2022091106  MIA       NE        REG             1
10     571 2022_01_~ 2022091106  MIA       NE        REG             1
# i 1,313 more rows
# i 365 more variables: posteam <chr>, posteam_type <chr>, ...
```

The output shows that the `posteam` variable contains only `NE`, which is the abbreviation for the New England Patriots in the `nflreadr` play-by-play data. In the code that produced `ne_offense`, **it is important to notice that:**

1. the "equal to" logical operator consist of *TWO* equal signs, not one.
2. the team abbreviation (`NE`) is in quotation marks.

In the first, a single equal sign (`=`) can be used as a synonym for assignment (`<-`) but is most often used when passing values into functions. To avoid confusion with this possibility, **the test for equality when using `filter()` is always `==`.**

In the second, you must use quotation marks around the character-based value you are placing into the `filter()` verb because, if not, R will interpret `posteam` incorrectly and ultimately generate an error. On the other hand, you *do not* need to include quotation marks if you are filtering out numeric-based variables. Below are incorrect and correct examples of both:

```
ne_offense <- pbp %>%
  filter(posteam == NE) #this is incorrect.
            ##Character values must be in quotation marks.

ne_offense <- pbp %>%
```

```
  filter(posteam == "NE") #this is correct for character
          values.

ne_offense <- pbp %>%
  filter(air_yards >= "5") #this is incorrect.
             #Numeric values do not need quotation marks.

ne_offense <- pbp %>%
  filter(air_yards >= 5) #this is correct for numeric values.
```

How do we approach the logical operators if we want to retrieve every offensive team in the `pbp` data *except* for New England? In that case, we can use the "not equal to" (`!=`) operator:

```
not_ne_offense <- pbp %>%
  filter(posteam != "NE")
```

The resulting data set titled `not_ne_offense` will still include all 372 variables housed within the `nflreadr` play-by-play data, but will not include any row in which New England is the offensive (`posteam`) team.

Continuing with examples, how do we use the `filter()` verb on multiple teams at once? For instance, let's use the above filtering process for offensive teams but only retrieve information from the play-by-play data for the four teams that comprise the AFC East (Buffalo Bills, Miami Dolphins, New England Patriots, and the New York Jets). There are, in fact, two logical operators that can produce the results we are looking for: the "or" logical operator (`|`) or by using the "includes" logical operator (`%in%`) combined with the "concatenate" operator (`c()`). Let's start with using the "or" operator.

```
afc_east <- pbp %>%
  filter(posteam == "NE" | posteam == "MIA"
         | posteam == "NYJ" | posteam == "BUF")
```

By using the `|` logical operator, which translates to the word "or," we can string together for separate filters for `posteam` within the play-by-platy data. That said, it probably seems odd to have to include the `posteam` argument four different times, rather than being able to do this:

```
afc_east <- pbp %>%
  filter(posteam == "NE" | "MIA" | "NYJ" | "BUF")
```

While the above example logically makes sense (verbally saying "**posteam** equals NE or MIA or NYJ or BUF"), it unfortunately results in an error. To that end, if you'd rather avoid the need to type **posteam** four different times, as in the above example, you can switch to using the **%in%** operator combined with **c()**. It is possible to combine just **filter()** and the **%in%** operator to retrieve one specific team. But, as in the above example where, we will receive an error if we try to do it for multiple teams without including the **c()** operator, as such:

```
afc_east <- pbp %>%
  filter(posteam %in% c("NE", "MIA", "NYJ", "BUF"))

afc_east
```

```
# A tibble: 5,625 x 372
   play_id game_id    old_game_id home_team away_team season_type  week
     <dbl> <chr>      <chr>       <chr>     <chr>     <chr>       <int>
 1      43 2022_01_~ 2022091107  NYJ       BAL       REG             1
 2      68 2022_01_~ 2022091107  NYJ       BAL       REG             1
 3      89 2022_01_~ 2022091107  NYJ       BAL       REG             1
 4     115 2022_01_~ 2022091107  NYJ       BAL       REG             1
 5     136 2022_01_~ 2022091107  NYJ       BAL       REG             1
 6     172 2022_01_~ 2022091107  NYJ       BAL       REG             1
 7     391 2022_01_~ 2022091107  NYJ       BAL       REG             1
 8     412 2022_01_~ 2022091107  NYJ       BAL       REG             1
 9     436 2022_01_~ 2022091107  NYJ       BAL       REG             1
10     469 2022_01_~ 2022091107  NYJ       BAL       REG             1
# i 5,615 more rows
# i 365 more variables: posteam <chr>, posteam_type <chr>, ...
```

The above approach is a simplified version of first creating a vector of the team abbreviations and then passing that into the **filter()** argument. For example, we can create the following vector that appears in the "Values" area of your RStudio environment, and then use that to retrieve the same results in did in the above code:

```
afc_east_teams_vector <- c("NE", "MIA", "NYJ", "BUF")

afc_east <- pbp %>%
  filter(posteam %in% afc_east_teams_vector)

afc_east_teams_vector
```

```
[1] "NE"  "MIA" "NYJ" "BUF"
```

In other cases, we may need to do the opposite of above – that is, select all teams *except* for specific ones. For example, using data from Pro Football Focus regarding a quarterback's average time to throw, let's determine how to *remove* all AFC North teams (Pittsburgh, Cleveland, Baltimore, and Cincinnati) from the data set while leaving all other teams intact. Such a procedure is necessary, for example, if we wanted to explore the average time to throw for the combined AFC North teams against the rest of the NFL.

We cannot leave the data as is and find the averages outright. We must remove the AFC North teams in order to avoid "baking them into" the league-wide average, thus skewing the results of our analysis. To complete this process, we turn to the `!` ("is not") logical operator.

To begin, we will use the `vroom` package to read in the data. Let's load the 'vroom' package before using it.

```
library(vroom)
```

```
time_in_pocket
        <- vroom("http://nfl-book.bradcongelio.com/qb-tip")
```

```
# A tibble: 30 x 5
   player       position team_name player_game_count avg_time_to_throw
   <chr>        <chr>    <chr>                 <dbl>             <dbl>
 1 Patrick Mah~ QB       KC                       20              2.85
 2 Tom Brady    QB       TB                       18              2.31
 3 Justin Herb~ QB       LAC                      18              2.75
 4 Joe Burrow   QB       CIN                      19              2.51
 5 Josh Allen   QB       BUF                      18              2.92
 6 Kirk Cousins QB       MIN                      18              2.7
 7 Trevor Lawr~ QB       JAX                      19              2.51
 8 Geno Smith   QB       SEA                      18              2.79
 9 Daniel Jones QB       NYG                      18              3.03
10 Jalen Hurts  QB       PHI                      18              2.86
# i 20 more rows
```

The structure and organization of the data is quite similar to the data sets we worked with in our above examples, despite coming from a different source (PFF vs. `nflreadr`). Our `time_in_pocket` data includes a quarterback's name on each row with his corresponding `team`, `player_game_count`, and `avg_time_to_throw`. Again, we are seeking to compare the average time to release – combined – for the AFC North against the rest of the NFL. In order to avoid skewing the results by "baking in data" (that is, not removing AFC North teams prior to finding the NFL average), we can use the `!` logical operator to remove all four teams at once (as oppose to structuring the `filter()` to include four separate `team_name !=` for all four different AFC North teams.

```
time_in_pocket_no_afcn <- time_in_pocket %>%
  filter(!team_name %in% c("PIT", "BAL", "CIN", "CLE"))

time_in_pocket_no_afcn
```

```
# A tibble: 26 x 5
   player        position team_name player_game_count avg_time_to_throw
   <chr>         <chr>    <chr>                 <dbl>             <dbl>
 1 Patrick Mah~ QB       KC                       20              2.85
 2 Tom Brady     QB       TB                       18              2.31
 3 Justin Herb~ QB       LAC                      18              2.75
 4 Josh Allen    QB       BUF                      18              2.92
 5 Kirk Cousins QB       MIN                      18              2.7
 6 Trevor Lawr~ QB       JAX                      19              2.51
 7 Geno Smith    QB       SEA                      18              2.79
 8 Daniel Jones QB       NYG                      18              3.03
 9 Jalen Hurts   QB       PHI                      18              2.86
10 Jared Goff    QB       DET                      17              2.7
# i 16 more rows
```

With Pittsburgh, Baltimore, Cincinnati, and Cleveland removed from the data set, we can – for the sake of completing the analysis – take our new `time_in_pocket_afcn` data set and find the league-wide average for `avg_time_to_throw`.

```
time_in_pocket_no_afcn %>%
  summarize(combined_average = mean(avg_time_to_throw))
```

```
# A tibble: 1 x 1
  combined_average
             <dbl>
1             2.76
```

```
time_in_pocket %>%
  filter(team_name %in% c("PIT", "BAL", "CIN", "CLE")) %>%
  summarize(combined_average = mean(avg_time_to_throw))
```

```
# A tibble: 1 x 1
  combined_average
             <dbl>
1             2.89
```

The AFC North quarterbacks had an average time to throw of 2.89 seconds while the rest of the NFL averaged 2.76 seconds, meaning the AFC North QBs released the ball, on average, 0.13 seconds slower than the rest of the quarterbacks in the league.

So far, all our `filter()` operations have include just one logical operator, or multiple values built into one operator using the `%in%` option. However, because of the structure of the logical operators, we are able to construct a single `filter()` combined with various operator requests. For example, let's gather only those rows that fit the following specifications:

1. that play occurs on 2nd down.
2. the play must be a pass play.
3. the play must include a charted QB hit.
4. the play must result in a complete pass.
5. the pass must have an air yards amount that is greater than or equal to the yards to go.

 Tip

Compiling the above `filter()` requires knowledge of the variable names within `nflreadr` play-by-play data. This is covered in Chapter 3.

That said, it is important to point out that many of the above variables have data values that are structured in binary – that is, the value is either 1 or 0.

In the above list, the `qb_hit` variable is binary. A play *with* a QB hit is denoted by the inclusion of a numeric `1` in the data, while the inclusion of a numeric `0` indicates that a QB hit did not occur.

Building the `filter()` based on the above specifications requires the use of both character and numeric values:

```
multiple_filters <- pbp %>%
  filter(down == 2 & play_type == "pass" &
           qb_hit == 1 &
           complete_pass == 1 &
           air_yards >= ydstogo)
```

The resulting data set, `multiple_filters`, indicates that exactly 105 such instances took place during the 2002 NFL season.

Lastly, in some instances, it is necessary to use `filter()` to remove rows that may provide you with incorrect information if include. To showcase this, it is necessary to also include the `summarize()` verb, which is covered in greater detail in the following two section. For example, let's gather information from the play-by-play data that will provide the total amount of yards per scrimmage – combined – for the 2022 regular season. Using our knowledge from the prior example, we can write code that first uses the `filter()` verb to retrieve just those rows that took place during the regular season and then uses the `summarize()` verb to distill the information down to the singular, combined total. In the example below, please note the use of `na.rm = TRUE`. When completing computation operations within the `summarize()` verb, it is often necessary to include `na.rm = TRUE` in order to drop any missing values in the data. This is especially important when using the `mean()` calculation, as any missing data will result in `NA` values being returned.

```
total_yards <- pbp %>%
  filter(season_type == "REG") %>%
  summarize(yards = sum(yards_gained, na.rm = TRUE))

total_yards
```

```
# A tibble: 1 x 1
   yards
   <dbl>
1 184444
```

How many total scrimmage yards did NFL teams earn during the 2022 NFL regular season? According to the results from our above code, the answer is 184,444. However, that is incorrect.

The right answer, in fact, is 184,332 total scrimmage yards. How did the results of our above coding come up 112 shorts of the correct answer (as provided by the 2022 Team Stats page on Pro Football Reference)?

The difference is the result of the above code including all plays, including 2-point conversion attempts. Because a 2-point conversion is technically considered an extra point attempt – not a play from scrimmage or an official down – no player stats are officially record for the play. We can verify this is the case by using the `filter()` verb to include only those rows that include regular-season games and those plays that do not include information pertaining to `down`:

```
two_points <- pbp %>%
  filter(season_type == "REG" & is.na(down)) %>%
  summarize(total = n(),
            total_yards = sum(yards_gained, na.rm = TRUE))

two_points
```

```
# A tibble: 1 x 2
  total total_yards
  <int>       <dbl>
1  8137         112
```

The results show that exactly 112 scrimmage yards were gained on places missing `down` information. There was, of course, not 8,137 2-point conversion attempts during the 2022 regular season. There are many other instances within the data that do not include information for `down`, including: kickoffs, extra points, time outs, the end of a quarter, etc. We can be more specific with our `filter()` argument in the above code to include only the binary argument for the `two_point_attempt` variable:

```
two_points_true <- pbp %>%
  filter(season_type == "REG" & two_point_attempt == 1) %>%
  summarize(total = n(),
            total_yards = sum(yards_gained, na.rm = TRUE))

two_points_true
```

```
# A tibble: 1 x 2
  total total_yards
  <int>       <dbl>
1   119         112
```

By altering the '`is.na(down)` argument to the binary indicator for `two_point_attempt`, we see that there were 119 2-point conversion attempts in the 2022 regular season while maintain the correct 112 yards that explains our difference from above. To account for our new-found knowledge that we must remove 2-point conversions to calculate the correct amount of total scrimmage yards, we can again use the `!is.na()` operator (where `!is.na()` means "is not missing this specific variable) to remove all rows that do not include information pertaining to `down`:

```
total_yards_correct <- pbp %>%
  filter(season_type == "REG" & !is.na(down)) %>%
  summarize(yards = sum(yards_gained, na.rm = TRUE))

total_yards_correct
```

```
# A tibble: 1 x 1
   yards
   <dbl>
1 184332
```

By filtering out any row that does not include the `down` for a given play, we correctly calculated the total amount of scrimmage yards – 184,332 – for the 2022 NFL regular season.

2.5.2 NFL Data and the `group_by()` Verb

As mentioned above, the `group_by()` verb allows you to group data by one or more specific variables in order to retrieve, among other actions, summary statistics. To showcase how `group_by()` is used within the `nflverse` data, let's first gather the 2022 regular season statistics and then use the `summarize()` verb to get the average success rate on rushing plays.

As well, we immediately make use of the `filter()` function to sort the data: (1) we first instruct to `filter()` the data to include just those instances where the `play_type` equals `run`, (2) we then say it must also be `play == 1`, meaning there was no penalty or other interruption that "cancelled out" the play, and (3) we lastly pass the argument that the `down` cannot be missing by using `!is.na` as a missing down is indicative of a two-point conversion attempt.

```
rushing_success_ungrouped <- pbp %>%
  filter(play_type == "run" & play == 1 & !is.na(down)) %>%
  summarize(success_rate = mean(success))

rushing_success_ungrouped
```

```
# A tibble: 1 x 1
  success_rate
         <dbl>
1        0.431
```

Without including the `group_by()` verb within the above code, the output is the average success rate for rushing plays for all 32 NFL teams, wherein success rate is the percentage of rushing plays that resulted in an EPA above zero. In this case, approximately 43% of NFL rushes had a positive success rate.

That said, we are interested in examining the success rate by team, not league-wide average. To do so, we add the `posteam` variable into the `group_by()` verb.

```
rushing_success_grouped <- pbp %>%
  filter(play_type == "run" & play == 1 & !is.na(down)) %>%
  group_by(posteam) %>%
  summarize(success_rate = mean(success)) %>%
  arrange(-success_rate)

rushing_success_grouped %>%
  slice(1:10)
```

```
# A tibble: 10 x 2
   posteam success_rate
   <chr>          <dbl>
 1 PHI            0.526
 2 BUF            0.488
 3 BAL            0.478
 4 GB             0.476
 5 PIT            0.470
 6 ATL            0.463
 7 KC             0.459
 8 NYG            0.454
 9 CIN            0.454
10 LV             0.449
```

In the above example, we have added the offensive team into the `group_by()` verb, while also arranging the data in descending order by `success_rate`, and then used `slice()` to gather just the ten teams with the highest rushing success rate. The Philadelphia Eagles led the NFL in rushing success rate during the 2022 NFL regular season at 52.3%. By removing the `slice()` function in the above example, we can see that Tampa Bay maintained the worst rushing success rate in the league at 37.3%.

While determining the rushing success rate of teams is interesting, we can also determine the same metric for individual running backs as well. To do so, we simply replace the variable in the `group_by()` verb. In the below example, we replace the `posteam` variable with the `rusher` variable to see which running backs have the highest success rate.

```
running_back_success <- pbp %>%
  filter(play_type == "run" & play == 1 & !is.na(down)) %>%
  group_by(rusher) %>%
  summarize(success_rate = mean(success)) %>%
  arrange(-success_rate)

running_back_success %>%
  slice(1:10)
```

```
# A tibble: 10 x 2
   rusher      success_rate
   <chr>              <dbl>
 1 A.Davis                1
 2 B.Aiyuk                1
 3 B.Allen                1
 4 B.Skowronek            1
 5 C.Kmet                 1
 6 C.Sutton               1
 7 C.Wilson               1
 8 D.Bellinger            1
 9 D.Brown                1
10 D.Gray                 1
```

The output, unfortunately, is not all that helpful. Because we did not use the `filter()` verb to stipulate a minimum number of rushing attempts, the output is saying that – for example, Daniel Bellinger, a tight end, was among the most successful rushers in the league with a 100% rushing success rate. To correct this, we must add a second metric to our `summarize()` verb (we will call it `n_rushes`) and then use the `filter()` verb afterward to include a minimum number of rushes required to be included in the final output.

As well, we will provide an additional argument in the first `filter()` verb that stops the output from including any rushing attempt that does not include the running back's name. The `n_rushes()` in the `summarize()` verb allows us to now include the number of attempts, per individual rusher, that fall within the first `filter()` parameter. Afterward, we include a second `filter()` argument to include just those rushers with at least 200 attempts.

```
running_back_success_min <- pbp %>%
  filter(play_type == "run" & play == 1 &
           !is.na(down) &
           !is.na(rusher)) %>%
  group_by(rusher) %>%
```

```
  summarize(success_rate = mean(success),
            n_rushes = n()) %>%
  filter(n_rushes >= 200) %>%
  arrange(-success_rate)

running_back_success_min %>%
  slice(1:10)
```

```
# A tibble: 10 x 3
   rusher      success_rate n_rushes
   <chr>              <dbl>    <int>
 1 M.Sanders          0.483      294
 2 I.Pacheco          0.469      207
 3 A.Jones            0.465      213
 4 J.Jacobs           0.444      340
 5 N.Chubb            0.434      302
 6 T.Allgeier         0.429      210
 7 Ja.Williams        0.427      262
 8 T.Etienne          0.424      250
 9 J.Mixon            0.422      249
10 B.Robinson         0.420      205
```

Unsurprisingly, Miles Sanders – a running back for the Eagles, who lead the NFL in team success rate – is the leader in rushing success among individual players with 49% of his attempts gaining positive EPA.

Much like the `filter()` verb, we are able to supply multiple arguments within one `group_by()` verb. This is necessary, for example, when you want to examine results over the course of multiple seasons. Let's explore this by determining each team's yards after catch between 2010 and 2022. To start, we will retrieve all NFL play-by-play data from those seasons.

> **Caution**
>
> Loading in multiple seasons of play-by-play data can be taxing on your computer's memory. RStudio's memory usage monitor, located in "Environment" toolbar tell you how much data is currently being stored in memory. As needed, you can click on the usage monitor and select "Free Unused Memory" to purge any data that is no longer also in your "Environment."

```
multiple_seasons <- nflreadr::load_pbp(2010:2022)
```

Once the data set is loaded, we can think ahead to what is needed in the `group_by()` verb. In this case, we are exploring team results on a season basis. Because of this, we will include *both* `posteam` and `season` in the `group_by()` verb.

```
yac_seasons <- multiple_seasons %>%
  filter(season_type == "REG" &
           !is.na(posteam) &
           !is.na(yards_after_catch)) %>%
  group_by(season, posteam) %>%
  summarize(total_yac = sum(yards_after_catch, na.rm = TRUE))
```

2.5.3 NFL Data and the `summarize()` Verb

As we've seen, the `summarize()` function can be used to find summary statistics based whichever option we pass to it via the `group_by()` verb. However, it can also be used to create new metrics built off data included in the `nflverse` play-by-play data.

Let's examine which teams were the most aggressive on 3rd and short passing attempts during the 2022 season. Of course, determining our definition of both what "short" is on 3rd down and "aggressive" is quite subjective. For the purposes of this example, however, let's assume that 3rd and short is considered 3rd down with five or less yards to go and that "aggressive" is a quarterback's air yards being to, at minimum, the first-down marker.

Must like our above examples working with rushing success rate, we begin constructing the metric with the `filter()` argument. In this case, we are filtering for just pass plays, we want the down to equal 3, the yards to go to be equal to or less than 5, we want it to be an official play, and we do not want it to be missing the down information. After the initial `filter()` process, we include the `posteam` variable within our `group_by()` verb.

In our `summarize()` section, we are first getting the total number of times each team passed the ball on 3rd down with no more than five yards to go. After, we are creating a new `aggressiveness` column that counts the number of times a quarterback's air yards were, at minimum, the required yards for a first down. Next, we create another new column titled `percentage` that takes `aggressiveness` and divides it by `total`.

```
team_aggressiveness <- pbp %>%
  filter(play_type == "pass" &
           down == 3 &
           ydstogo <= 5 &
           play == 1 &
           !is.na(down)) %>%
  group_by(posteam) %>%
  summarize(total = n(),
            aggressiveness = sum(air_yards >= ydstogo,
            na.rm = TRUE),
            percentage = aggressiveness / total) %>%
  arrange(-percentage)

team_aggressiveness %>%
  slice(1:10)
```

```
# A tibble: 10 x 4
   posteam total aggressiveness percentage
   <chr>   <int>          <int>      <dbl>
 1 LV         60             50      0.833
 2 BUF        61             47      0.770
 3 ARI        58             44      0.759
 4 SF         67             50      0.746
 5 PIT        79             58      0.734
 6 SEA        60             44      0.733
 7 NE         54             39      0.722
 8 TB         92             66      0.717
 9 MIA        62             43      0.694
10 CHI        42             29      0.690
```

The Las Vegas Raiders, based on our definitions, are the most aggressive passing team in the league on 3rd and short as just over 83% of their air yards were at – or past – the required yardage for a first down. On the other end of the spectrum, the New York Giants were the least aggressive team during the 2022 regular season, at 49.1%.

2.5.4 NFL Data and the `mutate()` Verb

In the our example above working with the `summarize()` verb, our output includes only the information contained in our `group_by()` and then whatever information we provided in the `summarize()` (such as `total`, `aggressiveness`, and `percentage`).

What if, however, you wanted to create new variables and then `summarize()` those? That is where the `mutate()` verb is used.

As an example, let's explore individual quarterback's average completion percentage over expected for specific air yard distances . To start, we can attempt to do this simply by including both `passer` and `air_yards` in the `group_by` verb.

```
airyards_cpoe <- pbp %>%
  group_by(passer, air_yards) %>%
  summarize(avg_cpoe = mean(cpoe, na.rm = TRUE))
```

Your output is going to include the `avg_cpoe` for each quarterback at each and every distance of `air_yards`. Not only is it difficult to find meaning in, but it would prove to be difficult – if not impossible – to visualize with `ggplot()`. To correct this issue, we must use the `mutate()` verb.

Rather than `summarize()` the completion percentage over expected for each distance of `air_yards`, we can use the `mutate()` verb to bundle together a grouping of distances. In the below example, we are using the `mutate()` verb to create a new variable titled `ay_distance` using the `case_when()` verb.

```
airyards_cpoe_mutate <- pbp %>%
  filter(!is.na(cpoe)) %>%
  mutate(
    ay_distance = case_when(
      air_yards < 0 ~ "Negative",
      air_yards >= 0 & air_yards < 10 ~ "Short",
      air_yards >= 10 & air_yards < 20 ~ "Medium",
      air_yards >= 20 ~ "Deep")) %>%
  group_by(passer, ay_distance) %>%
  summarize(avg_cpoe = mean(cpoe))
```

With the `air_yards` data now binned into four different groupings, we can examine quarterbacks at specific distances.

```
airyards_cpoe_mutate %>%
  filter(ay_distance == "Medium") %>%
  arrange(-avg_cpoe) %>%
  slice(1:10)
```

```
# A tibble: 82 x 3
# Groups:   passer [82]
   passer     ay_distance avg_cpoe
   <chr>      <chr>          <dbl>
 1 A.Brown    Medium        -6.09
 2 A.Cooper   Medium       -43.1
 3 A.Dalton   Medium         5.01
 4 A.Rodgers  Medium         2.02
 5 B.Allen    Medium        44.5
 6 B.Hoyer    Medium       -44.0
 7 B.Mayfield Medium       -15.4
 8 B.Perkins  Medium         0.266
 9 B.Purdy    Medium         9.72
10 B.Rypien   Medium        18.0
# i 72 more rows
```

2.6 Core Skills for Tidy Data

With your new understanding of the `tidyverse` flow in the R programming language, we are now going to hone these core skills by taking a dataset from ingestion through the cleaning and prepping process so that it is prepared for eventual data visualization (which will be done in Chapter 4 of this book). To do so, we are going to use data provided by Sports Info Solutions. Based out of Allentown, Pennsylvania, SIS collects its play-by-play and then does a manual hand-charting process to provide weekly data that covers a vast array of specific occurrences in football (defensive scheme, pre-snap motion, wide receivers formations, offensive personnel, drop type, and more).

In this example, we are going to compare the `Boom%` and the `Bust%` for each quarterback in the data, wherein the `Boom%` correlates to any pass attempt by the quarterback that resulted in an expected points add (EPA) of at least 1 and `Bust%` is any pass attempt that resulted in an EPA of at least -1. In order to prepare the data for visualization in `ggplot`, we will:

1. Import and examine the data.
2. Deal with missing values in the data.
3. Charge variable types to match what it needed for visualization.
4. Using the `mutate()` verb to correct and create new metrics within the data.
5. Merging the data with a secondary dataset.

2.6.1 Importing and Conducting Exploratory Data Analysis

To start, let's examine data from Sports Info Solutions by reading in the data from this book's Git repository using the `vroom` package.

```
sis_data <- vroom("http://nfl-book.bradcongelio.com/sis-bb")
```

Using `vroom`, we are reading in the `.csv` file into our environment and assigning it the name `sis_data`. With access to the data, we can begin our exploratory data analysis – the importance of which cannot be understated. The process of EDA results in several key items:

1. EDA helps to identify potential issues in the data set, such as missing or erroneous data, outliers, or inconsistent values – all of which must be addressed prior to further analysis or visualization.
2. Conducting an EDA allows you to gain a deeper understanding of the distribution, variability, and relationships between all the variables contained in the data set, thus allowing you to make informed decisions on appropriate statistical techniques, models, and visualizations that can best analyze and communicate the data effectively.
3. The EDA process can help you discover patterns or trends in the data that may be of interest or relevance to your research question, or help you discover to questions to answer.

To start the process of EDA on our newly created `sis_data` dataset, let's examine the current status fo the data set:

```
sis_data
```

```
# A tibble: 107 x 14
   Season Player            Team       Att `Points Earned` `PE Per Play`
    <dbl> <chr>             <chr>    <dbl>           <dbl>         <dbl>
 1   2022 Patrick Mahomes Chiefs     648            180.         0.267
 2   2022 Justin Herbert  Chargers   699            138.         0.187
 3   2022 Trevor Lawrence Jaguars    584            123.         0.202
 4   2022 Jared Goff      Lions      587            121.         0.199
 5   2022 Jalen Hurts     Eagles     460            106.         0.213
 6   2022 <NA>            <NA>       439            104.         0.12
 7   2022 Kirk Cousins    Vikings    643            104.         0.151
 8   2022 Tua Tagovailoa  Dolphins   400            102.         0.243
 9   2022 Joe Burrow      Bengals    606            102.         0.158
10   2022 Josh Allen      Bills      567            100.         0.167
# i 97 more rows
# i 8 more variables: `Points Above Avg` <dbl>, ...
```

The results immediately bring to light several issues in the data:

1. There are several instances of missing data (as indicated by `NA` values in the 6th row).
2. Many of the column names are not in `tidy` format. Specifically, the `Boom%` and `Bust%` columns, because of the inclusion of the percentage sign, require the small ticks at either end of the word. The same issue is apparent in `Points Earned` as it includes a space between the word and, finally, many of the variable names are in full caps.
3. The `Boom%` and `Bust%` columns are both listed as `<chr>` which indicates that they are currently character-based columns, rather than the needed numeric-based column.
4. The above issue is caused by all the values within `Boom%` and `Bust%` containing a percentage sign at the tail end.
5. Lastly, the `team` variable for Baker Mayfield's has a value of "2 teams" since he played for both the Carolina Panthers and Los Angeles Rams during the 2022 season.

To begin dealing with these issues, let's first tackle cleaning the variable names and then selecting just those columns relevant to our forthcoming data visualization in Chapter 4:

```
sis_data <- sis_data %>%
  janitor::clean_names() %>%
  select(player, team, att, boom_percent, bust_percent)

sis_data
```

```
# A tibble: 107 x 5
   player          team       att boom_percent bust_percent
   <chr>           <chr>    <dbl> <chr>        <chr>
 1 Patrick Mahomes Chiefs     648 25.10%       12.50%
 2 Justin Herbert  Chargers   699 21.20%       15.90%
 3 Trevor Lawrence Jaguars    584 23.70%       15.20%
 4 Jared Goff      Lions      587 23.60%       13.30%
 5 Jalen Hurts     Eagles     460 23.50%       16.10%
 6 <NA>            <NA>       439 <NA>         10.10%
 7 Kirk Cousins    Vikings    643 21.30%       17.40%
 8 Tua Tagovailoa  Dolphins   400 28.50%       15.70%
 9 Joe Burrow      Bengals    606 23.60%       14.80%
10 Josh Allen      Bills      567 23.20%       12.70%
# i 97 more rows
```

In the above code, we are using the `clean_names()` function within the `janitor` package to automatically rework the column names into a `tidy` format. After, we use the `select()` verb to keep only five of the included columns of data (`player`, `team`, `att`, `boom_percent`, and `bust_percent`). With the column names clean and easy to implement within our code, and with the relevant columns selected, we can move to the process of dealing with the missing data.

2.6.2 Dealing with Missing Data

Dealing with missing data is an important step in the exploratory data analysis process, as its inclusion can introduce bias, reduce statistical power, and ultimately influence the validity of your research findings. In any case, a missing value within an RStudio dataset is represented by `NA` values and there are several different approaches to handling the missing data in an effective manner.

A data set can present missing data for a variety of reasons, including participant non-responses (such in social science research), measurement error, or simple data processing issues. The process of handling missing data is ultimately the result of the end goal of the analysis. In some case, conducting a imputation (such as replacing all `NA` values with the mean of all existing data) is used. In other cases, a more complex method can be utilized such as using the `mice` package, which can impute missing data based on several different machine learning approaches (predictive mean matching, logistic regression, Bayesian polytomous regression, proportional odds model, etc.). Ultimately, the choice of method for handling missing data depends on the attributes of the data set, your research question, and your domain knowledge of what the data represents.

To begin exploring missing values in our `sis_data` dataset, we can run the following code:

```
sis_data_missing <- sis_data %>%
  filter_all(any_vars(is.na(.)))

sis_data_missing
```

```
# A tibble: 1 x 5
  player team    att boom_percent bust_percent
  <chr>  <chr> <dbl> <chr>        <chr>
1 <NA>   <NA>    439 <NA>         10.10%
```

By using `filter_all()` across all of our variables (`any_vars`), we can filter for any missing value, indicated by including `is.na` in the `filter()`. The output shows each row within the data set that includes at least one missing value. In this case, we only have one row with missing data, with values for `player`, `team`, and `boom_percent` all missing. Given the nature of our data, it is not prudent to impute a value in place of the missing data as we have no ability to accurately determine who the player is.

Instead, we will return to our original data set (`sis_data`) and use the `na.omit()` function to drop any row that includes missing data:

```
sis_data <- na.omit(sis_data)
```

In the above code, we are simply "recreating" our existing dataset but using `na.omit()` to drop rows that include `NA` values. Once complete, we can move onto the next step of the data preparation/cleaning process.

2.6.3 Changing Variable Types

As discovered in the exploratory data analysis process, we know that four columns (`player`, `team`, `boom_percent`, and `bust_percent`) are listed as a `character` data type while each quarterback's number of attempts (`att`) is listed as a numeric. While we do want both player and team names to be a character-based datatype, we need both the boom and bust percentage for each quarterback to be a numeric value rather than a character.

Moreover, the values in each `boom_percent` and `bust_percent` include a percentage sign at the tail-end.

Because of this, we have two issues we need to correct:

1. correctly changing the variable type for `boom_percent` and `bust_percent`
2. removing the `%` from the end of each value in `boom_percent` and `bust_percent`

Much like many things while working in the R programming language, there exists more than one way to tackle the above issues. In this specific case, we will first look at a method that uses the `stringr` package before switching the columns from `character` to `numeric` using an approach rooted in base R. After, we will use a much easier method to do both at one time.

> **Note**
>
> It is important to note here that I do not necessarily endorse using the first method below to change variable types and to drop the percentage sign, as the second option is ideal.
>
> However, it is important to see how both methods work as there could be a case, somewhere down the road when you are exploring and preparing data yourself, that the first option is the *only* or *best* option.
>
> As mentioned, there are typically multiple ways to get to the same endpoint in the R programming language. Showcasing both methods below simply provides you more options in your toolkit for tidying data.

2.6.3.1 Method #1: Using `stringr` and Changing Variables

In the below example, we are creating an "example" data frame titled `sis_data_stringr` from our current iteration of SIS data and then pipe into the `mutate()` verb (covered in more depth in the below `Creating New Variables` section). It is within this `mutate()` verb that we can drop the percentage sign.

We first indicate that we are going to `mutate()` our already existing columns (both `boom_percent` and `bust_percent`). After, we use an `=` sign to indicate that the following string is the argument to create or, in our case, edit the existing values in the column.

We then use the `str_remove()` function from the `stringr` package to locate the `%` sign in each value and remove it.

Afterward, we dive into base R (that is, not `tidyverse` structure) and use the `as.numeric()` function to change both of the boom and bust columns to the correct data type (using the `$` sign to notate that we are working on just one specific column in the data.

```
sis_data_stringr <- sis_data %>%
  mutate(boom_percent = str_remove(boom_percent, "%"),
         bust_percent = str_remove(bust_percent, "%"))

sis_data_stringr$boom_percent <-
  as.numeric(as.character(sis_data_stringr$boom_percent))
sis_data_stringr$bust_percent <-
```

```
  as.numeric(as.character(sis_data_stringr$bust_percent))

sis_data_stringr
```

```
# A tibble: 106 x 5
   player          team       att boom_percent bust_percent
   <chr>           <chr>    <dbl>        <dbl>        <dbl>
 1 Patrick Mahomes Chiefs     648         25.1         12.5
 2 Justin Herbert  Chargers   699         21.2         15.9
 3 Trevor Lawrence Jaguars    584         23.7         15.2
 4 Jared Goff      Lions      587         23.6         13.3
 5 Jalen Hurts     Eagles     460         23.5         16.1
 6 Kirk Cousins    Vikings    643         21.3         17.4
 7 Tua Tagovailoa  Dolphins   400         28.5         15.7
 8 Joe Burrow      Bengals    606         23.6         14.8
 9 Josh Allen      Bills      567         23.2         12.7
10 Geno Smith      Seahawks   572         20.6         17.6
# i 96 more rows
```

Our output shows that `att`, `boom_percent`, and `bust_percent` are now all three numeric and, moreover, the tailing percentage sign in each boom and bust value is now removed. At this point, the data is prepped and properly structured for the visualization process.

2.6.3.2 Method #2: Using the `parse_number` Function in `readr`

While the above example using `stringr` and base R is suitable for our needs, we can use the `readr` package – and its `parse_number()` function – to achieve the same results in a less verbose manner.

```
sis_data <- sis_data %>%
  mutate(boom_percent = readr::parse_number(boom_percent),
         bust_percent = readr::parse_number(bust_percent))

sis_data
```

```
# A tibble: 106 x 5
   player          team       att boom_percent bust_percent
   <chr>           <chr>    <dbl>        <dbl>        <dbl>
 1 Patrick Mahomes Chiefs     648         25.1         12.5
 2 Justin Herbert  Chargers   699         21.2         15.9
 3 Trevor Lawrence Jaguars    584         23.7         15.2
```

```
 4 Jared Goff      Lions      587          23.6          13.3
 5 Jalen Hurts     Eagles     460          23.5          16.1
 6 Kirk Cousins    Vikings    643          21.3          17.4
 7 Tua Tagovailoa  Dolphins   400          28.5          15.7
 8 Joe Burrow      Bengals    606          23.6          14.8
 9 Josh Allen      Bills      567          23.2          12.7
10 Geno Smith      Seahawks   572          20.6          17.6
# i 96 more rows
```

The above example, using `readr`, is quite similar to the first version that used `stringr` in that we are using the `mutate()` verb to edit our existing `boom_percent` and `bust_percent` columns. In this case, however, we are replacing our `stringer` argument with the `parse_number()` function from the `readr` package. When finished, you can see in the output that not only are both boom and bust correctly listed as numeric, but the `parse_number()` function automatically recognized and dropped the tailing percentage sign.

While both examples get us to the same endpoint, using the `readr` package allows us to get there with less work.

2.6.4 Correct and Create New Variables Using `mutate`

The `mutate()` verb is a powerful and widely used tool that allows us to create new variables (columns) based on existing ones within our data set. To that end, the `mutate()` verb applies a user-supplied argument to each row of the data set to create the new metric. Additionally, the `mutate()` verb – when combined with the `case_when()` argument – allows us to make correction to the data (such as the aforementioned issues with Mayfield's teams).

The syntax of the `mutate()` verb follows the `tidyverse` flow like many of the other `dplyr` verbs:

```
mutate(new_column_name = user_supplied_argument(existing_
            column_name))
```

The `mutate()` verb is highly flexible, allowing us to create a new column based on a wide-range of functions and arguments, including mathematical operations, logical operations, string operation, and many more – including applying functions from other packages.

Importantly, the `mutate()` verb allows us to create new variables **without changing the original data set**. While the `summarize()` verb can also create new columns based on various functions and argument, it also distills the data set down to just the columns created within the `summarize()` function (plus any column including in the `group_by()`). Conversely, the `mutate()` verb creates the new column and adds it on to the current iteration of the data set.

To showcase this, let's create a new column in our `sis_data` that calculate the difference between a quarterback's `boom_percent` and `bust_percent`. This is also a great time to limit the quarterbacks we include based on the number of attempts for each. Not doing so results in oddities in the data, such as Christian McCaffrey (a running back) having a 100% boom percentage on one passing attempt. To avoid such occurrences, let's add the `filter()` verb prior to our `mutate()` and limit the data to just those quarterbacks with at least 200 passing attempts:

```
sis_data <- sis_data %>%
  filter(att >= 200) %>%
  mutate(difference = boom_percent - bust_percent)

sis_data
```

```
# A tibble: 33 x 6
   player           team       att boom_percent bust_percent difference
   <chr>            <chr>    <dbl>        <dbl>        <dbl>      <dbl>
 1 Patrick Mahomes  Chiefs     648         25.1         12.5      12.6
 2 Justin Herbert   Chargers   699         21.2         15.9       5.3
 3 Trevor Lawrence  Jaguars    584         23.7         15.2       8.5
 4 Jared Goff       Lions      587         23.6         13.3      10.3
 5 Jalen Hurts      Eagles     460         23.5         16.1       7.4
 6 Kirk Cousins     Vikings    643         21.3         17.4       3.90
 7 Tua Tagovailoa   Dolphins   400         28.5         15.7      12.8
 8 Joe Burrow       Bengals    606         23.6         14.8       8.8
 9 Josh Allen       Bills      567         23.2         12.7      10.5
10 Geno Smith       Seahawks   572         20.6         17.6       3
# i 23 more rows
```

In the above example, by using the `mutate()` verb, we created a new column in our existing `sis_data` data set that calculates the difference between a quarterback's `boom_percent` and `bust_percent`. Importantly, as mentioned, this is done while keeping the rest of the dataset in place unlike using the `summarize()` verb.

As mentioned in the introduction to this section, it is possible to include other function within the `mutate()` verb. In fact, doing so is necessary if we wish to make the correction to Baker Mayfield's team. Currently, Mayfield's value for `team` is listed as `2 teams`. Instead, let's replace that with his most recent team in the 2022 NFL season (Rams).

To make this correction, we will begin by using the `mutate()` verb on the existing `team` column in our data set. However, where we calculated the different between boom and bust percent in the above example of `mutate()`, we will now use the `case_when()` function.

The `case_when()` function allows us to either correct values, or create new columns, based on multiple conditions and values. The syntax of `case_when()` is as follows:

```
mutate(column_name = case_when(
  condition_1 == value_1 ~ new_value,
  TRUE ~ default_value))
```

In the above code, we are essentially saying: "if condition_1 equals value_1, then replace the column_name's existing value with this new_value." The final argument, `TRUE ~ default_value`, specifies the default value to be used if none of the prior conditions are `TRUE`.

To make this more concrete, let's work through it using our issue with Baker Mayfield. In order to make the correction, we need to supply the `case_when()` argument three items from the data: the column name that we are applying the `mutate()` to (in this case, `team`), the first condition (in this case, `player`), and the value of that condition (in this case, `Baker Mayfield`). Next, we need to include the new value (in this case, `Rams`), and then supply what to do if the above condition is not `TRUE`. In other words, if the `player` is not `Baker Mayfield` then simply keep the existing `team` in place.

Following that verbal walk through of the process, we can place the conditions and values into the correct spot in the relevant code:

```
sis_data <- sis_data %>%
  mutate(team = case_when(
    player == "Baker Mayfield" ~ "Rams",
    TRUE ~ team))

sis_data %>%
  filter(player == "Baker Mayfield")
```

```
# A tibble: 1 x 6
  player         team    att boom_percent bust_percent difference
  <chr>          <chr> <dbl>        <dbl>        <dbl>      <dbl>
1 Baker Mayfield Rams    335           17         18.9      -1.90
```

By filtering the data down to include just Baker Mayfield, we can make sure that his team is now listed as the Rams. Further, a continued look at the data shows no other players' teams were changed in the process:

```
sis_data
```

```
# A tibble: 33 x 6
   player          team       att boom_percent bust_percent difference
   <chr>           <chr>    <dbl>        <dbl>        <dbl>      <dbl>
 1 Patrick Mahomes Chiefs     648         25.1         12.5      12.6
 2 Justin Herbert  Chargers   699         21.2         15.9       5.3
 3 Trevor Lawrence Jaguars    584         23.7         15.2       8.5
 4 Jared Goff      Lions      587         23.6         13.3      10.3
 5 Jalen Hurts     Eagles     460         23.5         16.1       7.4
 6 Kirk Cousins    Vikings    643         21.3         17.4       3.90
 7 Tua Tagovailoa  Dolphins   400         28.5         15.7      12.8
 8 Joe Burrow      Bengals    606         23.6         14.8       8.8
 9 Josh Allen      Bills      567         23.2         12.7      10.5
10 Geno Smith      Seahawks   572         20.6         17.6       3
# i 23 more rows
```

With that, our `sis_data` data set is nearly complete and ready for visualization. Before doing so, though, we can use the `load_teams` function within `nflreadr` to combine the appropriate team colors to each quarterback in the data, allowing us to customize the appearance of the data visualization.

2.6.5 Merging Datasets with `dplyr` and Mutating Joins

Mutating joins are a manipulation technique that allows us to add, modify, or remove columns in a data set based on a shared variable or a set of variables. The process is done by matching rows in the data sets based on an matching identifier between the two. Once completed, the matching data from the second data set is added to the first or, conversely, the data within the first dataset is modified as needed to include the incoming data from the second data set.

There are four common mutating joins in the `dplyr` package:

1. `left_join()`: a `left_join()` adds all the columns from the first dataset by only matching columns from the second data set.
2. `right_join()`: a `right_join()` is the opposite of the `left_join()` process in that it adds all the columns from the second data set and only matches those columns from the first.
3. `inner_join()`: a `inner_join()` will only match rows found in both data sets.
4. `full_join()`: a `full_join()` will include all the rows and columns from both data sets.

In this instance, using our sis_data data set, we will need to conduct a left_join() to bring in NFL team color information. To start, we can use the load_teams() function in nflreadr to load the team information and then use select() to keep just the information we needed for the merging process:

> **i Note**
>
> Please note that a much deeper dive into the nflverse functions is included in Chapter 3, including a more detailed approach to working with individual team data. The below code will be expanded upon in the next chapter but, for now, allows us to walk through the merging process.

```
teams <- nflreadr::load_teams(current = TRUE) %>%
  select(team_abbr, team_nick, team_name, team_color,
   team_color2)

teams
```

```
# A tibble: 32 x 5
   team_abbr team_nick team_name            team_color team_
    color2
   <chr>     <chr>     <chr>                <chr>      <chr>
 1 ARI       Cardinals Arizona Cardinals    #97233F    #000000
 2 ATL       Falcons   Atlanta Falcons      #A71930    #000000
 3 BAL       Ravens    Baltimore Ravens     #241773    #9E7C0C
 4 BUF       Bills     Buffalo Bills        #00338D    #C60C30
 5 CAR       Panthers  Carolina Panthers    #0085CA    #000000
 6 CHI       Bears     Chicago Bears        #0B162A    #C83803
 7 CIN       Bengals   Cincinnati Bengals   #FB4F14    #000000
 8 CLE       Browns    Cleveland Browns     #FF3C00    #311D00
 9 DAL       Cowboys   Dallas Cowboys       #002244    #B0B7BC
10 DEN       Broncos   Denver Broncos       #002244    #FB4F14
# i 22 more rows
```

For educational purposes, we use the above code to bring in each NFL's team abbreviation, nickname, and full name. Along with identifying information for each team, we also have selected team_color and team_color2 (which correspond to each team's primary and secondary colors).

As mentioned, we will use the left_join() function to merge this information into our sis_data. The basic syntax of a left_join includes a call to the original data set, the data set to be merged in, and then an argument to stipulate which variable the join should match on:

```
data_set1 <- data_set1 %>%
  left_join(data_set2, by = c("variable_1" = "variable_2"))
```

The above code is stating: "use **data_set1**, *and then* conduct a **left_join** with **data_set2**, by matching **variable_1** from **data_set1** to corresponding **variable_2** from **data_set2**."

To conduct the join, **we must determine which variable in teams matching a corresponding variable in sis_data**. We can use the `slice()` function to compare the first row of data from each:

```
sis_data %>%
  slice(1)
```

```
# A tibble: 1 x 6
  player          team     att boom_percent bust_percent difference
  <chr>           <chr>  <dbl>        <dbl>        <dbl>      <dbl>
1 Patrick Mahomes Chiefs   648         25.1         12.5       12.6
```

```
teams %>%
  slice(1)
```

```
# A tibble: 1 x 5
  team_abbr team_nick team_name         team_color team_color2
  <chr>     <chr>     <chr>             <chr>      <chr>
1 ARI       Cardinals Arizona Cardinals #97233F    #000000
```

In **sis_data**, the **team** column includes the nickname of the NFL team (in this case, the Chiefs). The corresponding variable in our **teams** data set is the column titled **team_nick**. Knowing that, we can conduct the **left_join** using **team** for **variable_1** from the above example syntax and **team_nick** for **variable_2**.

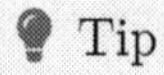

Tip

Notice that the columns containing the corresponding data do not have the same name (**team** vs. **team_nick**).

It is important to remember that the name of the columns *do not* have to match. It is the values contained within the column that must match.

```
sis_data <- sis_data %>%
  left_join(teams, by = c("team" = "team_nick"))

sis_data
```

```
# A tibble: 33 x 10
   player   team    att boom_percent bust_percent difference team_abbr
   <chr>    <chr> <dbl>        <dbl>        <dbl>      <dbl> <chr>
 1 Patrick~ Chie~   648         25.1         12.5      12.6  KC
 2 Justin ~ Char~   699         21.2         15.9       5.3  LAC
 3 Trevor ~ Jagu~   584         23.7         15.2       8.5  JAX
 4 Jared G~ Lions   587         23.6         13.3      10.3  DET
 5 Jalen H~ Eagl~   460         23.5         16.1       7.4  PHI
 6 Kirk Co~ Viki~   643         21.3         17.4       3.90 MIN
 7 Tua Tag~ Dolp~   400         28.5         15.7      12.8  MIA
 8 Joe Bur~ Beng~   606         23.6         14.8       8.8  CIN
 9 Josh Al~ Bills   567         23.2         12.7      10.5  BUF
10 Geno Sm~ Seah~   572         20.6         17.6       3    SEA
# i 23 more rows
# i 3 more variables: team_name <chr>, team_color <chr>, ...
```

You can see in the output that our two data sets correctly merged on the `team` and `team_nick` columns. Moreover, to illustrate what happens doing the joining process, please note these two items:

1. The `team_nick` column from our `teams` data set is not included in our merged `sis_data`, as it is dropped after matching/merging with the `team` information from our original data set.
2. While the `team_nick` column was dropped, the other team identifiers (`team_abbr` and `team_name`) were kept intact, as they were not used in the matching process. As well, `team_color` and `team_color2` are now correctly matched with each quarterback's team.

2.7 Exercises

The answers for the following answers can be found here: http://nfl-book.bradcongelio.com/ch2-answers.

Create a data frame titled `pbp` that contains the play-by-play information for the 2022 regular season by running the following code:

```
pbp <- nflreadr::load_pbp(2022) %>%
  filter(season_type == "REG")
```

After, using the data in pbp, complete the following:

2.7.1 Exercise 1

1. Use filter() to select only Patrick Mahomes and all passes that resulted in interceptions.
2. Add a summarize() to determine the average air yards (air_yards) for his combined interceptions.

2.7.2 Exercise 2

1. In a data frame titled wide_receivers, find the average yards after catch (yards_after_catch) and average yards after catch EPA (yac_epa) for the top-10 receivers in total receiving yards. In doing so, you will need to use group_by() on receiver.

2.7.3 Exercise 3

1. In a data frame titled qb_hits, find the total number of times each defensive unit (defteam) recorded a QB hit (qb_hit).

2.7.4 Exercise 4

1. After running the following code, add a column titled difference using that mutate() verb that calculates the difference between each quarterback's total_completions and total_incompletions.

```
cmp_inc <- pbp %>%
  filter(complete_pass == 1 |
           incomplete_pass == 1 |
           interception == 1, !is.na(down)) %>%
  group_by(passer) %>%
  summarize(total_completions = sum(complete_pass == 1,
                                    na.rm = TRUE),
```

```
                total_incompletions = sum(incomplete_pass == 1,
                                          na.rm = TRUE)) %>%
  filter(total_completions >= 180)
```

2.7.5 Exercise 5

1. After running the following code, use the `left_join()` function to merge the two data frames together into a single data frame titled `rb_success_combined`.

```
roster_information <- nflreadr::load_rosters(2022) %>%
  select(full_name, gsis_id, headshot_url)

rb_success <- pbp %>%
  group_by(rusher_player_id) %>%
  filter(!is.na(rusher) & !is.na(success)) %>%
  summarize(total_success = sum(success, na.rm = TRUE))
    %>%
  arrange(-total_success) %>%
  slice(1:10)
```

3

NFL Analytics with the `nflverse` *Family of Packages*

As mentioned in the Preface of this book, the `nflverse` has drastically expanded since the inception of `nflfastR` in April of 2020. In total, the current version of the `nflverse` is comprised of five separate R packages:

1. `nflfastR`
2. `nflseedR`
3. `nfl4th`
4. `nflreadr`
5. `nflplotR`

Installing the `nflverse` as a package in R will automatically install all five packages. However, the core focus of this book will be on `nflreadr`. It is understandable if you are confused by that, since the Preface of this book introduced the `nflfastR` package. The `nflreadr` package, as explained by its author (Tan Ho), is a "minimal package for downloading data from `nflverse` repositories." The data that *is* the `nflverse` is stored across five different GitHub repositories. Using `nflreadr` allows for easy access to any of these data sources. For lack of a better term, `nflreadr` acts as a shortcut of sorts while also operating with less dependencies.

As you will see in this chapter, using `nflreadr::` while coding provides nearly identical functions to those available when using `nflfastR::`. In fact, `nflfastR::`, in many instances, now calls, "under the hood," the equivalent function in `nflreadr::`. Because of the coalescing between the two, many of the new functions being developed are available only when using `nflreadr::`. For example, `nflreadr::` allows you to access data pertaining to the NFL Combine, draft picks, contracts, trades, injury information, and access to statistics on Pro Football Reference.

While `nflfastR` did initially serve as the foundation of the "amateur NFL analytics" movement, the `nflreadr` package has superseded it and now serves as the "catchall" package for all the various bits and pieces of the `nflverse`. Because of this, and to maintain consistency throughout, this book – nearly

DOI: 10.1201/9781003364320-3

exclusively – will use `nflreadr::` when calling functions housed within the `nflverse` rather than `nflfastR::`.

Figure 3.1 visualizes the relationship between `nflfastR` and `nflreadr`.

The purpose of this chapter is to explore `nflreadr` data in an introductory fashion using, what I believe, are the two most important functions in the `nflverse`: (1) `load_player_stats()` and (2) `load_pbp()`. It makes the assumption that you are versed in the R programming language. If you are not, please start with Chapter 2 where you can learn about R and the `tidyverse` language using examples from the `nflverse`.

3.1 nflreadr: An Introduction to the Data

The most important part of the `nflverse` is, of course, the data. To begin, we will examine the core data that underpins the `nflverse`: weekly player stats and the more detailed play-by-play data. Using `nflreadr`, the end user is able to collect weekly top-level stats via the `load_player_stats()` function or the much more robust play-by-play numbers by using the `load_pbp()` function.

As you may imagine, there is a **very important distinction between the `load_player_stats()` and `load_pbp()`**. As mentioned, `load_player_stats()` will provide you with weekly, pre-calculated statistics for either offense or kicking. Conversely, `load_pbp()` will provide over 350 metrics for every single play of every single game dating back to 1999.

The `load_player_stats()` function includes the following offensive information:

```
offensive.stats <- nflreadr::load_player_stats(2021)

ls(offensive.stats)
```

```
 [1] "air_yards_share"             "attempts"
 [3] "carries"                     "completions"
 [5] "dakota"                      "fantasy_points"
 [7] "fantasy_points_ppr"          "headshot_url"
 [9] "interceptions"               "pacr"
[11] "passing_2pt_conversions"     "passing_air_yards"
[13] "passing_epa"                 "passing_first_downs"
```

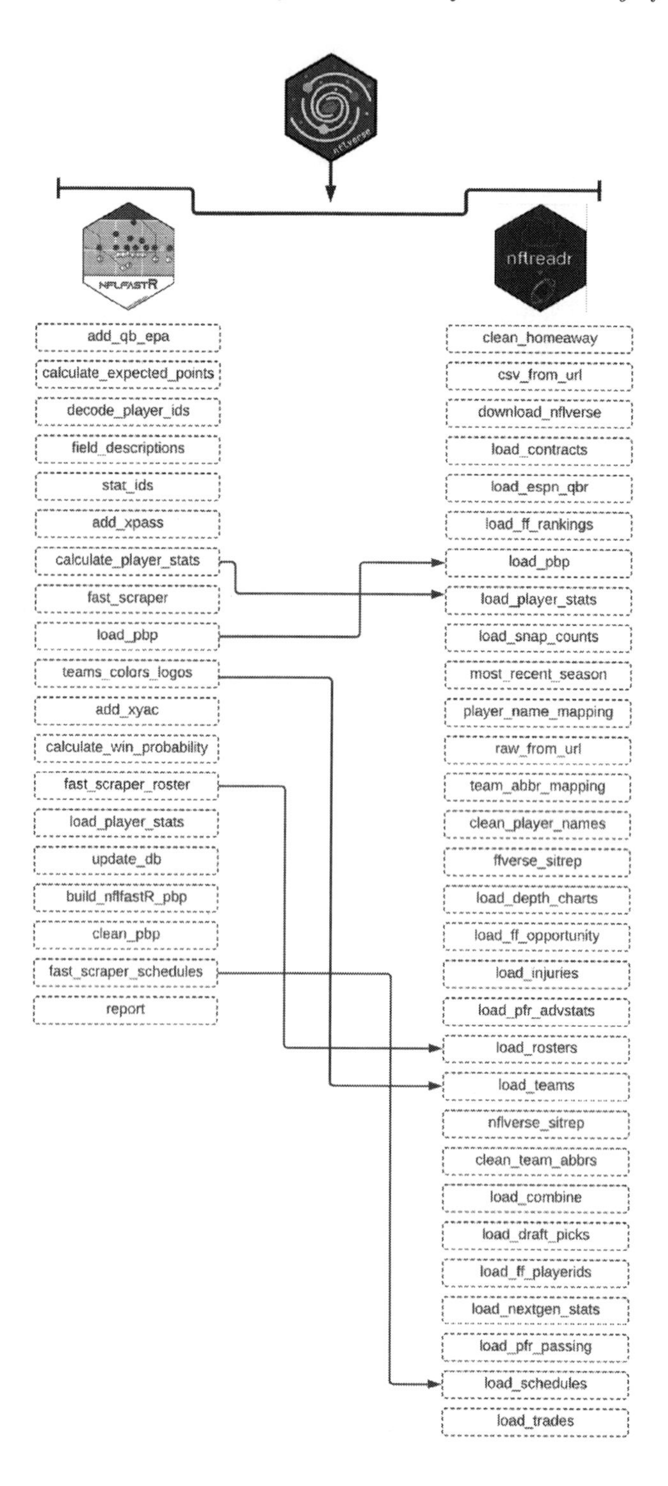

Figure 3.1: Comparison of nflreadr and nflfastR

```
[15] "passing_tds"                 "passing_yards"
[17] "passing_yards_after_catch"   "player_display_name"
[19] "player_id"                   "player_name"
[21] "position"                    "position_group"
[23] "racr"                        "receiving_2pt_conversions"
[25] "receiving_air_yards"         "receiving_epa"
[27] "receiving_first_downs"       "receiving_fumbles"
[29] "receiving_fumbles_lost"      "receiving_tds"
[31] "receiving_yards"             "receiving_yards_after_catch"
[33] "recent_team"                 "receptions"
[35] "rushing_2pt_conversions"     "rushing_epa"
[37] "rushing_first_downs"         "rushing_fumbles"
[39] "rushing_fumbles_lost"        "rushing_tds"
[41] "rushing_yards"               "sack_fumbles"
[43] "sack_fumbles_lost"           "sack_yards"
[45] "sacks"                       "season"
[47] "season_type"                 "special_teams_tds"
[49] "target_share"                "targets"
[51] "week"                        "wopr"
```

As well, switching the `stat_type` to "kicking" provides the following information:

```
kicking.stats <- nflreadr::load_player_stats(2021,
                                            stat_type = "kicking")

ls(kicking.stats)
```

```
 [1] "fg_att"                "fg_blocked"            "fg_blocked_distance"
 [4] "fg_blocked_list"       "fg_long"               "fg_made"
 [7] "fg_made_0_19"          "fg_made_20_29"         "fg_made_30_39"
[10] "fg_made_40_49"         "fg_made_50_59"         "fg_made_60_"
[13] "fg_made_distance"      "fg_made_list"          "fg_missed"
[16] "fg_missed_0_19"        "fg_missed_20_29"       "fg_missed_30_39"
[19] "fg_missed_40_49"       "fg_missed_50_59"       "fg_missed_60_"
[22] "fg_missed_distance"    "fg_missed_list"        "fg_pct"
[25] "gwfg_att"              "gwfg_blocked"          "gwfg_distance"
[28] "gwfg_made"             "gwfg_missed"           "pat_att"
[31] "pat_blocked"           "pat_made"              "pat_missed"
[34] "pat_pct"               "player_id"             "player_name"
[37] "season"                "season_type"           "team"
[40] "week"
```

While the data returned is not as rich as the play-by-play data we will covering next, the `load_player_stats()` function is extremely helpful when you need to quickly (and correctly!) recreate the official stats listed on either the NFL's website or on Pro Football Reference.

As an example, let's say you need to get Patrick Mahomes' total passing yard and attempts from the 2022 season. You could do so via `load_pbp()` but, if you do not need further context (such as down, distance, quarter, etc.), using `load_player_stats()` is much more efficient.

3.1.1 Getting Player Stats via `load_player_stats()`

We can generate a data frame titled `mahomes_pyards` but running the following code:

```
mahomes_pyards <- nflreadr::load_player_stats(seasons = 2022)
```

In the above example, we specified that the returned data be from the 2022 season. It needs to be noted that running `load_player_stats()` (without a specific season) will return the newest season in the data. Moreover, separating two seasons with a colon (`:`) will provide multiple seasons of data. In our working example, the `mahomes_pyards` data frame contains statistics for every player from every week of the 2022 season.

It is important to note that the structure of the data returned by `load_player_stats()` includes passing yards, rushing yards, receiving yards, etc. for each player. As a result, the data returns – for example – the week-by-week statistics for rushing and receiving for Peyton Manning, not just his passing numbers. Because of this, we can calculate Peyton's rushing statistics from 2000 to 2022.

```
# A tibble: 1 x 4
  carries rushing_yards rushing_tds rush_epa
    <int>         <dbl>       <int>    <dbl>
1     409           531          18   -0.740
```

Knowing that Peyton's average rushing EPA from 2000 to the end of his career was -0.740 is not going to win you final Jeopardy nor is it going to be helpful in any sort of Twitter football discourse (as opposed to knowing the rushing EPA of Patrick Mahomes who is, for lack of a better term, a bit more nimble on the run). When working with `load_player_stats()`, it is important that you know the variable names that are useful to the research you are doing.

> **Tip**
>
> The relevant passing statistics housed in `load_player_stats()` includes:

```
 [1] "completions"                 "attempts"
 [3] "passing_yards"               "passing_tds"
 [5] "interceptions"               "sacks"
 [7] "sack_yards"                  "sack_fumbles"
 [9] "sack_fumbles_lost"           "passing_air_yards"
[11] "passing_yards_after_catch"   "passing_first_downs"
[13] "passing_epa"                 "passing_2pt_conversions"
[15] "pacr"                        "dakota"
```

> The columns relevant to rushing statistics in `load_player_stats()` are:

```
[1] "carries"                 "rushing_yards"
[3] "rushing_tds"             "rushing_fumbles"
[5] "rushing_fumbles_lost"    "rushing_first_downs"
[7] "rushing_epa"             "rushing_2pt_conversions"
```

> Lastly, the columns for receiving statistics in `load_player_stats()` include:

```
 [1] "receptions"                  "targets"
 [3] "receiving_yards"             "receiving_tds"
 [5] "receiving_fumbles"           "receiving_fumbles_lost"
 [7] "receiving_air_yards"         "receiving_yards_after_catch"
 [9] "receiving_first_downs"       "receiving_epa"
[11] "receiving_2pt_conversions"   "racr"
[13] "target_share"                "air_yards_share"
```

Continuing with our above example working with `load_player_stats()` and Patrick Mahomes, we can find his total passing yards during the 2022 regular season by using the `filter()` function to sort the data for either `player_name` or `player_display_name` as well the `season_type` and then use `summarize()` to get the total of his `passing_yards` over the course of the regular season.

```
mahomes_pyards <- mahomes_pyards %>%
  filter(player_name == "P.Mahomes" & season_type == "REG") %>%
  summarize(passing_yards = sum(passing_yards))

mahomes_pyards
```

```
# A tibble: 1 x 1
  passing_yards
          <dbl>
1          5250
```

The above code returns a result of 5,250 passing yards which is an exact match from the data provided by Pro Football Reference. In the above example, we used the `filter()` function on the `player_name` column in the data. You will get the same result if you replace that with `player_display_name == "Patrick Mahomes"`. While there is no difference in how the data is collected, there will be a difference if you were to visualize the data frame and wished to display the player names along with their respective statistics. In most cases, in order to maintain a clean design, using the `player_name` option is best as it allows you to plot just the player's first initial and last name.

We can build upon the passing yards example to replicate the vast majority of statistics found in the"Passing" data table for Mahomes from Pro Football Reference.

```
mahomes_pfr <- nflreadr::load_player_stats(seasons = 2022) %>%
  filter(player_name == "P.Mahomes" & season_type == "REG") %>%
  summarize(
    completions = sum(completions),
    attempts = sum(attempts),
    cmp_pct = completions / attempts,
    yards = sum(passing_yards),
    touchdowns = sum(passing_tds),
    td_pct = touchdowns / attempts * 100,
    interceptions = sum(interceptions),
    int_pct = interceptions / attempts * 100,
    first_down = sum(passing_first_downs),
    yards_attempt = yards / attempts,
    adj_yards_attempt = (yards + 20 * touchdowns
      - 45 * interceptions) / attempts,
    yards_completions = yards / completions,
    yards_game = yards / 17,
    sacks = sum(sacks),
    sack_yards = sum(sack_yards),
    sack_pct = sacks / (attempts + sacks) * 100)

mahomes_pfr
```

```
# A tibble: 1 x 16
  completions attempts cmp_pct yards touchdowns td_pct interceptions
        <int>    <int>   <dbl> <dbl>      <int>  <dbl>         <dbl>
1         435      648   0.671  5250         41   6.33            12
# i 9 more variables: int_pct <dbl>, first_down <dbl>,
#   yards_attempt <dbl>, adj_yards_attempt <dbl>, ...
```

What if we wanted to find Mahomes' adjusted yards gained per pass attempt for every one of his seasons between 2018 and 2022? It is as simple as gathering all the data from years (`seasons = 2018:2022`), using the same `filter()` from above, and then using `group_by()` on the `season` variable.

```
mahomes_adjusted_yards
   <- nflreadr::load_player_stats(seasons = 2018:2022) %>%
  filter(player_name == "P.Mahomes" &
           season_type == "REG") %>%
  group_by(season) %>%
  summarize(
    adj_yards_attempt = (sum(passing_yards) + 20 *
                           sum(passing_tds) - 45 *
                           sum(interceptions)) /
sum(attempts))

mahomes_adjusted_yards
```

```
# A tibble: 5 x 2
  season adj_yards_attempt
   <int>             <dbl>
1   2018              9.58
2   2019              8.94
3   2020              8.89
4   2021              7.59
5   2022              8.53
```

Based on the data output, Mahomes' best year for adjusted yards gained per pass attempt was 2018 with 9.58 adjusted yards. How does this compare to other quarterbacks from the same season? We can find the answer by doing slight modifications to our above code. Rather than filtering the information out to just Patrick Mahomes, we will add the `player_name` variable as the argument in the `group_by()` function.

```
all_qbs_adjusted
        <- nflreadr::load_player_stats(seasons = 2018) %>%
  filter(season_type == "REG" & position == "QB") %>%
  group_by(player_name) %>%
  summarize(
    adj_yards_attempt = (sum(passing_yards) + 20 *
                           sum(passing_tds) - 45 *
                           sum(interceptions)) /
                           sum(attempts)) %>%
```

```
  arrange(-adj_yards_attempt)

all_qbs_adjusted
```

```
# A tibble: 73 x 2
   player_name    adj_yards_attempt
   <chr>                      <dbl>
 1 N.Sudfeld                  21
 2 G.Gilbert                  13.3
 3 M.Barkley                  11.7
 4 K.Allen                     9.87
 5 C.Henne                     9.67
 6 P.Mahomes                   9.58
 7 M.Glennon                   9.24
 8 D.Brees                     9.01
 9 R.Wilson                    8.98
10 R.Fitzpatrick               8.80
# i 63 more rows
```

Mahomes, with an adjusted yards per pass attempt of 9.58, finishes in sixth place in the 2018 season behind C. Henne (9.67), K. Allen (9.87), and then three others that have results between 10 and 20. This is a situation where I tell my students the results do not pass the "eye test." Why? It is unlikely that an adjusted yards per pass attempt of 21 and 13.3 for Nate Sudfeld and Garrett Gilbert, respectively, are from a season worth of data. The results for Kyle Allen and Matt Barkley are questionable, as well.

> **! Important**
>
> You will often discover artificially-inflated statistics like above when the players have limited number of pass attempts/rushing attempts/receptions/etc. compared to the other players on the list. To confirm this, we can add `pass_attempts = sum(attempts)` to the above code to compare the number of attempts Sudfeld and Gilbert had compared to the rest of the list.

```
all_qbs_adjusted_with_attempts <-
  nflreadr::load_player_stats(seasons = 2018) %>%
  filter(season_type == "REG" & position == "QB") %>%
  group_by(player_name) %>%
  summarize(
    pass_attempts = sum(attempts),
    adj_yards_attempt = (sum(passing_yards) + 20 *
                         sum(passing_tds) - 45 *
                         sum(interceptions)) /
                         sum(attempts)) %>%
  arrange(-adj_yards_attempt)

all_qbs_adjusted_with_attempts
```

```
# A tibble: 73 x 3
   player_name    pass_attempts adj_yards_attempt
   <chr>                  <int>             <dbl>
 1 N.Sudfeld                  2             21
 2 G.Gilbert                  3             13.3
 3 M.Barkley                 25             11.7
 4 K.Allen                   31              9.87
 5 C.Henne                    3              9.67
 6 P.Mahomes                580              9.58
 7 M.Glennon                 21              9.24
 8 D.Brees                  489              9.01
 9 R.Wilson                 427              8.98
10 R.Fitzpatrick            246              8.80
# i 63 more rows
```

The results are even worse than initially thought. There are a total of six players with an inflated adjusted yards per pass attempt:

1. Nate Sudfeld (2 attempts, 21 adjusted yards).
2. Garrett Gilbert (3 attempts, 13.3 adjusted yards).
3. Matt Barkley (25 attempts, 11.7 adjusted yards).
4. Kyle Allen (31 attempts, 9.87 adjusted yards).
5. Chris Henne (3 attempts, 9.67 adjusted yards).

To remove those players with inflated statistics resulting from a lack of attempts, we can apply a second `filter()` at the end of our above code to limit the results to just those quarterbacks with no less than 100 pass attempts in the 2018 season.

```
all_qbs_attempts_100 <-
  nflreadr::load_player_stats(seasons = 2018) %>%
  filter(season_type == "REG" & position == "QB") %>%
  group_by(player_name) %>%
  summarize(
    pass_attempts = sum(attempts),
    adj_yards_attempt = (sum(passing_yards) + 20 *
                         sum(passing_tds) - 45 *
                         sum(interceptions)) /
                         sum(attempts)) %>%
  filter(pass_attempts >= 100) %>%
  arrange(-adj_yards_attempt)

all_qbs_attempts_100
```

After including a `filter()` to the code to remove those quarterbacks with less than 100 attempts, it is clear that Mahomes has the best adjusted pass yards per attempt (9.58) in the 2018 season with Drew Brees in second place with 9.01. Based on our previous examples, we know that a Mahomes 9.58 adjusted pass yards per attempt in 2018 was his career high (and is also the highest in the 2018 season). How do his other seasons stack up? Is his lowest adjusted pass yards (7.59) in 2021 still the best among NFL quarterbacks in that specific season?

To find the answer, we can make slight adjustments to our existing code.

```
best_adjusted_yards <-
  nflreadr::load_player_stats(seasons = 2018:2022) %>%
  filter(season_type == "REG" & position == "QB") %>%
  group_by(season, player_name) %>%
  summarize(
    pass_attempts = sum(attempts),
    adj_yards_attempt = (sum(passing_yards) + 20 *
                         sum(passing_tds) - 45 *
                         sum(interceptions)) /
                         sum(attempts)) %>%
  filter(pass_attempts >= 100) %>%
  ungroup() %>%
  group_by(season) %>%
  filter(adj_yards_attempt == max(adj_yards_attempt))


best_adjusted_yards
```

```
# A tibble: 5 x 4
# Groups:   season [5]
  season player_name  pass_attempts adj_yards_attempt
   <int> <chr>                <int>             <dbl>
1   2018 P.Mahomes              580              9.58
2   2019 R.Tannehill            286             10.2 
3   2020 A.Rodgers              526              9.57
4   2021 J.Burrow               520              8.96
5   2022 T.Tagovailoa           400              9.22
```

To get the answer, we've created code that follows along with our prior examples. However, after using `filter()` to limit the number of attempts each quarterback must have, we use `ungroup()` to remove the previous grouping between `season` and `player_name` and then use `group_by()` again on just the `season` information in the data. After, we use `filter()` to select just the highest adjusted yards per attempt for each individual season. As a result, we can see that Mahomes only led the NFL in this specific metric in the 2018 season. In fact, between the 2018 and 2022 seasons, Ryan Tannehill had the highest adjusted yards per attempt with 10.2 (but notice he had just 286 passing attempts).

As mentioned, the `load_player_stats()` information is provided on a week-by-week basis. In our above example, we are aggregating all regular season weeks into a season-long metric. Given the structure of the data, we can take a similar approach but to determine the leaders on a week-by-week basis. To showcase this, we can determine the leader in rushing EPA for every week during the 2022 regular season.

```
rushing_epa_leader
        <- nflreadr::load_player_stats(season = 2022) %>%
  filter(season_type == "REG" & position == "RB") %>%
  filter(!is.na(rushing_epa)) %>%
  group_by(week, player_name) %>%
  summarize(
    carries = carries,
    rush_epa = rushing_epa) %>%
  filter(carries >= 10) %>%
  ungroup() %>%
  group_by(week) %>%
  filter(rush_epa == max(rush_epa))
```

```
# A tibble: 18 x 4
# Groups:   week [18]
    week player_name carries rush_epa
   <int> <chr>         <int>    <dbl>
 1     1 D.Swift          15    10.0
 2     2 A.Jones          15     8.08
 3     3 C.Patterson      17     5.76
 4     4 R.Penny          17     9.73
 5     5 A.Ekeler         16    11.6
 6     6 K.Drake          10     6.72
 7     7 J.Jacobs         20     6.93
 8     8 T.Etienne        24     9.68
 9     9 J.Mixon          22    11.3
10    10 A.Jones          24     5.47
11    11 J.Cook           11     2.92
12    12 M.Sanders        21     6.72
13    13 T.Pollard        12     3.97
14    14 M.Sanders        17     8.95
15    15 T.Allgeier       17    10.5
16    16 D.Foreman        21     7.22
17    17 A.Ekeler         10    10.0
18    18 A.Mattison       10     5.85
```

Rather than using the `season` variable in the `group_by()` function, we instead use `week` to calculate the leader in rushing EPA. The results show that the weekly leader is varied, with only Austin Ekeler and Aaron Jones appearing on the list more than once. The Bengals' Joe Mixon produced the largest rushing EPA in the 2022 season, recording 11.3 in week 9 while James Cook's 2.92 output in week 11 was the "least" of the best weekly performances.

Much like we did with the passing statistics, we can use the rushing data in `load_player_stats()` to replicate the majority found on Pro Football Reference. To do so, let's use Joe Mixon's week 9 performance from the 2022 season.

```
mixon_week_9
        <- nflreadr::load_player_stats(seasons = 2022) %>%
  filter(player_name == "J.Mixon" & week == 9) %>%
  summarize(
    rushes = sum(carries),
    yards = sum(rushing_yards),
    yards_att = yards / rushes,
    tds = sum(rushing_tds),
    first_down = sum(rushing_first_downs))

mixon_week_9
```

```
# A tibble: 1 x 5
  rushes yards yards_att   tds first_down
   <int> <dbl>     <dbl> <int>      <dbl>
1     22   153      6.95     4         12
```

Mixon gained a total of 153 yards on the ground on 22 carries, which is just under 7 yards per attempt. Four of those carries resulting in touchdowns, while 12 of them kept drives alive with first downs. Other contextual statistics for this performance, such as yards after contact, are not included in `load_player_stats()` but can be retrieved using other functions within the `nflverse()` (which are covered later in this chapter).

Finally, we can use the `load_player_stats()` function to explore wide receiver performances. As listed above, the wide receivers statistics housed within `load_player_stats()` include those to be expected: yards, touchdowns, air yards, yards after catch, etc. Rather than working with those in an example, let's use `wopr` which stands for `Weighted Opportunity Rating`, which is a weighted average that contextualizes how much value any one wide receivers bring to a fantasy football team. The equation is already built into the data, but for clarity it is as follows:

$$1.5 * targetshare + 0.7 * airyardsshare$$

We can use `wopr` to determine which wide receiver was the most valuable to fantasy football owners during the 2022 season.

```
receivers_wopr
        <- nflreadr::load_player_stats(seasons = 2022) %>%
  filter(season_type == "REG" & position == "WR") %>%
  group_by(player_name) %>%
  summarize(
    total_wopr = sum(wopr)) %>%
  arrange(-total_wopr)

receivers_wopr
```

```
# A tibble: 225 x 2
   player_name total_wopr
   <chr>            <dbl>
 1 D.Moore           13.9
 2 D.Adams           13.4
 3 T.Hill            12.4
 4 A.Brown           12.2
 5 J.Jefferson       11.9
```

```
 6 C.Lamb              11.8
 7 A.Cooper            11.8
 8 D.Johnson           11.2
 9 D.London            10.9
10 D.Metcalf           10.7
# i 215 more rows
```

D.J. Moore led the NFL in the 2022 season with a 13.9 Weighted Opportunity Rating, followed closely by Davante Adams with 13.4. However, before deciding that D.J. Moore is worth the number one pick in your upcoming fantasy football draft, it is worth exploring if there is a relationship between a high Weight Opportunity Rating and a player's total fantasy points. To do so, we can take our above code that was used to gather each player's `WOPR` over the course of the season and add `total_ff = sum(fantasy_points)` and also calculate a new metric called "Fantasy Points per Weighted Opportunity" which is simply the total of a player's points divided by the `wopr`.

```
receivers_wopr_ff_context <-
  nflreadr::load_player_stats(seasons = 2022) %>%
  filter(season_type == "REG" & position == "WR") %>%
  group_by(player_name) %>%
  summarize(
    total_wopr = sum(wopr),
    total_ff = sum(fantasy_points),
    ff_per_wopr = total_ff / total_wopr) %>%
  arrange(-total_wopr)

receivers_wopr_ff_context
```

```
# A tibble: 225 x 4
   player_name total_wopr total_ff ff_per_wopr
   <chr>            <dbl>    <dbl>       <dbl>
 1 D.Moore           13.9     136.        9.76
 2 D.Adams           13.4     236.       17.6
 3 T.Hill            12.4     222.       17.9
 4 A.Brown           12.2     212.       17.4
 5 J.Jefferson       11.9     241.       20.3
 6 C.Lamb            11.8     195.       16.5
 7 A.Cooper          11.8     168        14.2
 8 D.Johnson         11.2      94.7       8.48
 9 D.London          10.9     107.        9.77
10 D.Metcalf         10.7     137.       12.7
# i 215 more rows
```

There are several insights evident in the results. Of the top `wopr` earners in the 2022 season. only Diontae Johnson scored fewer fantasy points than D.J. Moore (94.7 to 136). Given the relationship between `total_wopr` and `total_ff`, a lower `ff_per_wopr` is indicative of less stellar play. In this case, Moore maintained a large amount of both the team's `target_share` and `air_yards_share`, but was not able to translate that into a higher amount of fantasy points (as evidenced by a lower `ff_per_wopr` score). On the other hand, Justin Jefferson's `ff_per_wopr` score of 20.3 argues that he used his amount of `target_share` and `air_yards_share` to increase the amount of fantasy points he produced.

Based on the above examples, it is clear that the `load_player_stats()` function is useful when needing to aggregate statistics on a weekly or season-long basis. While this does allow for easy matching of official statistics, it does not provide the ability to add context to the findings. For example, we know that Patrick Mahomes had the highest adjusted pass yards per attempt during the 2018 regular season with 9.58. But, what if we wanted to explore that same metric but only on passes that took place on 3rd down with 10 or less yards to go?

Unfortunately, the `load_player_stats()` function does not provide ability to distill the statistics into specific situations. Because of this, we must turn to the `load_pbp()` function.

3.2 Using `load_pbp()` to Add Context to Statistics

Using the `load_pbp()` function is preferable when you are looking to add context to a player's statistics, as the `load_player_stats()` function is, for all intents and purposes, aggregated statistics that limit your ability to find deeper meaning.

The `load_pbp()` function provides over 350 various metrics, as listed below:

```
pbp.data <- nflreadr::load_pbp(2022)
ls(pbp.data)
```

```
[1] "aborted_play"
[2] "air_epa"
[3] "air_wpa"
[4] "air_yards"
```

```
 [5] "assist_tackle"
 [6] "assist_tackle_1_player_id"
 [7] "assist_tackle_1_player_name"
 [8] "assist_tackle_1_team"
 [9] "assist_tackle_2_player_id"
[10] "assist_tackle_2_player_name"
[11] "assist_tackle_2_team"
[12] "assist_tackle_3_player_id"
[13] "assist_tackle_3_player_name"
[14] "assist_tackle_3_team"
[15] "assist_tackle_4_player_id"
[16] "assist_tackle_4_player_name"
[17] "assist_tackle_4_team"
[18] "away_coach"
[19] "away_score"
[20] "away_team"
[21] "away_timeouts_remaining"
[22] "away_wp"
[23] "away_wp_post"
[24] "blocked_player_id"
[25] "blocked_player_name"
[26] "comp_air_epa"
[27] "comp_air_wpa"
[28] "comp_yac_epa"
[29] "comp_yac_wpa"
[30] "complete_pass"
[31] "cp"
[32] "cpoe"
[33] "def_wp"
[34] "defensive_extra_point_attempt"
[35] "defensive_extra_point_conv"
[36] "defensive_two_point_attempt"
[37] "defensive_two_point_conv"
[38] "defteam"
[39] "defteam_score"
[40] "defteam_score_post"
[41] "defteam_timeouts_remaining"
[42] "desc"
[43] "div_game"
[44] "down"
[45] "drive"
[46] "drive_end_transition"
[47] "drive_end_yard_line"
[48] "drive_ended_with_score"
[49] "drive_first_downs"
```

```
[50] "drive_game_clock_end"
[51] "drive_game_clock_start"
[52] "drive_inside20"
[53] "drive_play_count"
[54] "drive_play_id_ended"
[55] "drive_play_id_started"
[56] "drive_quarter_end"
[57] "drive_quarter_start"
[58] "drive_real_start_time"
[59] "drive_start_transition"
[60] "drive_start_yard_line"
[61] "drive_time_of_possession"
[62] "drive_yards_penalized"
[63] "end_clock_time"
[64] "end_yard_line"
[65] "ep"
[66] "epa"
[67] "extra_point_attempt"
[68] "extra_point_prob"
[69] "extra_point_result"
[70] "fantasy"
[71] "fantasy_id"
[72] "fantasy_player_id"
[73] "fantasy_player_name"
[74] "fg_prob"
[75] "field_goal_attempt"
[76] "field_goal_result"
[77] "first_down"
[78] "first_down_pass"
[79] "first_down_penalty"
[80] "first_down_rush"
[81] "fixed_drive"
[82] "fixed_drive_result"
[83] "forced_fumble_player_1_player_id"
[84] "forced_fumble_player_1_player_name"
[85] "forced_fumble_player_1_team"
[86] "forced_fumble_player_2_player_id"
[87] "forced_fumble_player_2_player_name"
[88] "forced_fumble_player_2_team"
[89] "fourth_down_converted"
[90] "fourth_down_failed"
[91] "fumble"
[92] "fumble_forced"
[93] "fumble_lost"
[94] "fumble_not_forced"
```

```
 [95] "fumble_out_of_bounds"
 [96] "fumble_recovery_1_player_id"
 [97] "fumble_recovery_1_player_name"
 [98] "fumble_recovery_1_team"
 [99] "fumble_recovery_1_yards"
[100] "fumble_recovery_2_player_id"
[101] "fumble_recovery_2_player_name"
[102] "fumble_recovery_2_team"
[103] "fumble_recovery_2_yards"
[104] "fumbled_1_player_id"
[105] "fumbled_1_player_name"
[106] "fumbled_1_team"
[107] "fumbled_2_player_id"
[108] "fumbled_2_player_name"
[109] "fumbled_2_team"
[110] "game_date"
[111] "game_half"
[112] "game_id"
[113] "game_seconds_remaining"
[114] "game_stadium"
[115] "goal_to_go"
[116] "half_sack_1_player_id"
[117] "half_sack_1_player_name"
[118] "half_sack_2_player_id"
[119] "half_sack_2_player_name"
[120] "half_seconds_remaining"
[121] "home_coach"
[122] "home_opening_kickoff"
[123] "home_score"
[124] "home_team"
[125] "home_timeouts_remaining"
[126] "home_wp"
[127] "home_wp_post"
[128] "id"
[129] "incomplete_pass"
[130] "interception"
[131] "interception_player_id"
[132] "interception_player_name"
[133] "jersey_number"
[134] "kick_distance"
[135] "kicker_player_id"
[136] "kicker_player_name"
[137] "kickoff_attempt"
[138] "kickoff_downed"
[139] "kickoff_fair_catch"
```

```
[140] "kickoff_in_endzone"
[141] "kickoff_inside_twenty"
[142] "kickoff_out_of_bounds"
[143] "kickoff_returner_player_id"
[144] "kickoff_returner_player_name"
[145] "lateral_interception_player_id"
[146] "lateral_interception_player_name"
[147] "lateral_kickoff_returner_player_id"
[148] "lateral_kickoff_returner_player_name"
[149] "lateral_punt_returner_player_id"
[150] "lateral_punt_returner_player_name"
[151] "lateral_receiver_player_id"
[152] "lateral_receiver_player_name"
[153] "lateral_receiving_yards"
[154] "lateral_reception"
[155] "lateral_recovery"
[156] "lateral_return"
[157] "lateral_rush"
[158] "lateral_rusher_player_id"
[159] "lateral_rusher_player_name"
[160] "lateral_rushing_yards"
[161] "lateral_sack_player_id"
[162] "lateral_sack_player_name"
[163] "location"
[164] "name"
[165] "nfl_api_id"
[166] "no_huddle"
[167] "no_score_prob"
[168] "old_game_id"
[169] "opp_fg_prob"
[170] "opp_safety_prob"
[171] "opp_td_prob"
[172] "order_sequence"
[173] "out_of_bounds"
[174] "own_kickoff_recovery"
[175] "own_kickoff_recovery_player_id"
[176] "own_kickoff_recovery_player_name"
[177] "own_kickoff_recovery_td"
[178] "pass"
[179] "pass_attempt"
[180] "pass_defense_1_player_id"
[181] "pass_defense_1_player_name"
[182] "pass_defense_2_player_id"
[183] "pass_defense_2_player_name"
[184] "pass_length"
```

```
[185] "pass_location"
[186] "pass_oe"
[187] "pass_touchdown"
[188] "passer"
[189] "passer_id"
[190] "passer_jersey_number"
[191] "passer_player_id"
[192] "passer_player_name"
[193] "passing_yards"
[194] "penalty"
[195] "penalty_player_id"
[196] "penalty_player_name"
[197] "penalty_team"
[198] "penalty_type"
[199] "penalty_yards"
[200] "play"
[201] "play_clock"
[202] "play_deleted"
[203] "play_id"
[204] "play_type"
[205] "play_type_nfl"
[206] "posteam"
[207] "posteam_score"
[208] "posteam_score_post"
[209] "posteam_timeouts_remaining"
[210] "posteam_type"
[211] "punt_attempt"
[212] "punt_blocked"
[213] "punt_downed"
[214] "punt_fair_catch"
[215] "punt_in_endzone"
[216] "punt_inside_twenty"
[217] "punt_out_of_bounds"
[218] "punt_returner_player_id"
[219] "punt_returner_player_name"
[220] "punter_player_id"
[221] "punter_player_name"
[222] "qb_dropback"
[223] "qb_epa"
[224] "qb_hit"
[225] "qb_hit_1_player_id"
[226] "qb_hit_1_player_name"
[227] "qb_hit_2_player_id"
[228] "qb_hit_2_player_name"
[229] "qb_kneel"
```

```
[230] "qb_scramble"
[231] "qb_spike"
[232] "qtr"
[233] "quarter_end"
[234] "quarter_seconds_remaining"
[235] "receiver"
[236] "receiver_id"
[237] "receiver_jersey_number"
[238] "receiver_player_id"
[239] "receiver_player_name"
[240] "receiving_yards"
[241] "replay_or_challenge"
[242] "replay_or_challenge_result"
[243] "result"
[244] "return_team"
[245] "return_touchdown"
[246] "return_yards"
[247] "roof"
[248] "run_gap"
[249] "run_location"
[250] "rush"
[251] "rush_attempt"
[252] "rush_touchdown"
[253] "rusher"
[254] "rusher_id"
[255] "rusher_jersey_number"
[256] "rusher_player_id"
[257] "rusher_player_name"
[258] "rushing_yards"
[259] "sack"
[260] "sack_player_id"
[261] "sack_player_name"
[262] "safety"
[263] "safety_player_id"
[264] "safety_player_name"
[265] "safety_prob"
[266] "score_differential"
[267] "score_differential_post"
[268] "season"
[269] "season_type"
[270] "series"
[271] "series_result"
[272] "series_success"
[273] "shotgun"
[274] "side_of_field"
```

```
[275] "solo_tackle"
[276] "solo_tackle_1_player_id"
[277] "solo_tackle_1_player_name"
[278] "solo_tackle_1_team"
[279] "solo_tackle_2_player_id"
[280] "solo_tackle_2_player_name"
[281] "solo_tackle_2_team"
[282] "sp"
[283] "special"
[284] "special_teams_play"
[285] "spread_line"
[286] "st_play_type"
[287] "stadium"
[288] "stadium_id"
[289] "start_time"
[290] "success"
[291] "surface"
[292] "tackle_for_loss_1_player_id"
[293] "tackle_for_loss_1_player_name"
[294] "tackle_for_loss_2_player_id"
[295] "tackle_for_loss_2_player_name"
[296] "tackle_with_assist"
[297] "tackle_with_assist_1_player_id"
[298] "tackle_with_assist_1_player_name"
[299] "tackle_with_assist_1_team"
[300] "tackle_with_assist_2_player_id"
[301] "tackle_with_assist_2_player_name"
[302] "tackle_with_assist_2_team"
[303] "tackled_for_loss"
[304] "td_player_id"
[305] "td_player_name"
[306] "td_prob"
[307] "td_team"
[308] "temp"
[309] "third_down_converted"
[310] "third_down_failed"
[311] "time"
[312] "time_of_day"
[313] "timeout"
[314] "timeout_team"
[315] "total"
[316] "total_away_comp_air_epa"
[317] "total_away_comp_air_wpa"
[318] "total_away_comp_yac_epa"
[319] "total_away_comp_yac_wpa"
```

```
[320] "total_away_epa"
[321] "total_away_pass_epa"
[322] "total_away_pass_wpa"
[323] "total_away_raw_air_epa"
[324] "total_away_raw_air_wpa"
[325] "total_away_raw_yac_epa"
[326] "total_away_raw_yac_wpa"
[327] "total_away_rush_epa"
[328] "total_away_rush_wpa"
[329] "total_away_score"
[330] "total_home_comp_air_epa"
[331] "total_home_comp_air_wpa"
[332] "total_home_comp_yac_epa"
[333] "total_home_comp_yac_wpa"
[334] "total_home_epa"
[335] "total_home_pass_epa"
[336] "total_home_pass_wpa"
[337] "total_home_raw_air_epa"
[338] "total_home_raw_air_wpa"
[339] "total_home_raw_yac_epa"
[340] "total_home_raw_yac_wpa"
[341] "total_home_rush_epa"
[342] "total_home_rush_wpa"
[343] "total_home_score"
[344] "total_line"
[345] "touchback"
[346] "touchdown"
[347] "two_point_attempt"
[348] "two_point_conv_result"
[349] "two_point_conversion_prob"
[350] "vegas_home_wp"
[351] "vegas_home_wpa"
[352] "vegas_wp"
[353] "vegas_wpa"
[354] "weather"
[355] "week"
[356] "wind"
[357] "wp"
[358] "wpa"
[359] "xpass"
[360] "xyac_epa"
[361] "xyac_fd"
[362] "xyac_mean_yardage"
[363] "xyac_median_yardage"
[364] "xyac_success"
```

```
[365] "yac_epa"
[366] "yac_wpa"
[367] "yardline_100"
[368] "yards_after_catch"
[369] "yards_gained"
[370] "ydsnet"
[371] "ydstogo"
[372] "yrdln"
```

The amount of information contained in the **nflverse** play-by-play data can be overwhelming. Luckily, the **nflreadr** website includes a searchable directory of all the variables with a brief description of what each one means. You can visit that here: nflreadr Field Descriptions.

We can recreate our examination of 2018 adjusted pass yards per attempt to just those passes on 3rd down with 10 or less yards to go by running the following code:

```
pbp <- nflreadr::load_pbp(2018) %>%
  filter(season_type == "REG")

adjusted_yards <- pbp %>%
  group_by(passer_player_name) %>%
  filter(down == 3 & ydstogo <= 10) %>%
  filter(complete_pass == 1 | incomplete_pass == 1 |
           interception == 1 &
           !is.na(down)) %>%
  summarize(
    total_attempts = n(),
    adj_yards = (sum(yards_gained) + 20 *
                   sum(touchdown == 1) - 45 *
                   sum(interception == 1)) /
                    total_attempts) %>%
  filter(total_attempts >= 50) %>%
  arrange(-adj_yards)

adjusted_yards
```

```
# A tibble: 32 x 3
   passer_player_name total_attempts adj_yards
   <chr>                       <int>     <dbl>
 1 A.Rodgers                      98     11.3
 2 P.Mahomes                     102     10.3
 3 R.Wilson                      104     10.3
 4 E.Manning                     119      9.18
```

```
 5 M.Mariota                      79        8.43
 6 M.Ryan                        117        8.26
 7 J.Winston                        66        8.05
 8 D.Brees                        97        7.99
 9 D.Prescott                    104        7.87
10 J.Flacco                       74        7.84
# i 22 more rows
```

Because the `load_pbp()` data is not pre-aggregated, we must do a bit of the work ourselves before we are able to achieve our answer. To start, we load the 2018 play-by-play data into a data frame titled `pbp` but using the `load_pbp()` function with the `season` argument set to "2018." As well, we use the `filter()` function to make sure the data gathered into our `pbp` data frame is only the regular season statistics. Once the play-by-play is loaded, we create a data frame off of it titled `adjusted_yards`, wherein we use `group_by()` to calculate the metric for each individual quarterback, then ensure that we are collecting the play-by-play data for only those plays that took place on 3rd down with ten or less yards to go.

We then use a second `filter()` to gather those plays, where `complete_pass == 1` *or* `incomplete_pass == 1` *or* `interception == 1` *and* the `down` is not missing (which usually indicates a 2-point conversion attempt). As a result, each quarterback has an exact match to the number of complete passes, attempts, and interceptions as found in the official statistics of the season.

Despite leading the league in adjusted pass yard per attempt in 2018, Mahomes finished a whole yard behind Aaron Rodgers when exploring the same metric on 3rd down with 10 or less yards to go (and was tied for second place with Russel Wilson).

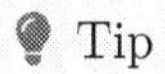 Tip

Why is it necessary to use such a long list of specifications in the `filter()` function in order to match the official QB statistics found elsewhere?

That is a good question, especially since there are other variables within the play-by-play data that would seemingly return the correct results (such as `pass == 1` or `play_type == "pass"`. We can view the differences in all three methods by doing the following.

```
pbp <- nflreadr::load_pbp(2018) %>%
  filter(season_type == "REG")

pbp %>%
  filter(complete_pass == 1 | incomplete_pass == 1 |
           interception == 1 &
           !is.na(down)) %>%
  filter(passer_player_name == "P.Mahomes") %>%
  summarize(total_attempts = n())
```

```
# A tibble: 1 x 1
  total_attempts
           <int>
1            580
```

```
pass_numbers <- pbp %>%
  filter(passer_player_name == "P.Mahomes") %>%
  summarize(
    using_pass = sum(pass == 1, na.rm = TRUE),
    using_playtype = sum(play_type == "pass"),
    using_filter = 580)

pass_numbers
```

```
# A tibble: 1 x 3
  using_pass using_playtype using_filter
       <int>          <int>        <dbl>
1        606            606          580
```

We know that Patrick Mahomes attempted 580 passes during the 2022 regular season. We can match that number exactly using the `filter()` method for `complete_pass`, `incomplete_pass`, `interception`, and removing plays with a missing `down` observation. When trying to replicate this number using either the `pass == 1` variable in the play-by-play data or `play_type == "pass"`, we get just over 20 more passes than expected.

The main reason for this is the inclusion of the `qb_spike`, `qb_scramble`, and `sack` variables. We can actually use the code above that was used to create the `pass_numbers` data frame, add an additional `filter()` to account for these, and get the correct number of passing attempts for each method.

```
pass_numbers_correct <- pbp %>%
  filter(passer_player_name == "P.Mahomes") %>%
  filter(qb_spike == 0 & qb_scramble == 0 & sack == 0) %>%
  summarize(
    using_pass = sum(pass == 1, na.rm = TRUE),
    using_playtype = sum(play_type == "pass"),
    using_filter = 580)

pass_numbers_correct
```

```
# A tibble: 1 x 3
  using_pass using_playtype using_filter
       <int>          <int>        <dbl>
1        580            580          580
```

By using `filter()` to remove any play that included a QB spike, a QB scramble, or a sack, we get 580 attempts when using both `pass == 1` and `play_type == "pass"` which replicates the official statistics from the 2018 season for Patrick Mahomes.

What about the use of `passer` vs. `passer_player_name`?

It is also important to remember that you will receive differing numbers based on your use of `passer` and `passer_player_name`. The `passer` variable is created internally by the `nflverse` system and is used to "mimic" the statistics without spikes, scrambles, or sacks in the data. The `passer_player_name` variable comes from the official statistics and inherently includes this information.

The following, with the first method using `group_by(passer)` and the second using `group_by(passer_player_name)` returns differing results.

```
passer_grouping <- pbp %>%
  filter(complete_pass == 1 |
           incomplete_pass == 1 |
           interception == 1 & !is.na(down)) %>%
  group_by(passer) %>%
  summarize(total_attempts = n()) %>%
  arrange(-total_attempts) %>%
  slice(1:5)

passer_player_grouping <- pbp %>%
  filter(complete_pass == 1 |
           incomplete_pass == 1 |
           interception == 1 & !is.na(down)) %>%
  group_by(passer_player_name) %>%
  summarize(total_attempts = n()) %>%
  arrange(-total_attempts) %>%
  slice(1:5)

passer_grouping
```

```
# A tibble: 5 x 2
  passer             total_attempts
  <chr>                       <int>
1 B.Roethlisberger              672
2 A.Luck                        637
3 M.Ryan                        607
4 K.Cousins                     603
5 A.Rodgers                     595
```

```
passer_player_grouping
```

```
# A tibble: 5 x 2
  passer_player_name total_attempts
  <chr>                       <int>
1 B.Roethlisberger              675
2 A.Luck                        639
3 M.Ryan                        608
4 K.Cousins                     606
5 A.Rodgers                     597
```

When grouping by `passer`, we get a total attempts of 672 for Ben Roethlisberger. That number increases to 675 when grouping by `passer_player_name`. Again, this is because the `passer` variable is created by the `nflverse` and automatically removes spikes, scrambles, and sacks while `passer_player_name` includes all three.

We can run the following code to determine where the difference of three passing attempts is coming from between `passer` and `passer_player_name`.

```
roethlisberger_difference <- pbp %>%
  filter(complete_pass == 1 |
           incomplete_pass == 1 |
           interception == 1 & !is.na(down)) %>%
  group_by(passer_player_name) %>%
  summarize(total_attempts = n(),
            spikes = sum(qb_spike == 1),
            scramble = sum(qb_scramble == 1),
            sacks = sum(sack == 1)) %>%
  arrange(-total_attempts) %>%
  slice(1:5)

roethlisberger_difference
```

```
# A tibble: 5 x 5
  passer_player_name total_attempts spikes scramble sacks
  <chr>                       <int>  <int>    <int> <int>
1 B.Roethlisberger              675      3        0     0
2 A.Luck                        639      2        0     0
3 M.Ryan                        608      1        0     0
4 K.Cousins                     606      3        0     0
5 A.Rodgers                     597      2        0     0
```

In the case of Roethlisberger, the three pass attempt difference was the result of the `passer_player_name` grouping including three QB spikes in the data. **Aside from when attempting to replicate the official statistics, it is better to use just `passer` as it removes those instances where a QB spike, scramble, or sack my skew the results of your data.**

To continue working with `load_pbp()` data, let's create a metric that examines a QB's aggressiveness on 3rd down passing attempts. The metric is designed to determine which QBs in the NFL are most aggressive in 3rd down situations by gauging how often they throw the ball to, or pass, the first down line. It is an interesting metric to explore as, just like many metrics in the NFL, not all air yards are created equal. For example, eight air yards on 1st and 10 are less valuable than the same eight air yards on 3rd and 5.

```
pbp <- nflreadr::load_pbp(2022)

aggressiveness <- pbp %>%
  filter(complete_pass == 1 |
           incomplete_pass == 1 |
           interception == 1 &
           !is.na(down)) %>%
  filter(down == 3 & ydstogo >= 5 & ydstogo <= 10) %>%
  group_by(passer) %>%
  summarize(
    team = last(posteam),
    total = n(),
    aggressive = sum(air_yards >= ydstogo, na.rm = TRUE),
    percentage = aggressive / total) %>%
  filter(total >= 50) %>%
  arrange(-percentage) %>%
  slice(1:10)

aggressiveness
```

```
# A tibble: 10 x 5
   passer     team  total aggressive percentage
   <chr>      <chr> <int>      <int>      <dbl>
 1 P.Mahomes  KC       75         55      0.733
 2 K.Pickett  PIT      54         36      0.667
 3 M.Jones    NE       56         36      0.643
 4 D.Carr     LV       74         47      0.635
 5 R.Wilson   DEN      63         40      0.635
 6 G.Smith    SEA      65         41      0.631
 7 A.Dalton   NO       55         34      0.618
 8 J.Hurts    PHI      65         40      0.615
 9 K.Cousins  MIN      84         51      0.607
10 J.Allen    BUF      61         35      0.574
```

Following our above example, we use `filter()` on the play-by-play data to calculate the total number of attempts for each quarterback on 3rd down when there were between 5 and 10 yards to go for a first down. Because we want to make sure we are not gathering attempts with spikes, scrambles, etc., we use `passer` in the `group_by()` function and then calculate the `total` attempts for each quarterback. After, we find the total number of times each quarterback was `aggressive` by finding the sum of attempts, where the `air_yards` of the pass were greater than or equal to the required `ydstogo`. The two numbers are then divided (`aggressive / total`) to get each quarterback's percentage.

As you can see in the output of `aggressiveness`, Mahomes was the most aggressive quarterback in 3rd down passing situation in the 2022 season, passing to, or beyond, the line of gain just over 73% of the time.

One item to consider, however, is the concept of "garbage time." Are the above results, perhaps, skewed by including quarterbacks that were forced to work the ball downfield while attempting a game-winning drive?

3.2.1 QB Aggressiveness: Filtering for "Garbage Time?"

To examine the impact of "garbage time" statistics, we can add an additional `filter()` that removes those plays that take place within the last 2-minutes of the half and when the probability of winning for either team is over 95% or under 5%.

Let's add the "garbage time" filter to the code we've already prepared:

```
aggressiveness_garbage <- pbp %>%
  filter(complete_pass == 1 |
           incomplete_pass == 1 |
           interception == 1 &
           !is.na(down)) %>%
  filter(down == 3 & ydstogo >= 5 & ydstogo <= 10) %>%
  filter(wp > .05 & wp < .95 &
           half_seconds_remaining > 120) %>%
  group_by(passer) %>%
  summarize(
    team = last(posteam),
    total = n(),
    aggressive = sum(air_yards >= ydstogo, na.rm = TRUE),
    percentage = aggressive / total) %>%
  filter(total >= 50) %>%
  arrange(-percentage) %>%
  slice(1:10)

aggressiveness_garbage
```

```
# A tibble: 10 x 5
   passer     team  total aggressive percentage
   <chr>      <chr> <int>      <int>      <dbl>
 1 P.Mahomes  KC       67         48      0.716
 2 K.Cousins  MIN      63         41      0.651
 3 G.Smith    SEA      50         32      0.64
```

```
 4 D.Carr      LV          61          38          0.623
 5 D.Prescott DAL          52          28          0.538
 6 J.Herbert  LAC          71          38          0.535
 7 J.Burrow   CIN          67          35          0.522
 8 T.Lawrence JAX          69          35          0.507
 9 T.Brady    TB           70          35          0.5
10 A.Rodgers  GB           53          26          0.491
```

We are now using the same code, but have included three new items to the `filter()`. First, we are stipulating that, aside from the down and distance inclusion, we only want those plays that occurred when the offense's win probability (the `wp` variable) was between 5% and 95%, as well as ensuring that the plays did not happen after the two-minute warning of either half.

The decision on range of the win probability numbers is very much a personal preference. When `nflfastR` was first released, analyst often used a 20–80% range for the win probability cutoff point. However, Sebastian Carl – one of the creators of the `nflverse` explained in the package's Discord:

> Sebastian Carl: "I am generally very conservative with filtering plays using wp. Especially the vegas wp model can reach >85% probs early in the game because it incorporates market lines. I never understood the 20% <= wp <= 80% "garbage time" filter. This is removing a ton of plays. My general advice is a lower boundary of something around 5% (i.e., 5% <= wp <= 95%).

Ben Baldwin followed up on Carl's thoughts:

> Ben Baldwin: "agree with this. 20–80% should only be used as a filter for looking at how run-heavy a team is (because outside of this range is when teams change behavior a lot). and possibly how teams behave on 4th downs. but not for team or player performance."

Based on that advice, I typically stick to the 5–95% range when filtering for win probability in the `nflverse` play-by-play data. And, in this case, it did have an impact. In fact, when accounting for "garbage time," Kenny Pickett went from being the second most aggressive QB in the league to not even being in the top ten. Conversely, Kirk Cousin went from being the ninth most aggressive quarterback to the second when accounting for win probability and the time left in the half.

That said, I more often than not do not concern myself with removing "garbage time" statistics. Despite the robust amount of data provided by the `nflverse` play-by-play function, the information still lacks great amount of granularity and, because of this, I believe removing "garbage time" plays often does more harm than good in the data analysis process.

3.3 More Examples Using load_pbp() Data

3.3.1 QB Air Yards EPA by Down

To continue working with quarterback information using the `load_pbp()` function, we can explore each quarterback's air yards EPA for 1st, 2nd, and 3rd down.

```
pbp <- nflreadr::load_pbp(2022) %>%
  filter(season_type == "REG")

qb_ay_by_down <- pbp %>%
  filter(complete_pass == 1 |
           incomplete_pass == 1 |
           interception == 1 &
           !is.na(down)) %>%
  filter(down <= 3) %>%
  group_by(passer, down) %>%
  summarize(
    attempts = n(),
    team = last(posteam),
    mean_airepa = mean(air_epa, na.rm = TRUE)) %>%
  filter(attempts >= 80) %>%
  arrange(-mean_airepa)

qb_ay_by_down
```

Josh Allen, Dak Prescott, and Tua Tagovailoa were all outstanding on 3rd down during the 2022 season, all recording over 1.00 in mean air yards EPA.

3.3.2 Measuring Impact of Offensive Line

Brian Burke – prior to moving to ESPN – used to host his own blog where he showcased his analytics work. In November of 2014, he wrote a post that detailed his process in determining how to value an offensive line's performance over the course of an NFL season.

The process of making such a valuation is understandably difficult as, as Burke pointed out, an offensive line's performance on the field is often characterized by the overall absence of stats. The less sacks, short yardage plays, tackles for loss, and quarterback scrambles – for example – the better. But how can the performance of an offensive line be quantified based on statistics that are absent?

To do so, Burke devised a rather ingenious method that is simple at its core: to measure the value of an offensive line's play, we must determine the impact in which the opposing defensive line had on the game.

In his 2014 work on the topic, Burke used quarterback sacks, tackles for losses, short gains, tipped passes, and quarterback hits to provide a quantifiable valuation to a defensive line to then calculate the opposing valuation of the offensive line. Building upon this philosophy, we can build the same sort of study using the publicly available data in the `nflverse` play-by-play data by first using the following standard metrics:

1. sacks
2. QB hits
3. tackles for loss
4. yards gained less than/equal to 2
5. forcing a QB scramble

To add more context, we can also create two new variables in the data that dictate whether a `qb_hit` and `incomplete_pass` are present in the same play, and the same for `qb_hit` and `interception`.

```
pbp <- nflreadr::load_pbp(2022) %>%
  filter(season_type == "REG") %>%
  mutate(qbh_inc = ifelse(qb_hit == 1 &
          incomplete_pass == 1, 1,0),
         qb_int = ifelse(qb_hit == 1 &
          interception == 1, 1,0))

pitt_plays <- pbp %>%
  filter(posteam == "PIT") %>%
  group_by(week) %>%
  summarize(
    total_plays = sum(play_type == "run" |
                        play_type == "pass", na.rm = TRUE))

pitt_line_value <- pbp %>%
  filter(sack == 1 |
           tackled_for_loss == 1 |
           yards_gained <= 2 |
           qb_scramble == 1 |
```

```
            qb_hit == 1 |
            qbh_inc == 1 |
            qb_int == 1) %>%
  filter(posteam == "PIT") %>%
  group_by(posteam, week) %>%
  left_join(pitt_plays, by = c("week" = "week")) %>%
  summarize(opponent = unique(defteam),
            sum_wpa = sum(wpa, na.rm = TRUE),
            avg_wpa = (sum_wpa / unique(total_plays) * 100))
```

However, as Burke pointed out in his initial study of the topic, calculating the number for just Pittsburgh, as above, does not provide enough information to draw a meaningful conclusion because even the most elite offensive lines in the NFL cannot avoid negative-pointed plays. In order to build this assumption into the data, we can gather the same information as above, but for the entire NFL minus the Steelers, and then calculate the difference in the Steelers' weekly WPA to the league-wide average WPA.

```
nfl_plays <- pbp %>%
  filter(posteam != "PIT") %>%
  group_by(week, posteam) %>%
  summarize(total_plays = sum(play_type == "run" |
                               play_type == "pass",
                             na.rm = TRUE))

nfl_line_value <- pbp %>%
  filter(posteam != "PIT") %>%
  filter(sack == 1 |
           tackled_for_loss == 1 |
           yards_gained <= 2 |
           qb_scramble == 1 |
           qb_hit == 1 |
           qbh_inc == 1 |
           qb_int == 1) %>%
  left_join(nfl_plays,
          by = c("posteam" = "posteam", "week" = "week")) %>%
  group_by(week) %>%
  mutate(nfl_weekly_plays = sum(unique(total_plays))) %>%
  summarize(nfl_sum_wpa = sum(wpa, na.rm = TRUE),
            nfl_avg_wpa = (nfl_sum_wpa /
            unique(nfl_weekly_plays) * 100))
```

We can now merge the NFL data back into the Steelers data to conduct the final calculation.

```
pitt_line_value <- pitt_line_value %>%
  left_join(nfl_line_value, by = c("week" = "week"))

pitt_line_value <- pitt_line_value %>%
  mutate(final_value = avg_wpa - nfl_avg_wpa) %>%
  select(week, opponent, final_value)

pitt_line_value %>%
  print(n = 17)
```

The output provides a quantitative value of the Steelers' offensive line performance over the course of the season in the `final_value` column. We've structured the data so that we can make the argument that the Steelers' offensive line, in week 1, provided 1.55% more WPA than the rest of the NFL. In week 11, against New Orleans, the offensive line added a 1.11% chance of winning the game compared to all the other offensive line performances that specific week.

Conversely, negative numbers indicate that the play of the offensive line took away that percentage amount toward the probability of winning.

Unsurprisingly, the numbers across the board for the Steelers' offensive line remain small, indicated that the group's performance provided little to no help in earning victories for Pittsburgh.

3.4 Retrieving and Working with Data for Multiple Seasons

In the case of both `load_pbp()` and `load_player_stats()`, it is possible to load data over multiple seasons.

In our above example calculating average air yard per attempt, it is important to note that Russell Wilson's league-leading average of 9.89 air yards per attempt is calculated using *all* passing attempts, meaning pass attempts that were both complete and incomplete.

In our first example of working with data across multiple seasons, let's examine average air yards for only completed passes. To begin, we will retrieve the play-by-play data for the last five seasons:

```
ay_five_years <- nflreadr::load_pbp(2017:2022)
```

To retrieve multiple seasons of data, a colon : is placed between the years that you want. When you run the code, **nflreadr** will output the data to include the play-by-play data starting with the oldest season (in this case, the 2017 NFL season).

Once you have the data collected, we can run code that looks quite similar to our code above that explored 2021's air yards per attempt leaders using `load_player_stats()`. In this case, however, we are including an additional `filter()` to gather those passing attempts that resulted *only* in complete passes:

```
average_airyards <- ay_five_years %>%
  group_by(passer_id) %>%
  filter(season_type == "REG" & complete_pass == 1) %>%
  summarize(player = first(passer_player_name),
            completions = sum(complete_pass),
            air.yards = sum(air_yards),
            average = air.yards / completions) %>%
  filter(completions >= 1000) %>%
  arrange(-average)

average_airyards
```

```
# A tibble: 28 x 5
   passer_id  player      completions air.yards average
   <chr>      <chr>             <dbl>     <dbl>   <dbl>
 1 00-0031503 J.Winston          1081      8830    8.17
 2 00-0033537 D.Watson           1285      9019    7.02
 3 00-0034857 J.Allen            1604     10912    6.80
 4 00-0029263 R.Wilson           1895     12742    6.72
 5 00-0034796 L.Jackson          1055      7046    6.68
 6 00-0026143 M.Ryan             2263     14694    6.49
 7 00-0033077 D.Prescott         1874     12105    6.46
 8 00-0034855 B.Mayfield         1386      8889    6.41
 9 00-0029701 R.Tannehill        1261      8043    6.38
10 00-0026498 M.Stafford         1874     11763    6.28
# i 18 more rows
```

Of those QBs with at least 1,000 complete passes since the 2017 season, Jameis Winston has the highest average air yards per complete pass at 8.17.

3.5 Working with the Various nflreadr Functions

The `nflreadr` package comes with a multitude of "under the hood" functions designed to provide you with supplemental data, items for data visualization, and utilities for efficiently collecting and storing the data on your system. You can view the entire list of these options using the `ls` to output all the objects in the package.

```
ls("package:nflreadr")
```

```
 [1] "clean_homeaway"             "clean_player_names"
 [3] "clean_team_abbrs"           "clear_cache"
 [5] "csv_from_url"               "dictionary_combine"
 [7] "dictionary_contracts"       "dictionary_depth_charts"
 [9] "dictionary_draft_picks"     "dictionary_espn_qbr"
[11] "dictionary_ff_opportunity"  "dictionary_ff_playerids"
[13] "dictionary_ff_rankings"     "dictionary_injuries"
[15] "dictionary_nextgen_stats"   "dictionary_participation"
[17] "dictionary_pbp"             "dictionary_pfr_passing"
[19] "dictionary_player_stats"    "dictionary_rosters"
[21] "dictionary_schedules"       "dictionary_snap_counts"
[23] "dictionary_trades"          "ffverse_sitrep"
[25] "get_current_season"         "get_current_week"
[27] "get_latest_season"          "join_coalesce"
[29] "load_combine"               "load_contracts"
[31] "load_depth_charts"          "load_draft_picks"
[33] "load_espn_qbr"              "load_ff_opportunity"
[35] "load_ff_playerids"          "load_ff_rankings"
[37] "load_from_url"              "load_injuries"
[39] "load_nextgen_stats"         "load_officials"
[41] "load_participation"         "load_pbp"
[43] "load_pfr_advstats"          "load_pfr_passing"
[45] "load_player_stats"          "load_players"
[47] "load_rosters"               "load_rosters_weekly"
[49] "load_schedules"             "load_snap_counts"
[51] "load_teams"                 "load_trades"
[53] "most_recent_season"         "nflverse_download"
```

```
[55] "nflverse_game_id"          "nflverse_releases"
[57] "nflverse_sitrep"           "parquet_from_url"
[59] "player_name_mapping"       "progressively"
[61] "qs_from_url"               "raw_from_url"
[63] "rbindlist_with_attrs"      "rds_from_url"
[65] "team_abbr_mapping"         "team_abbr_mapping_norelocate"
```

Going forward in this chapter, we will be exploring specific use cases for the functions provided by `nflreadr` – but not all of them. For example, the `dictionary_` functions can more easily be used directly on the `nflreadr` website, where the package's maintainers keep a copy of each. Many, like the dictionary for play-by-play data, includes a search feature to allow you to quickly find how the variables you are looking for is provided in the column name. Others, like the `clear_cache()` function is used only when you want to wipe any memoized data stored by `nflreadr` – often needed if you are troubleshooting a pesky error message – or `join_coalesce()` which is an experimental function that is only used internally to help build player IDs into the data. The `load_pbp()` and `load_player_stats()` function will also not be covered here, as the first portion of this chapter explored the use of both in great detail. We will briefly discuss the use of `load_players()` and `load_rosters()` but a more detailed discussion of each is provided in Chapter 4: Data Visualization with NFL Analytics. The remaining functions will be presented in the order provided on the `nflreadr` function reference website.

3.5.1 The `load_participation()` Function

The `load_participation()` function allows us to create a data frame of player participation data dating back to 2016 with an option to infuse that information into the existing `nflreadr` play-by-play data. The resulting data frame, when not including play-by-play data, includes information pertaining to: the individual game ID, the play ID, the possession team, what formation the offense was in, the layout of the offensive personnel, how many defenders were in the box, the defensive personnel, the number of rushers on pass players, and the unique ID number for each player on the field for that specific play.

Despite the `load_participation()` function being one of the newest editions to the `nflreadr` package, it is already being used to create contextual analysis regarding a team's use of its players out out formations. For example, Joseph Hefner uses the data to create tables (built with the `gt` and `gtExtras` packages) that calculates not only each player's rate of usage out of different personnel packages, but how the team's EPA per play and pass rate fluctuate with each. In the spirit of R's open-source nature, he also created a Team Formation ShinyApp that allows anybody to explore the data and output the results in .png format.

To build a data frame of 2022 participation data, that includes complete play-by-play, you must pass both the `season` and `include_pbp` argument with the `load_participation()` function.

```
participation <-
  nflreadr::load_participation(season = 2022,
        include_pbp = TRUE)

participation
```

```
# A tibble: 49,969 x 383
   nflverse_game_id play_id possession_team offense_formation
   <chr>              <int> <chr>           <chr>
 1 2022_01_BAL_NYJ        1 ""              <NA>
 2 2022_01_BAL_NYJ       43 "BAL"           <NA>
 3 2022_01_BAL_NYJ       68 "NYJ"           SINGLEBACK
 4 2022_01_BAL_NYJ       89 "NYJ"           SHOTGUN
 5 2022_01_BAL_NYJ      115 "NYJ"           SINGLEBACK
 6 2022_01_BAL_NYJ      136 "NYJ"           SHOTGUN
 7 2022_01_BAL_NYJ      172 "NYJ"           <NA>
 8 2022_01_BAL_NYJ      202 "BAL"           SINGLEBACK
 9 2022_01_BAL_NYJ      230 "BAL"           SHOTGUN
10 2022_01_BAL_NYJ      254 "BAL"           EMPTY
# i 49,959 more rows
# i 379 more variables: offense_personnel <chr>, ...
```

Rather than exploring the data at the league-level, let's take a micro-approach to the participation data and examine the 2022 Pittsburgh Steelers. It is important to note that the participation data, as gathered using the `load_participation()` function, is not fully prepped for immediate analysis as the `players_on_play`, `offense_players`, and `defense_players` information (which contains the unique player identification numbers) are separated by a delimiter (in this case, a semicolon) in what are called "concatenated strings" or "delimited strings." While this is a compact way to store the values in the core data frame, it does require us to clean the information through the "splitting" or "tokenizing" process. As you will see in the below code, we are going to place the unique identifiers into separate rows, based on formation, by utilizing the `separate_rows()` function in `tidyr`.

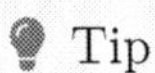 Tip

If you use `load_participation()` but set `include_pbp = FALSE` it is important to remember that the `posteam` variable that is part of the play-by-play data will not exist. Instead, the offensive team is indicated by the `possession_team` column.

If you gather the participation data *with* play-by-play information, you can use either `possession_team` or `posteam` for any filtering.

The same is not true when looking at defensive participation data. A defensive team variable is not provided without including the play-by-play in the data. Once set to `TRUE`, the defensive team is housed under the typical `defteam` column.

```
participation_split <- participation %>%
  filter(!is.na(offense_formation)) %>%
  filter(posteam == "PIT") %>%
  tidyr::separate_rows(offense_players, sep = ";") %>%
  group_by(offense_personnel, offense_players) %>%
  summarize(total = n()) %>%
  ungroup()

participation_split
```

```
# A tibble: 157 x 3
   offense_personnel offense_players total
   <chr>             <chr>           <int>
 1 1 RB, 1 TE, 3 WR  00-0032897          3
 2 1 RB, 1 TE, 3 WR  00-0033869        262
 3 1 RB, 1 TE, 3 WR  00-0034142         37
 4 1 RB, 1 TE, 3 WR  00-0034347        781
 5 1 RB, 1 TE, 3 WR  00-0034768        781
 6 1 RB, 1 TE, 3 WR  00-0034785        744
 7 1 RB, 1 TE, 3 WR  00-0034928        235
 8 1 RB, 1 TE, 3 WR  00-0035216        745
 9 1 RB, 1 TE, 3 WR  00-0035217         25
10 1 RB, 1 TE, 3 WR  00-0035222        258
# i 147 more rows
```

We are doing quite a few things in the above example:

1. **Those rows where the is an "NA"' value for offense_formation are removed.** The participation data includes information for kickoffs, extra points, field goals, punts, QB kneels, no plays, and more. It is possible to remove items using `play_type == "run" | play_type == "pass"` but such an approach, if done with the Steelers, results in one fake punt being included in the data (as the `play_type` was listed as "run"). If you wish to include plays such as fake punts, you do so by filtering for only specific play types.
2. **Only those rows that includes "PIT" as the posteam are included.**
3. **The separate_rows() function from tidyr is used to conduct the splitting of the concatenated players identifiers in the offense_players column.** The `separate_rows()` function needs just two arguments to complete the process – the column name and the delimiter (provided in the argument as `sep =`).
4. **The newly split data is then grouped by offense_personnel and offense_players in order to calculate the number of times each specific ID was associated with personnel package.** The resulting data frame lists the offense formation type for each player that participated in it, along with the player's respective participation count.

As is, the `load_participation()` data does not include any way to further identify players outside of the unique players IDs. Because of this, it is necessary to use the `load_rosters()` function to include this information (this process is included below in the `load_rosters()` section).

3.5.2 The load_rosters() and load_rosters_weekly() Functions

The `load_rosters()` function houses a multitude of useful tools for use in the `nflverse`. With player information dating back to 1920, `load_rosters()` will provide you with basic demographic information about each player in the NFL such as their name, birth date, height, weight, the college and high school they attended, and items regarding primary position and depth chart position.

Importantly, `load_rosters()` also provides the unique identifier for each player over nine different sources: the `gsis_id` (which is the core ID used in `nflverse` data), `sleeper_id`, `espn_id`, `yahoo_id`, `rotowire_id`, `pff_id`, `fantasy_data_id`, `sportradar_id`, and `pfr_id`. The different IDs become extremely useful when using data collected from outside the `nflverse`, or when combining data from two outside sources, as matching the information can be done by matching the various IDs as needed. As well, as outlined in Chapter 4: Data Visualization with NFL Analytics, the URLs for player headshots are included in `load_rosters()` data.

Let's return to our `participation_split` data frame that we just built while working with the `load_participation()` function. While the data is formatted and prepared for analysis, there is no information that allows for easy identification of each player (aside from unique IDs, which is not helpful). To correct this, we can bring in the 2022 Pittsburgh Steelers roster information.

```
rosters <- nflreadr::load_rosters(2022)
```

The only argument you must provide the `load_rosters()` function is the years in which you want to collect roster information (in this specific case, 2022). While we know we only want the Steelers' roster, it is not suggested to use `filter()` to gather any specific team at this point. Case in point: doing so with this example will result in the `gsis_id` *00-0036326* having an "NA" value for a name, despite having over 380 snaps in the 1 RB, 1TE, 3 WR offensive formation. The player associated with that `gsis_id` is Chase Claypool, who the Steelers traded to the Bears on November 1 of 2022. Because of this, no player name will be associated with that specific `gsis_id` on the Steelers' roster (with the end result being a missing value).

> **i Note**
>
> You may notice that the resulting data frame, titled `pit_roster`, has 80 players on the roster despite NFL teams only being permitted to have 53 active players at a time.
>
> This is because the information collected with `load_rosters()` includes not only active players, but those listed on the practice squad and injured reserve. Additionally, you may notice players listed as `R/Retired` (such as Stephon Tuitt in our data frame, who retired at the end of the 2022 season). In these situations, this `status` indicates that the player is listed as "reserved-retired" and that the team continues to hold the player's rights until the official expiration of their contract. In this case, Tuitt could not come out of retirement prior to his contract ending and play with another team unless first released or traded by Pittsburgh.

Despite the wealth of information in the roster information, only the `gsis_id` and the `full_name` is required to complete our `participation_split` data frame. Rather than bringing unnecessary information over during the merge, we can select just the two columns we need and then use `left_join()` to merge each player's name on the matching ID in the `offense_players` column and `gsis_id` from the roster information.

```
rosters <- nflreadr::load_rosters(2022)

rosters <- rosters %>%
  select(gsis_id, full_name)

participation_split <- participation_split %>%
  left_join(rosters, by = c("offense_players" = "gsis_id"))

participation_split
```

```
# A tibble: 157 x 4
   offense_personnel offense_players total full_name
   <chr>             <chr>           <int> <chr>
 1 1 RB, 1 TE, 3 WR  00-0032897          3 Derek Watt
 2 1 RB, 1 TE, 3 WR  00-0033869        262 Mitchell Trubisky
 3 1 RB, 1 TE, 3 WR  00-0034142         37 J.C. Hassenauer
 4 1 RB, 1 TE, 3 WR  00-0034347        781 James Daniels
 5 1 RB, 1 TE, 3 WR  00-0034768        781 Chukwuma Okorafor
 6 1 RB, 1 TE, 3 WR  00-0034785        744 Mason Cole
 7 1 RB, 1 TE, 3 WR  00-0034928        235 Steven Sims
 8 1 RB, 1 TE, 3 WR  00-0035216        745 Diontae Johnson
 9 1 RB, 1 TE, 3 WR  00-0035217         25 Benny Snell
10 1 RB, 1 TE, 3 WR  00-0035222        258 Zach Gentry
# i 147 more rows
```

Tip

The `load_rosters()` data is capable of providing data for multiple seasons at once. Through each season, a player's `gsis_id` will remain static. Despite this, when merging multiple years of participation data with multiple years of roster information, the data must be matched on the `season` variable was well. This process involves including `season` in both the participation data `group_by()` and the roster information, as seen below.

```
participation_multi_years <-
  load_participation(season = 2018:2022,
  include_pbp = TRUE)

participation_2018_2022 <- participation_multi_years %>%
  filter(!is.na(offense_formation)) %>%
  filter(posteam == "PIT") %>%
  tidyr::separate_rows(offense_players, sep = ";") %>%
  group_by(season, offense_personnel, offense_players) %>%
  summarize(total = n())

rosters_2018_2022 <- nflreadr::load_rosters(2018:2022) %>%
  select(season, gsis_id, full_name)

participation_2018_2022 <- participation_2018_2022 %>%
  left_join(rosters_2018_2022,
            by = c("season", "offense_players" = "gsis_id"))
```

If your analysis requires providing great granularity, the `load_rosters_weekly()` function provides the same information as `load_rosters()` but structures it in weekly format by season (or over multiple seasons, if needed).

3.5.3 The `load_teams()` Function

A tool heavily used in the data visualization process, the `load_teams()` function provides information pertaining to each NFL team's colors and logos, as well as providing a way to merge data frames that have differing values for teams.

More often than not, we will be using `left_join` to bring in team information (colors, logos, etc.) into a another data frame. We will utilize this function heavily in the example below.

However, it is important to note the one argument that `load_teams()` accepts: `current`. Depending on the time span of your data, it is possible that you have teams included prior to relation (like the St. Louis Rams). If you are sure that you do not have such a scenario, you can use `load_teams(current = TRUE)` which will bring in just the current 32 NFL teams.

However, if you need to include teams before expansion, you can use `load_teams(current = FALSE)` which will result in a data frame with 36 NFL teams.

3.5.4 The load_officials() Function

The `load_officials()` data will return data, from 2015 to present, outlining which officials were assigned to which game. The data also includes information regarding each referee's position, jersey number, and their official NFL ID number. Importantly, the data is structured to also include both a `game_id` and `game_key` that are sorted by `season` and `week`, allowing you to merge the information other data frames.

With the data, we can – for example – examine which NFL officiating crews called the most penalties during the 2022 NFL season. Doing so requires a bit of work in order to assign a unique `crew_id` to each stable of officials "working" under a lead referee.

As an example, we can create data frame called `nfl_officials` that contains the official and referee information for each game during the 2022 season. After, in order to get the referee for each crew, use `filter()` to select those crew members with the "Referee" position and then create a new column titled `crew_id` that takes the numeric official ID for each distinct Referee.

```
nfl_officials <- nflreadr::load_officials(seasons = 2022)

referees <- nfl_officials %>%
  filter(position == "Referee") %>%
  mutate(crew_id = match(official_id, unique(official_id)))
```

With a unique `crew_id` associated with each Referee, we can now use `left_join()` to bring that information back into our `nfl_officials` data by merging the `game_id` and `crew_id` to the corresponding `game_id`.

```
nfl_officials <- nfl_officials %>%
  left_join(referees %>% select(game_id, crew_id),
           by = "game_id")
```

The resulting `nfl_officials` data frame now includes all the original information and now a consistent `crew_id` for each team that the referee works with.

```
nfl_officials
```

```
# A tibble: 2,065 x 10
   game_id   game_key official_name position jersey_number official_id
   <chr>     <chr>    <chr>         <chr>            <int> <chr>
 1 20220908~ 58838    Nathan Jones  Field J~            42 174
 2 20220908~ 58838    Matt Edwards  Back Ju~            96 178
 3 20220908~ 58838    Mike Carr     Down Ju~            63 168
 4 20220908~ 58838    Eugene Hall   Side Ju~           103 108
 5 20220908~ 58838    Jeff Seeman   Line Ju~            45 23
 6 20220908~ 58838    Carl Cheffers Referee             51 3
 7 20220908~ 58838    Brandon Cruse Umpire               0 201
 8 20220911~ 58839    Mike Morton   Umpire               0 206
 9 20220911~ 58839    John Jenkins  Field J~           117 86
10 20220911~ 58839    Danny Short   Down Ju~           113 172
# i 2,055 more rows
# i 4 more variables: season <int>, season_type <chr>, ...
```

To calculate the total number of penalties flagged by each crew, we must first bring in the play-by-play from the 2022 season. After, we will use `filter()` to gather only those plays where a flag was thrown, and then `select()` the relevant columns, and finally `summarize()` by the `old_game_id` to get the total penalties called and the total penalty yards as a result of those penalties.

After, we conduct another `left_join()` to bring in the specific penalty information into our `nfl_officials` data frame.

```
penalty_pbp <- nflreadr::load_pbp(seasons = 2022)

penalties <- penalty_pbp %>%
  filter(penalty == 1) %>%
  select(game_id, old_game_id, season, play_id,
         desc, home_team, away_team,
         posteam, defteam, week, penalty_team,
         penalty_type, penalty, penalty_player_id,
         penalty_player_name, penalty_yards) %>%
  group_by(old_game_id) %>%
  summarize(total_called = sum(penalty == 1, na.rm = TRUE),
            total_yards = sum(penalty_yards, na.rm = TRUE))

nfl_officials <- nfl_officials %>%
  left_join(penalties, by = c("game_id" = "old_game_id")) %>%
  select(game_id, official_name, position,
         crew_id, total_called, total_yards)
```

With the data now combined, we can `group_by()` each game's unique `game_id` and then use `summarize()` to sum the penalties called and the penalty yardage. After arranging the results in descending order by the total number of penalties, we can see that Referee Carl Cheffers, and his crew, called the most penalties during the 2022 NFL season with 223 and, unsurprisingly, also had the highest amount of penalty yards with 1,916.

```
nfl_officials %>%
  group_by(game_id) %>%
  filter(position == "Referee") %>%
  ungroup() %>%
  group_by(crew_id) %>%
  summarize(referee = unique(official_name),
            all_pen = sum(total_called, na.rm = TRUE),
            all_yards = sum(total_yards, na.rm = TRUE)) %>%
  arrange(desc(all_pen)) %>%
  slice(1:10)
```

```
# A tibble: 10 x 4
   crew_id referee         all_pen all_yards
     <int> <chr>             <int>     <dbl>
 1       1 Carl Cheffers       223      1916
 2      11 Scott Novak         204      1584
 3       6 Brad Allen          203      1511
 4      16 Clete Blakeman      203      1562
 5       5 Shawn Hochuli       199      1780
 6       4 Clay Martin         198      1684
 7      17 Adrian Hill         193      1564
 8       2 Alex Kemp           192      1602
 9      10 Tra Blake           189      1545
10      15 Ronald Torbert      185      1549
```

3.5.5 The `load_trades()` Function

The `load_trades()` function returns a data frames that includes all trades in the NFL on a season-by-season basis with information pertaining to: `trade_id`, `season`, `trade_date`, `gave`, `received`, `pick_season`, `pick_round`, `pick_number`, `conditional`, `pfr_id`, `pfr_name`.

For example, we can gather every trade involving the New England Patriots with the following code:

```
ne_trades <- nflreadr::load_trades(seasons = 2000:2022) %>%
  filter(gave == "NE" | received == "NE")

ne_trades
```

```
# A tibble: 506 x 11
   trade_id season trade_date gave  received pick_season pick_round
      <dbl>  <dbl> <date>     <chr> <chr>          <dbl>      <dbl>
 1      704   2002 2002-03-11 GB    NE              2002          4
 2      704   2002 2002-03-11 NE    GB                NA         NA
 3      716   2002 2002-04-20 NE    WAS             2002          1
 4      716   2002 2002-04-20 NE    WAS             2002          3
 5      716   2002 2002-04-20 NE    WAS             2002          7
 6      716   2002 2002-04-20 WAS   NE              2002          1
 7      725   2002 2002-04-21 DEN   NE              2002          4
 8      725   2002 2002-04-21 NE    DEN             2002          4
 9      725   2002 2002-04-21 NE    DEN             2002          5
10      727   2002 2002-04-21 DAL   NE              2002          7
# i 496 more rows
# i 4 more variables: pick_number <dbl>, conditional <dbl>, ...
```

If you want to view a trade that involves a specific player, you can do the same as above but `filter()` for a specific player. As an example, we do search for the trade that resulted in the New England Patriots sending Drew Bledsoe to the Buffalo Bills.

```
bledsoe_trade <- nflreadr::load_trades() %>%
  filter(trade_id == trade_id[pfr_name %in% c("Drew Bledsoe")])

bledsoe_trade
```

```
# A tibble: 2 x 11
  trade_id season trade_date gave  received pick_season pick_round
     <dbl>  <dbl> <date>     <chr> <chr>          <dbl>      <dbl>
1      728   2002 2002-04-22 BUF   NE              2003          1
2      728   2002 2002-04-22 NE    BUF               NA         NA
# i 4 more variables: pick_number <dbl>, conditional <dbl>,
#   pfr_id <chr>, pfr_name <chr>
```

Since the `load_trades()` function also includes NFL Draft round and pick numbers (if included in a trade), we can also – for example – determine all trades that involved a top ten pick switching hands.

```
top_ten_picks <- nflreadr::load_trades() %>%
  filter(pick_round == 1 & pick_number <= 10)

top_ten_picks
```

```
# A tibble: 57 x 11
   trade_id season trade_date gave  received pick_season pick_round
      <dbl>  <dbl> <date>     <chr> <chr>          <dbl>      <dbl>
 1      711   2002 2002-04-20 DAL   KC              2002          1
 2      711   2002 2002-04-20 KC    DAL             2002          1
 3      661   2003 2003-04-26 CHI   NYJ             2003          1
 4      662   2003 2003-04-26 ARI   NO              2003          1
 5       16   2004 2004-04-24 CLE   DET             2004          1
 6       16   2004 2004-04-24 DET   CLE             2004          1
 7       64   2005 2005-03-03 OAK   MIN             2005          1
 8      189   2007 2007-03-22 ATL   HOU             2007          1
 9      189   2007 2007-03-22 HOU   ATL             2007          1
10      197   2007 2007-04-28 SF    NE              2008          1
# i 47 more rows
# i 4 more variables: pick_number <dbl>, conditional <dbl>, ...
```

3.5.6 The `load_draft_picks()` Function

The `load_draft_picks()` function will load information pertaining to every draft pick dating back to 1980. Aside from the information you would expect (the player's name, the team that draft, round, pick number, position, etc.), the `load_draft_picks()` function also includes data regarding how many seasons the player played, the amount of times they were named to Pro Bowls, and top-level statistics regarding rushing, passing, receiving, and defensive metrics.

The `load_draft_picks()` function can be used to explore multiple different facets of the NFL Draft. For example, there is a belief in the analytics community to never draft a running back in the first round. Without getting into the reasoning behind that belief, we can quickly create a visualization of career rushing yards per running back compared to their draft position.

```
draft_picks <- nflreadr::load_draft_picks()
teams <- nflreadr::load_teams()

rb_picks <- draft_picks %>%
  filter(position == "RB") %>%
  select(pick, team, rush_atts, rush_yards) %>%
  filter(pick <= 100)
```

```
rb_picks <- rb_picks %>%
  left_join(teams, by = c("team" = "team_abbr"))


ggplot(data = rb_picks, aes(x = rush_yards, y = pick)) +
  geom_point(color = rb_picks$team_color, size =
           rb_picks$rush_atts / 500) +
  scale_y_continuous(breaks = scales::pretty_breaks()) +
  scale_x_continuous(breaks = scales::pretty_breaks(),
                     labels = scales::comma_format()) +
  geom_smooth(se = FALSE) +
  xlab("Career Rushing Yards") +
  ylab("Pick Number") +
  nfl_analytics_theme() +
  labs(title = "**Career Rushing Yards vs. Pick Number**",
     subtitle = "1980 to 2022",
     caption = "*An Introduction to NFL Analytics with R*<br>
     **Brad J. Congelio**")
```

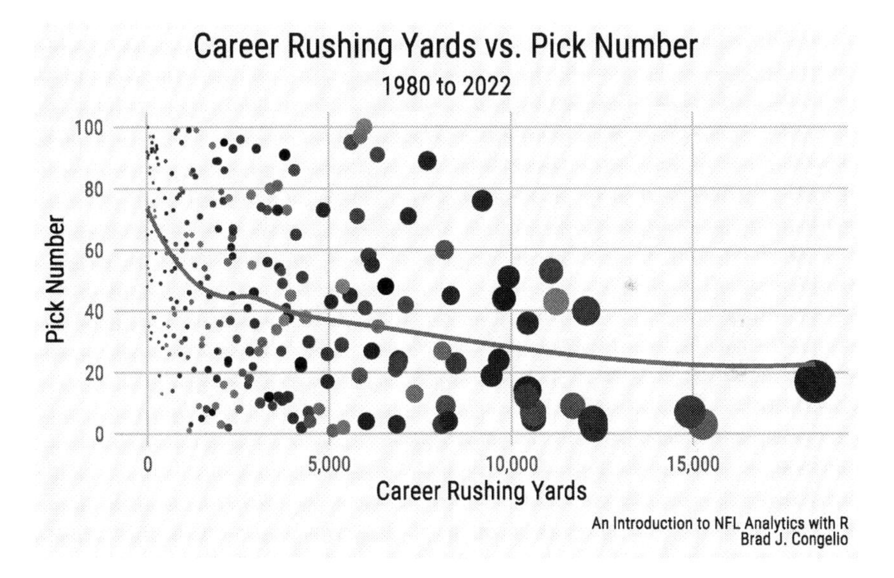

Figure 3.2: Career rushing yards vs. draft pick

According to the plot, there has been a limited number of running backs draft in the top 20 to go over 10,000 career yards. However, there are more running backs – some still active – drafted in the mid-range of the draft that are approaching, or have eclipsed, the 10,000 yard mark.

3.5.7 The load_combine() Function

The `load_combine()` function provides NFL Combine data dating back to 2000. Aside from biographical information for each player (including eventual draft position), the data include the player's scores in the 40-yard dash, the bench press, the vertical, the broad jump, the cone drill, and the shuttle drill.

We can join information from `load_combine()` with outside information to determine if there is any correlation between a running back's 40-yard dash time in the combine and the total number of rushing yard accumulated during his career.

```
combine_data <- nflreadr::load_combine() %>%
  select(pfr_id, forty) %>%
  filter(!is.na(pfr_id) & !is.na(forty))

rosters <- nflreadr::load_rosters(2000:2022) %>%
  select(gsis_id, pfr_id) %>%
  distinct(gsis_id, .keep_all = TRUE)

player_stats <- nflreadr::load_player_stats(seasons = TRUE,
                                            stat_type = "offense") %>%
  filter(position == "RB" & !is.na(player_name) &
           season_type == "REG") %>%
  group_by(player_name, player_id) %>%
  summarize(total_yards = sum(rushing_yards, na.rm = TRUE),
            team = last(recent_team))

player_stats <- player_stats %>%
  left_join(rosters, by = c("player_id" = "gsis_id"))

player_stats <- player_stats %>%
  filter(!is.na(pfr_id))

player_stats <- player_stats %>%
  left_join(combine_data, by = c("pfr_id" = "pfr_id")) %>%
  filter(!is.na(forty))

player_stats <- player_stats %>%
  left_join(teams, by = c("team" = "team_abbr"))
```

```
ggplot(data = player_stats, aes(x = forty, y = total_yards)) +
  geom_point(color = player_stats$team_color, size = 3.5) +
```

```
  geom_smooth(method = lm, se = FALSE,
              color = "black",
              linetype = "dashed",
              size = 0.8) +
  scale_x_continuous(breaks = scales::pretty_breaks()) +
  scale_y_continuous(breaks = scales::pretty_breaks(),
                     labels = scales::comma_format()) +
  nfl_analytics_theme() +
  xlab("Forty-Yard Dash Time") +
  ylab("Career Rushing Yards") +
  labs(title =
     "**Forty-Yard Dash Time vs. Career Rushing Yards**",
       subtitle = "2000 to 2022",
       caption =
          "*An Introduction to NFL Analytics with R*<br>
       **Brad J. Congelio**")
```

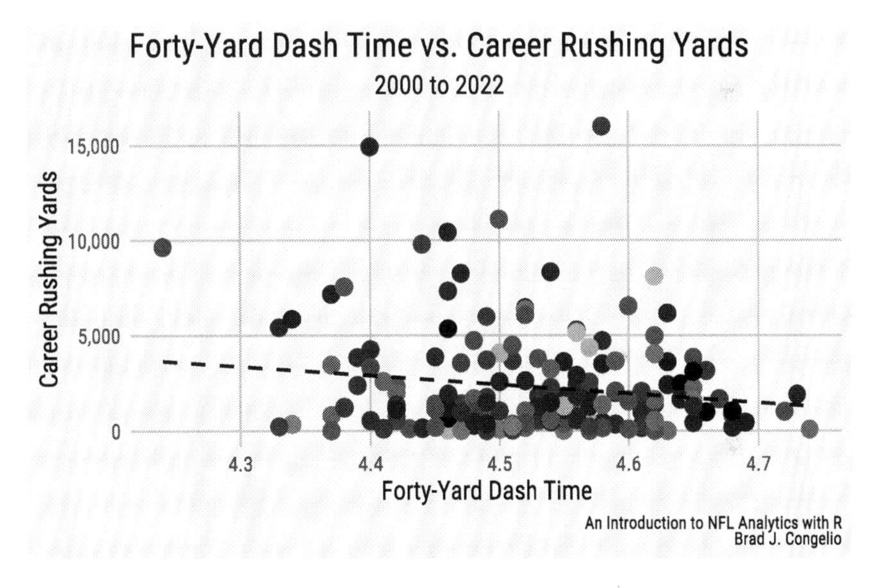

Figure 3.3: Comparing RBs forty-yard dash time vs. career total yards

The visualization indicates that there is little, if any, direct correlation between a running back's 40-yard dash time and his career total rushing yards. There are likely other factors that contribute to total rushing yards, such as a running back's agility and vision and – perhaps most important – the quality of the offensive line over the duration of the running back's career.

3.5.8 The load_nextgen_stats() Function

The load_nextgen_stats() function retrieves player-level weekly statistics as provided by NFL Next Gen Stats dating back to the 2016 season. While three different stat types are provided (passing, receiving, and rushing), it is important to note that the data will only contain those players above a minimum number of attempts as determined by the NFL Next Gen Stats team.

To illustrate what can be done with the load_nextgen_stats() function, we will gather information to create two different plots. First, we can plot each quarterback's average time to throw against their average completed air yards. Second, we will construct a graph to highlight which running backs had more actual rushing yards than the NGS "expected rushing yards" model.

```
ngs_data_passing
        <- nflreadr::load_nextgen_stats(seasons = 2022,
                stat_type = "passing") %>%
  filter(week == 0) %>%
  select(player_display_name, team_abbr,
         avg_time_to_throw, avg_completed_air_yards)

ngs_data_passing <- ngs_data_passing %>%
  left_join(teams, b = c("team_abbr" = "team_abbr"))
```

```
ggplot(data = ngs_data_passing,
        aes(x = avg_time_to_throw,
                y = avg_completed_air_yards)) +
  geom_hline(yintercept =
          mean(ngs_data_passing$avg_completed_air_yards),
             color = "black", size = 0.8,
                     linetype = "dashed") +
  geom_vline(xintercept =
          mean(ngs_data_passing$avg_time_to_throw),
             color = "black", size = 0.8,
                     linetype = "dashed") +
  geom_point(size = 3.5, color =
           ngs_data_passing$team_color) +
  scale_x_continuous(breaks = scales::pretty_breaks(),
                     labels = scales::comma_format()) +
  scale_y_continuous(breaks = scales::pretty_breaks(),
                     labels = scales::comma_format()) +
  geom_text_repel(aes(label = player_display_name),
```

```
                family = "Roboto", fontface = "bold",
                        size = 3.5) +
nfl_analytics_theme() +
xlab("Average Time to Throw") +
ylab("Average Completed Air Yards") +
labs(title =
      "**Avgerage Time to Throw vs. Average Air Yards**",
     subtitle = "2022 Regular Season",
     caption =
          "*An Introduction to NFL Analytics with R*<br>
     **Brad J. Congelio**")
```

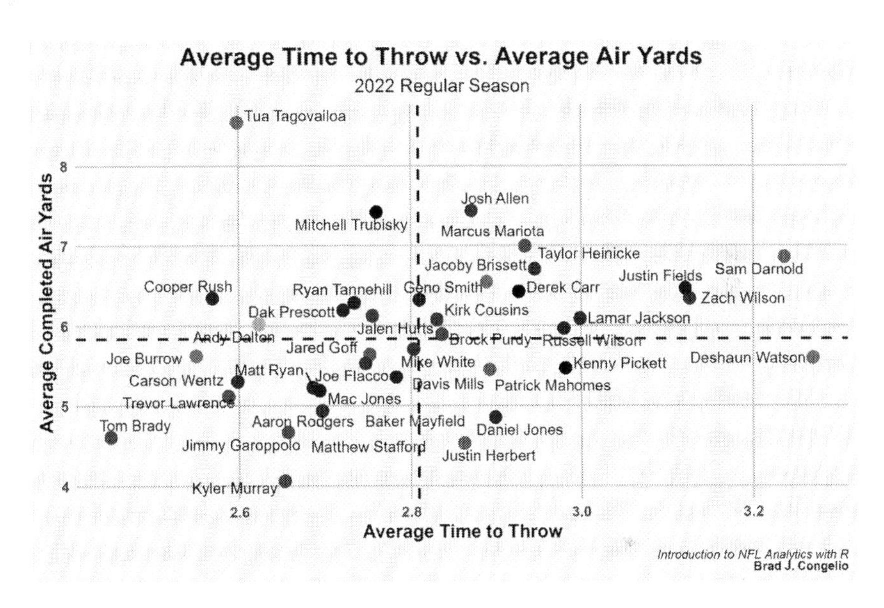

Figure 3.4: Comparing a QB's average time to throw to their average completed air yards

The resulting plot shows that Dak Prescott, Jalen Hurts, and Ryan Tannehill had an average completed air yard above the NFL average despite getting rid of the ball quicker than most other quarterbacks in the league. Conversely, Patrick Mahomes, Daniel Jones, and Justin Herbert has under average completed air yards despite holding on to the ball longer than the rest of the league (on average).

```
ngs_data_rushing
       <- nflreadr::load_nextgen_stats(seasons = 2022,
                 stat_type = "rushing") %>%
  filter(week == 0) %>%
  select(player_display_name, team_abbr, expected_rush_yards,
         rush_yards_over_expected) %>%
  mutate(actual_rush_yards = expected_rush_yards +
           rush_yards_over_expected)

ngs_data_rushing <- ngs_data_rushing %>%
  left_join(teams, by = c("team_abbr" = "team_abbr"))
```

```
ggplot(data = ngs_data_rushing, aes(x = expected_rush_yards,
                                    y = actual_rush_yards)) +
  geom_smooth(method = lm, se = FALSE,
              color = "black",
              size = 0.8,
              linetype = "dashed") +
  geom_point(size = 3.5, color =
           ngs_data_rushing$team_color) +
  scale_x_continuous(breaks = scales::pretty_breaks(),
                     labels = scales::comma_format()) +
  scale_y_continuous(breaks = scales::pretty_breaks(),
                     labels = scales::comma_format()) +
  geom_text_repel(aes(label = player_display_name),
                  family = "Roboto",
                  fontface = "bold", size = 3.5) +
  nfl_analytics_theme() +
  xlab("Expected Rush Yards") +
  ylab("Actual Rush Yards") +
  labs(title = "Expected Rush Yards vs. Actual Rush Yards",
       subtitle = "2022 Regular Season",
       caption =
               "*An Introduction to NFL Analytics with R*<br>
       **Brad J. Congelio**")
```

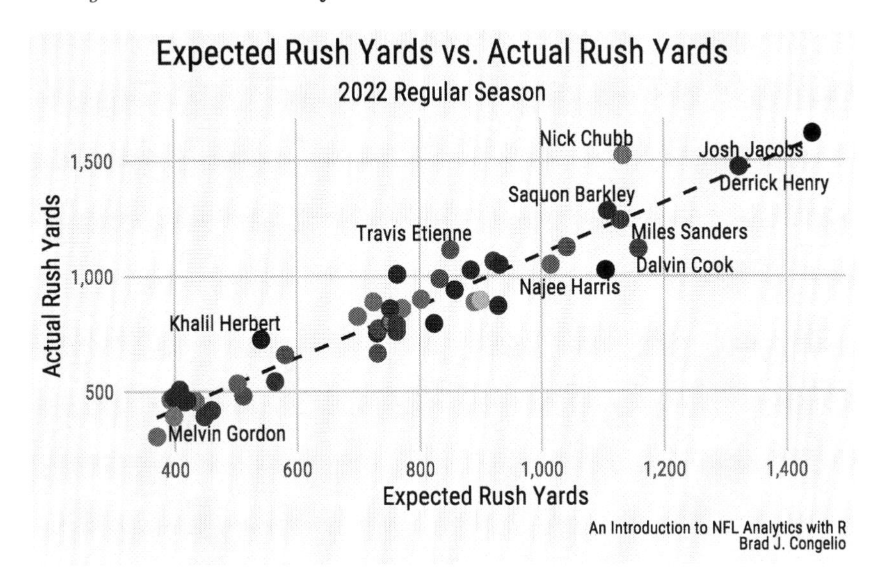

Figure 3.5: Expected rushing yards vs. actual rushing yards using Next Gen Stats

After plotting the rushing data from Next Gen Stats, we can see that Nick Chubb was well above the expected yardage metric while Ezekiel Elliott, Alvin Kamara, Najee Harris, and Dalvin Cook all had actual rushing yards that fell below what was expected by the model.

3.5.9 The `load_espn_qbr()` Function

With information dating back to the 2006 season, the `load_espn_qbr()` function provides a multitude of data points, including:

1. `qbr_total` – the adjusted total QBR which is calculated on a scale of 0–100 based on the strength of the defenses played.
2. `pts_added` – the number of total points added by a quarterback adjusted above the average level of all other QBs.
3. `epa_total` – the total number of expected points added for each quarterback as calculated by ESPN's win probability model.

4. `pass` – the total of expected points added for just pass plays.
5. `run` – the total of expected points added for just run plays.
6. `qbr_raw` – the QBR without adjustment for strength of defense.
7. `sack` – the adjustment for expected points added for sacks.

While the QBR metric has fallen out of grace among the analytics community in recent years, we can still compare the difference between the adjusted and unadjusted QBR scores to visualize the impact of defensive strength on the total.

```
espn_qbr <- nflreadr::load_espn_qbr(seasons = 2022) %>%
  select(name_short, team_abb, qbr_total, qbr_raw)

espn_qbr <- espn_qbr %>%
  left_join(teams, by = c("team_abb" = "team_abbr"))
```

```
ggplot(data = espn_qbr, aes(x = qbr_total, y = qbr_raw)) +
  geom_smooth(method = lm, se = FALSE,
              color = "black",
              linetype = "dashed",
              size = 0.8) +
  geom_point(color = espn_qbr$team_color, size = 3.5) +
  scale_x_continuous(breaks = scales::pretty_breaks()) +
  scale_y_continuous(breaks = scales::pretty_breaks()) +
  nfl_analytics_theme() +
  xlab("QBR - Adjusted") +
  ylab("QBR - Unadjusted") +
  labs(title = "QBR: Adjusted vs. Adjusted Scores",
    subtitle = "Based on ESPN's Model: 2022 Regular Season",
    caption = "*An Introduction to NFL Analytics with R*<br>
    **Brad J. Congelio**")
```

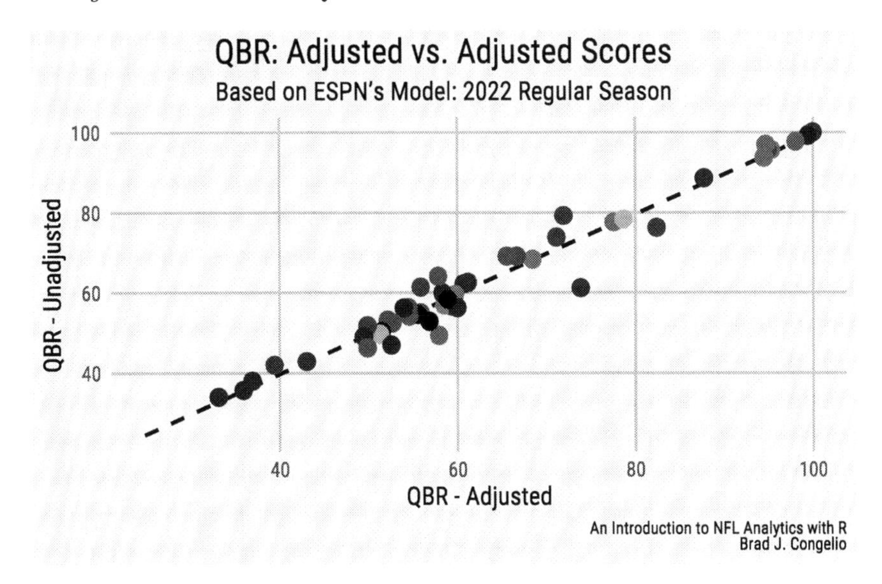

Figure 3.6: Comparing adjusted and adjusted quarterback ratings

3.5.10 The `load_pfr_advstats()` Function

The `load_pfr_advstats()` function provides statistics from Pro Football Reference starting with the 2018 season for passing, rushing, receiving, and defense. The function allows for collecting the data at either the weekly or season level with the `summary_level` argument.

To begin exploring the data, let's examine the relationship between the number of dropped passes each quarterback endured during the 2022 regular season compared to the total number of `bad_throws` charted by Pro Football Reference. An argument can be made that a dropped pass without the pass being considered "bad" is the fault of the wide receiver, while a dropped pass that *is* charted as "bad" falls on the quarterback.

> **! Important**
>
> When working with data from `load_pfr_advstats()`, it is important to remember that you must use the `clean_team_abbrs()` function from `nflreadr` as there is a difference between the abbreviations used by PFR and the abbreviations used within the `nflverse`.

```
pfr_stats_pass <- nflreadr::load_pfr_advstats(seasons = 2022,
                stat_type = "pass",
                summary_level = "season") %>%
  select(player, pass_attempts, team, drops, bad_throws) %>%
  filter(pass_attempts >= 400)

pfr_stats_pass$team <- clean_team_abbrs(pfr_stats_pass$team)

pfr_stats_pass <- pfr_stats_pass %>%
  left_join(teams, by = c("team" = "team_abbr"))
```

```
ggplot(data = pfr_stats_pass, aes(x = bad_throws, y = drops)) +
  geom_hline(yintercept = mean(pfr_stats_pass$drops),
             color = "black", linetype = "dashed", size = 0.8) +
  geom_vline(xintercept = mean(pfr_stats_pass$bad_throws),
             color = "black", linetype = "dashed", size = 0.8) +
  geom_point(color = pfr_stats_pass$team_color, size = 3.5) +
  scale_x_continuous(breaks = scales::pretty_breaks()) +
  scale_y_continuous(breaks = scales::pretty_breaks()) +
  nfl_analytics_theme() +
  geom_text_repel(aes(label = player),
                  family = "Roboto", fontface = "bold",
                  size = 3.5) +
  xlab("Bad Throws") +
  ylab("Drops") +
  labs(title = "**QB Bad Throws vs. WR Drops**",
       subtitle = "2022 Regular Season",
       caption = "*An Introduction to NFL Analytics with R*<br>
       **Brad J. Congelio**")
```

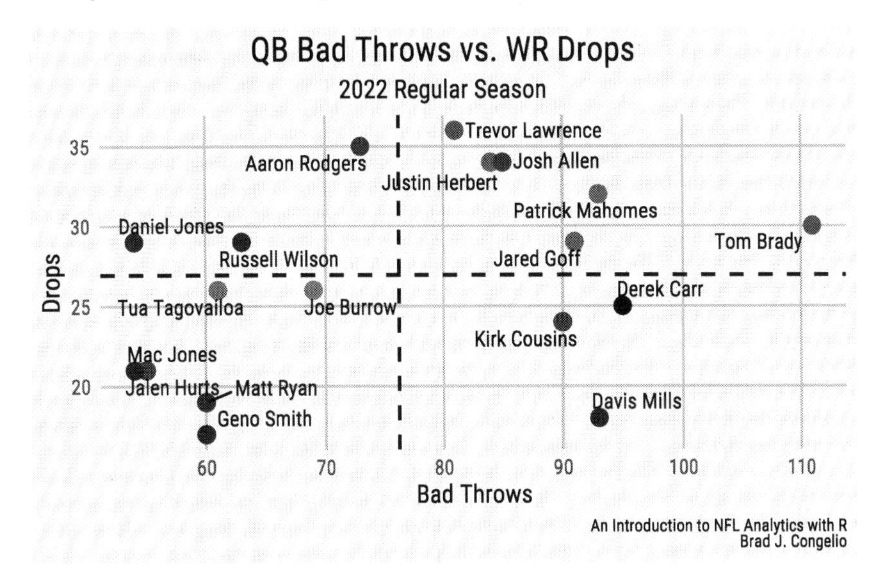

Figure 3.7: QB Bad Throws vs. WR Drops

Quarterbacks such as Patrick Mahomes, Josh Allen, and Tom Brady had some of the highest drop numbers in the league but were also charted as having the highest number of bad throws, as well. This indicates that – without additional context – the wide receivers may not be as much at fault for the drops compared to the wide receivers that dropped passes from Daniel Jones and Russell Wilson, both of whom had a lower number of bad throws.

By using the rushing data provided by `load_pfr_advstats()`, we can examine how "hard" it is to tackle a running back by examining the relationship between their yards before contact and yards after contact.

> **! Important**
>
> Because the list of running backs with more than 200 attempts includes Christian McCaffrey, it is necessary to use the `case_when()` function to switch his `tm` from `2TMS` to the team that he finished the season with (`SF`). Otherwise, using `left_join()` for team colors would not work for McCafffrey since `2TMS` is not a recognized team abbreviation.

```
pfr_stats_rush <- nflreadr::load_pfr_advstats(seasons = 2022,
               stat_type = "rush",
               summary_level = "season") %>%
  select(player, tm, att, ybc, yac) %>%
  filter(att >= 200) %>%
  mutate(tm = case_when(
    player == "Christian McCaffrey" ~ "SF",
    TRUE ~ tm))

pfr_stats_rush$tm <- clean_team_abbrs(pfr_stats_rush$tm)

pfr_stats_rush <- pfr_stats_rush %>%
  left_join(teams, by = c("tm" = "team_abbr"))


ggplot(data = pfr_stats_rush, aes(x = ybc, y = yac)) +
  geom_hline(yintercept = mean(pfr_stats_rush$yac),
             color = "black", linetype = "dashed",
             size = 0.8) +
  geom_vline(xintercept = mean(pfr_stats_rush$ybc),
             color = "black", linetype = "dashed",
             size = 0.8) +
  geom_point(color = pfr_stats_rush$team_color,
          size = 3.5) +
  scale_x_continuous(breaks = scales::pretty_breaks()) +
  scale_y_continuous(breaks = scales::pretty_breaks()) +
  nfl_analytics_theme() +
  geom_text_repel(aes(label = player),
                  family = "Roboto", fontface = "bold",
                  size = 3.5) +
  xlab("Yards Before Contact") +
  ylab("Yards After Contact") +
  labs(title =
         "**Running Backs: Yards Before and After Contact**",
       subtitle = "2022 Regular Season",
       caption =
              "*An Introduction to NFL Analytics with R*<br>
       **Brad J. Congelio**")
```

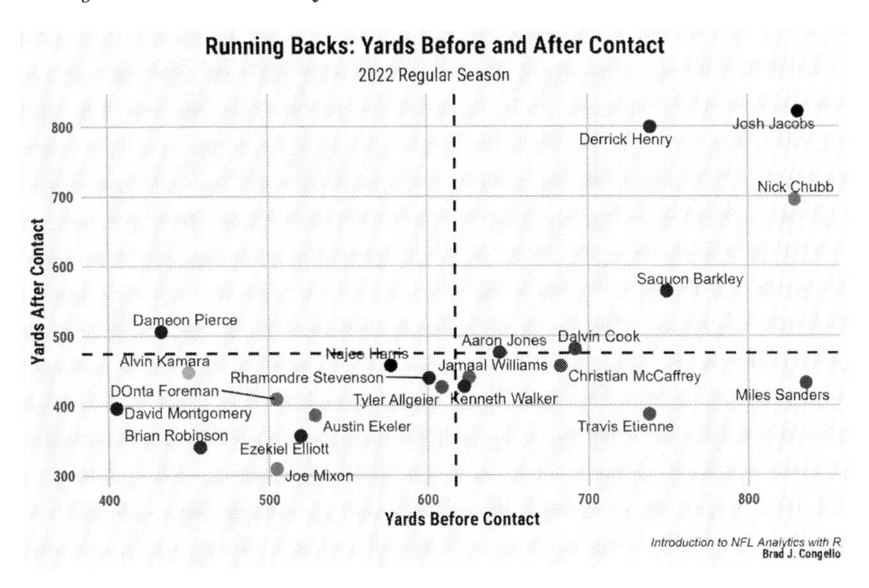

Figure 3.8: Exploring which running backs get more yards before and after contact

In the same vein as the running back statistics, we can use `load_pfr_advstats()` to explore which wide receivers gained the most yardage after catching the ball.

```
pfr_stats_rec <- nflreadr::load_pfr_advstats(seasons = 2022,
        stat_type = "rec",
                summary_level = "season") %>%
  filter(rec >= 80 & pos == "WR") %>%
  select(player, tm, rec, ybc, yac)

pfr_stats_rec <- pfr_stats_rec %>%
  left_join(teams, by = c("tm" = "team_abbr"))


ggplot(data = pfr_stats_rec, aes(x = ybc, y = yac)) +
  geom_hline(yintercept = mean(pfr_stats_rec$yac),
             color = "black", linetype = "dashed",
                     size = 0.8) +
  geom_vline(xintercept = mean(pfr_stats_rec$ybc),
             color = "black", linetype = "dashed",
                     size = 0.8) +
  geom_point(color = pfr_stats_rec$team_color, size = 3.5) +
```

```
scale_x_continuous(breaks = scales::pretty_breaks(),
                   labels = scales::comma_format()) +
scale_y_continuous(breaks = scales::pretty_breaks()) +
nfl_analytics_theme() +
geom_text_repel(aes(label = player),
                family = "Roboto", fontface = "bold",
                        size = 3.5) +
xlab("Yards At Point of Catch") +
ylab("Yards After Catch") +
labs(title =
        "**Yards at Point of Catch vs. Yards After Catch**",
     subtitle = "2022 Regular Season",
     caption =
             "*An Introduction to NFL Analytics with R*<br>
     **Brad J. Congelio**")
```

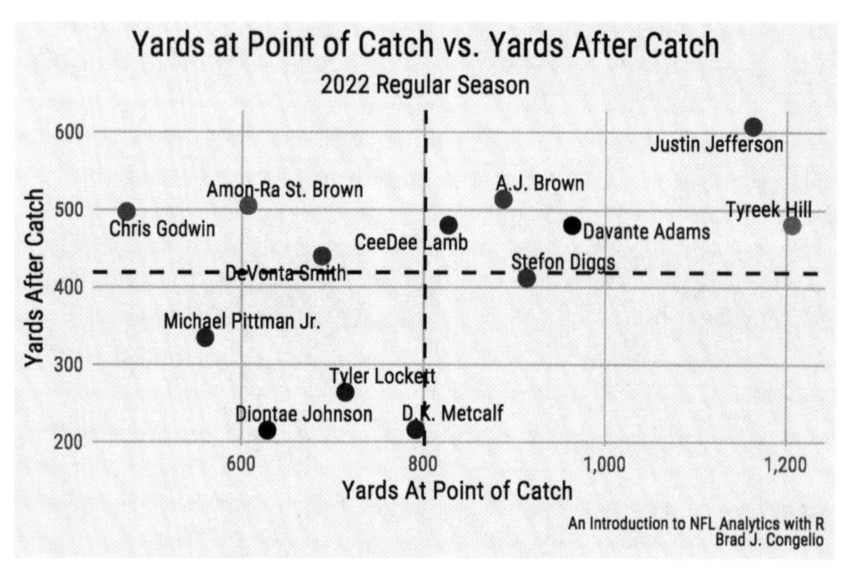

Figure 3.9: Which wide receivers gain the most yardage after catching the ball?

Given the outstanding season that Justin Jefferson had in 2022, it is unsurprising to see him in the upper-right quadrant of the plot. Of his 1,809 receiving yards in the season, just under 1,200 of them came through air yards while he gained just over an additional 600 yards on the ground after the catch. On the other hand, D.K. Metcalf and Diontae Johnson were more often than not tackled almost immediately after catching the football.

Given the structure of the data collected from `load_pfr_advstats()`, the information can also be aggregated to define team total. For example, we can use the defensive statistics for blitzes and sacks from individual players to calculate the relationship between a team's total number of blitzes against the number of sacks those blitzes produce.

```
pfr_stats_def <- nflreadr::load_pfr_advstats(seasons = 2022,
                stat_type = "def",
                summary_level = "season") %>%
  filter(!tm %in% c("2TM", "3TM")) %>%
  select(tm, bltz, sk) %>%
  group_by(tm) %>%
  summarize(total_blitz = sum(bltz, na.rm = TRUE),
            total_sack = sum(sk, na.rm = TRUE))

pfr_stats_def <- pfr_stats_def %>%
  left_join(teams, by = c("tm" = "team_abbr"))


ggplot(data = pfr_stats_def, aes(x = total_blitz,
       y = total_sack)) +
  geom_hline(yintercept = mean(pfr_stats_def$total_sack),
             color = "black", linetype = "dashed",
             size = 0.8) +
  geom_vline(xintercept = mean(pfr_stats_def$total_blitz),
             color = "black", linetype = "dashed",
             size = 0.8) +
  geom_image(aes(image = team_logo_wikipedia), asp = 16/9) +
  scale_x_continuous(breaks = scales::pretty_breaks()) +
  scale_y_continuous(breaks = scales::pretty_breaks()) +
  nfl_analytics_theme() +
  xlab("Total Blitzes") +
  ylab("Total Sacks") +
  labs(title = "**Total Blitzes vs. Total Sacks**",
       subtitle = "2022 Regular Season",
       caption =
               "*An Introduction to NFL Analytics with R*<br>
       **Brad J. Congelio**")
```

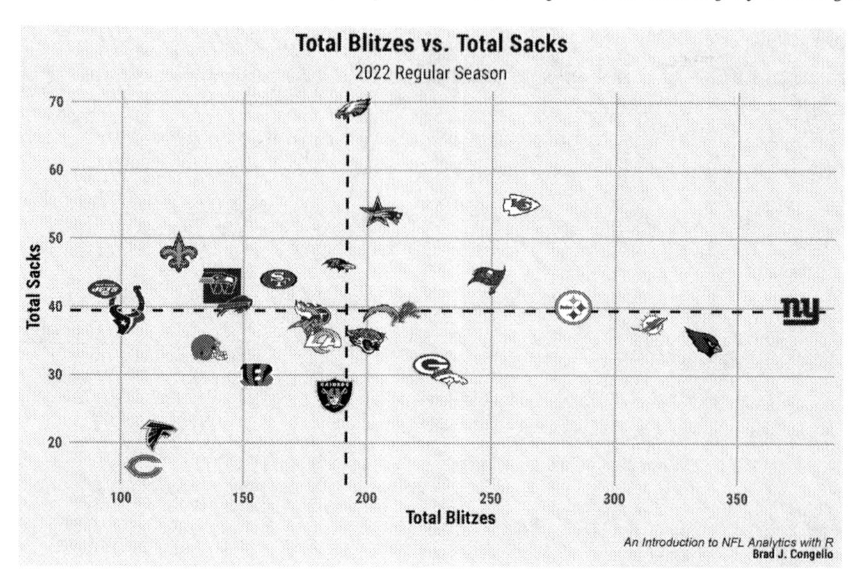

Figure 3.10: Which defensive units are most efficient at creating sacks from blitzes?

The Philadelphia Eagles were absolutely dominate during the 2022 regular season when it came to turning blitzes into sacks. On the other hand, the Arizona Cardinals, New York Giants, and Pittsburgh Steelers produced high amounts of pressure with blitzes but were not able to convert them into sacks as a consistent basis.

3.5.11 The `load_snap_counts()` Function

The `load_snap_counts()` function returns week-level snap counts for all players from Pro Football Reference dating back to the 2012 season.

3.5.12 The `load_contracts()` Function

The `load_contracts()` function brings in data from OverTheCap.com. It is important to remember that much of the information is "nested" and, if you wish to see the yearly information, you must use the `unnest()` function from `tidyr`. To highlight this, we can run the following code to bring in the data as it is stored in the `nflverse`.

```
nfl_contracts <- nflreadr::load_contracts()

colnames(nfl_contracts)
```

```
 [1] "player"               "position"          "team"
 [4] "is_active"            "year_signed"       "years"
 [7] "value"                "apy"               "guaranteed"
[10] "apy_cap_pct"          "inflated_value"    "inflated_apy"
[13] "inflated_guaranteed"  "player_page"       "otc_id"
[16] "date_of_birth"        "height"            "weight"
[19] "college"              "draft_year"        "draft_round"
[22] "draft_overall"        "draft_team"        "cols"
```

The returned data includes the contract information for the year in which it was signed, but does not include a year-by-year breakdown of money paid and other relevant information. This data is stored in the `cols` information and needs to be "opened" up in order to view, like below.

```
nfl_contracts_unnested <- nfl_contracts %>%
  tidyr::unnest(cols, name_sep = "_") %>%
  select(player, year, cash_paid) %>%
  filter(year != "Total") %>%
  mutate(cash_paid = as.numeric(as.character(cash_paid)))
```

After using the `unnest` function, we use `select()` to gather just the player's name, the year, and the cash paid for each year. After, we use `filter()` to remove the column that tallies the total of the player's contact and then `mutate()` the `cash_paid` column in order to turn it into a number rather than a character.

We can now bring in other information to compare a player's cash paid to their performance on the field. I am going to use an example from the final research project from one of my former Sport Analytics students, Matt Dougherty.[1] We will compare a player's yearly pay against their DAKOTA score (which is the adjusted EPA + CPOE composite based on the coefficients which best predict adjusted EPA/play). In order to do so, we must merge each quarterback's DAKOTA composite based on the year.

[1] Matt was in my SPT 313 class during the Spring 2023 semester and asked more questions and showed more willingness to learn coding and analytics than all of my prior classes combined. If he is not yet hired by an NFL team, it is a complete injustice.

```
nfl_contracts_unnested <- nfl_contracts %>%
  tidyr::unnest(cols, name_sep = "_") %>%
  select(player, year, cash_paid) %>%
  filter(year != "Total") %>%
  mutate(cash_paid = as.numeric(as.character(cash_paid)),
         year = as.numeric(as.character(year))) %>%
  filter(year == 2022)

dakota_composite <- nflreadr::load_player_stats(2022) %>%
  filter(position == "QB") %>%
  group_by(player_display_name, season) %>%
  summarize(attempts = sum(attempts, na.rm = TRUE),
            mean_dakota = mean(dakota, na.rm = TRUE)) %>%
  filter(attempts >= 200)

teams <- nflreadr::load_teams(current = TRUE)

nfl_contracts_unnested <- nfl_contracts_unnested %>%
  left_join(dakota_composite,
            by = c("player" = "player_display_name"))

nfl_contracts_unnested <- na.omit(nfl_contracts_unnested)

nfl_contracts_unnested <- nfl_contracts_unnested %>%
  distinct(player, year, .keep_all = TRUE)


ggplot(data = nfl_contracts_unnested,
       aes(x = cash_paid, y = mean_dakota)) +
  geom_hline(yintercept =
          mean(nfl_contracts_unnested$mean_dakota),
             color = "black", linetype = "dashed",
                     size = 0.8) +
  geom_vline(xintercept =
          mean(nfl_contracts_unnested$cash_paid),
             color = "black", linetype = "dashed",
                     size = 0.8) +
  geom_point(size = 3.5) +
  scale_x_continuous(breaks = scales::pretty_breaks(),
                     labels = scales::dollar_format()) +
  scale_y_continuous(breaks = scales::pretty_breaks()) +
  xlab("Cash Paid (in millions)") +
  ylab("Mean DAKOTA Composite") +
  geom_text_repel(aes(label = player),
```

```
                  family = "Roboto", fontface = "bold",
                          size = 3.5) +
  nfl_analytics_theme() +
  labs(title = "Cash Paid vs. Mean DAKOTA Composite",
       subtitle = "2022 Regular Season",
       caption =
               "*An Introduction to NFL Analytics with R*<br>
       **Brad J. Congelio**")
```

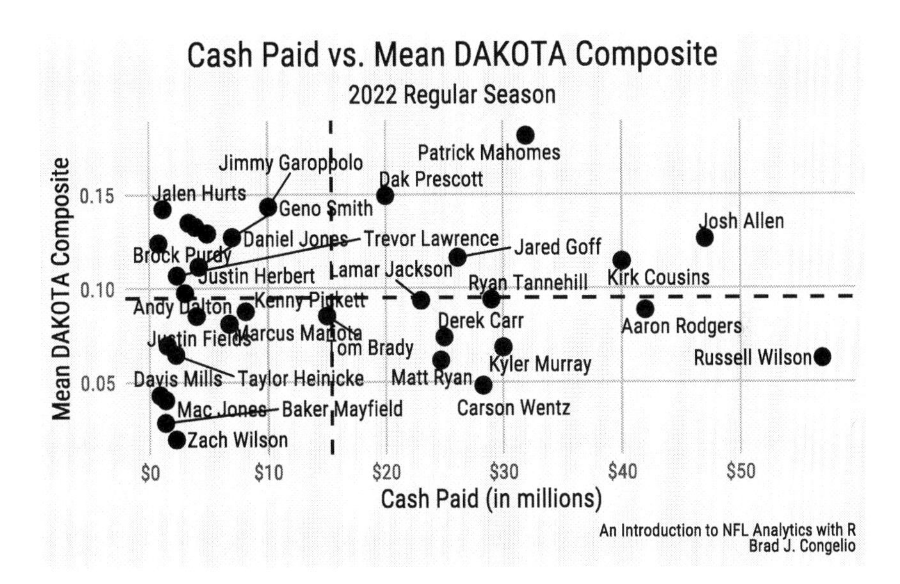

Figure 3.11: The relationship between a QB's average DAKOTA and annual cash paid

Those players in the upper-right quadrant (Mahomes, Allen, Cousins, etc.) are among the highest paid quarterbacks in the league, but also are the highest performing players based on the DAKOTA composite. On the other hand, those QBs in the lower-right quadrant are – based on the DAKOTA composite – overpaid relevant to their performance on the field. The lower-left QBs are not highly paid, but also do not perform well. Those players in the upper-left performed extremely well (some better than those in the upper-right) but come with a team-friendly contract (in terms of pure cash paid).

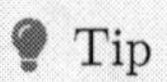

As mentioned in the book's Preface, you can feel free to reach out to me with questions regarding the R programming language and NFL analytics.

That said, there is a **fantastic `nflverse`** Discord channel that was created and currently maintained by the creators of the packages (Ben Baldwin, Sebastian Carl, Tan Ho, Thomas Mock, etc.).

You can become a member here: http://nfl-book.bradcongelio.com/discord-invite

However, before creating a thread and seeking assistance, be sure to visit the "#How To" section in the General Help area to learn to create a `reprex` (reproducible example). Providing a `reprex` allows other users to copy and paste your code in order to recreate the issue you are experiencing.

3.6 Exercises

The answers for the following answers can be found here: http://nfl-book.bradcongelio.com/ch3-answers.

3.6.1 Exercise 1

1. Load data from the 2010 to 2022 regular seasons into a data frame titled `pbp`.
2. In a data frame titled `rushing_success`, determine how many rushing attempts each offensive team had on 1st down.
3. Determine how many of those attempts resulted in a positive EPA (`success`).
4. Calculate the percentage of the results into a column titled `success_pct`.
5. Arrange the results in descending order by `success_pct`.

3.6.2 Exercise 2

Load data from the 2022 regular season into a data frame titled `pbp_2022`.

1. In a data frame titled `qb_short_third`, determine how many 3rd down passing attempts each QB had.
2. Determine the number of times the QB's air yards was less than the required yards to go.
3. In a column titled `ay_percent`, calculate the percentage of these results.
4. Filter the results to those QBs with 100 or more attempts.
5. Arrange the results in descending order by `ay_percent`.

3.6.3 Exercise 3

Tom Brady had a long and storied career, serving as a full-time started in the NFL from 2001 to 2022. The `qb_epa` metric gives a quarterback credit for EPA up to the point, where a receivers lost a fumble after a completed catch. For this question, create data frame titled `tom_brady` and find Brady's average `qb_epa` per season, from 2001 to 2022. After, arrange in descending order by his average QB EPA.

3.6.4 Exercise 4

Create a data frame titled `made_field_goals` and find, between the 2000 and 2022 season, the number of field goal attempts and percentage made on all kicks greater than 40-yards in distance.

3.6.5 Exercise 5

On December 11 of the 2005 season, Pittsburgh's Jerome Bettis trucked Bears linebacker Brian Urlacher. You can view the play here: Bettis Trucks Urlacher. Using the `load_pbp` function, find this specific play (using the video for contextual clues). After, determine how much win probability this individual play added (that is: the differnece between `home_wp` and `home_wp_post`.

4

Data Visualization with NFL Data

Effective data visualization is an important part of any data analysis project as it helps in highlighting key insights into the data, identifying trends, patterns, and anomalies, as while as allowing you to communicate results to the outside world.

Jim Stikeleather, writing for the *Harvard Business Review*, outlined three key elements that make a successful data visualization (albeit, leaving *us* to decide the definition of what a "successful" data visualization is). Despite that philosophical gap, the three elements provided by Stikeleather are succinct enough to allow us to build a framework in this chapter for how to successfully craft an NFL analytics data visualization. In his piece, Stikeleather outlines the following three characteristics of a successful data visualization: it understands the audience, it sets up a clear framework, and it tells a story (Stikeleather, 2013).

To illustrate the importance of these three elements, let's take a look at example visualizations using NFL data to further contextualize each one.

4.1 Data Viz Must Understand the Audience

As explained by Stikeleather, the core purpose of a data visualization is to take "great quantities of information" and then convey that information in such a way that it is "easily assimilated by the consumers of the information." In other words, the process of data visualization should allow for a great quantity of data to be distilled into an easily consumable (and understandable!) format.

Speaking specifically to NFL analytics, when doing visualizations we must be conscious about whether or not the intended audience will understand the terminology and concepts we use in the plot. For example, most all NFL fans understand the "non-advanced" statistics in the sport. But, when plots start using metrics such as EPA or completion percentage over expected, for example,

DOI: 10.1201/9781003364320-4

the audience looking at the plot may very well have little understanding of what is being conveyed.

Because of this, most of my data visualizations include "directables" within the plot. These "directables" may be arrows that indicate which trends on the plot are "good" or they may be text within a scatterplot that explains what each quadrant means. Or, for example, I sometimes include a textual explanation of the "equation" used to develop a metric as seen in Figure 4.1.

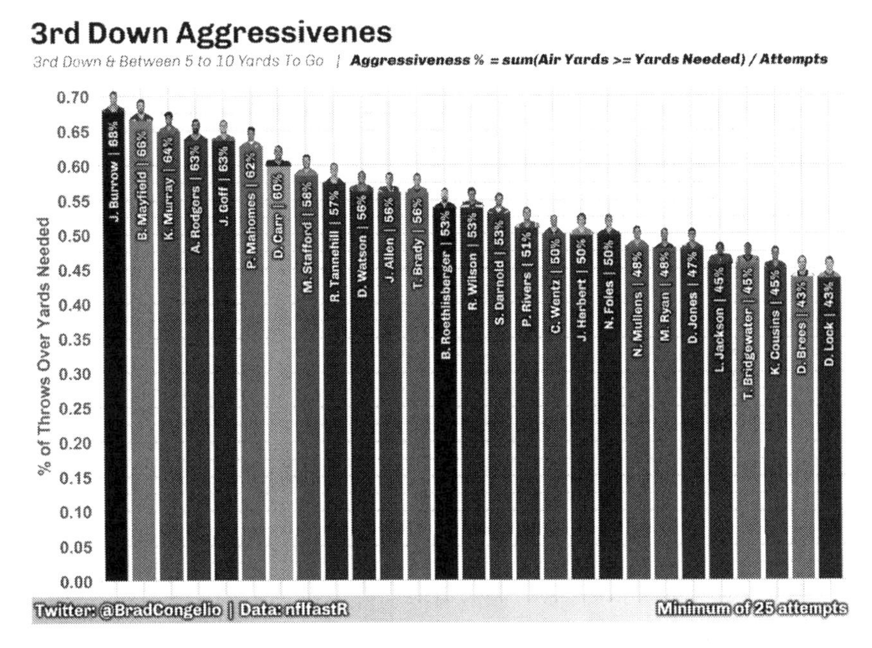

Figure 4.1: Joe Burrow was the most aggressive QB in the league when throwing on 3rd down

The above plot explores which QBs, from the 2020 season, were most aggressive on 3rd down with between 5 to 10 yards to go. Since "aggressiveness" is not a typical, day-to-day metric discussed by NFL fans, I included a "directable" within the subtitle of the plot that explained that the plot, first, was examining just 3rd down pass attempts within a specific yard range. And, second, I made the decision to include how "aggressiveness" was calculated by including the simple equation within the subtitle as well. Doing so allows even the most casual of NFL fans to easily understand what the plot is showing – in this case, that Joe Burrow's 3rd down pass attempts with between 5 and 10 yards to go made it to the line of gain, or more, on 68% of his attempts. On the other hand, Drew Lock and Drew Brees were the least aggressive QBs in the line based on the same metric.

4.2 Setting Up for Data Viz

While most of your journey through NFL analytics in this book required you to use the `tidyverse` and a handful of other packages, the process of creating compelling and meaningful data visualizations will require you to utilize multitudes of other packages. Of course, the most important is `ggplot2` which is already installed via the `tidyverse`. However, in order to recreate the visualizations included in this chapter, it is required that you install other R packages. To install the necessary packages, you can run the following code in RStudio:

```
install.packages(c("extrafont",
                   "ggrepel",
                   "ggimage",
                   "ggridges",
                   "ggtext",
                   "ggfx",
                   "geomtextpath",
                   "cropcircles",
                   "magick",
                   "glue",
                   "gt",
                   "gtExtras"))
```

4.3 The Basics of Using `ggplot2`

The basics of any `ggplot` visualization involves three basic calls to information in a data set as well as stipulation which type of geom you would like to use:

1. the data set to be used in the visualization
2. an aesthetic call for the x-axis
3. an aesthetic call for the y-axis
4. your desired geom type

```
ggplot(data = 'dataset_name', aes(x = 'x_axsis',
       y = 'y_axis')) +
  geom_type()
```

To showcase this, let's use data from Sports Info Solutions regarding quarterback statistics when using play action versus when not using play action. To start, collect the data using the `vroom` function.

```
play_action_data
       <- vroom("http://nfl-book.bradcongelio.com/pa-data")
```

To provide an easy-to-understand example of building a visualization with `ggplot`, let's use each QB's total yardage when using play action and when not. In this case, our two variable names are `yds` and `pa_yds` with the `yds` variable being placed on the x-axis and the `pa_yds` variable being placed on the y-axis.

> **Tip**
>
> It is important to remember which axis is which as you begin to learn using `ggplot`:
>
> The x-axis is the horizontal axis that runs left-to-right.
>
> The y-axis is the vertical axis that runs top-to-bottom.

```
ggplot(data = play_action_data, aes(x = yds, y = pa_yds)) +
  geom_point()
```

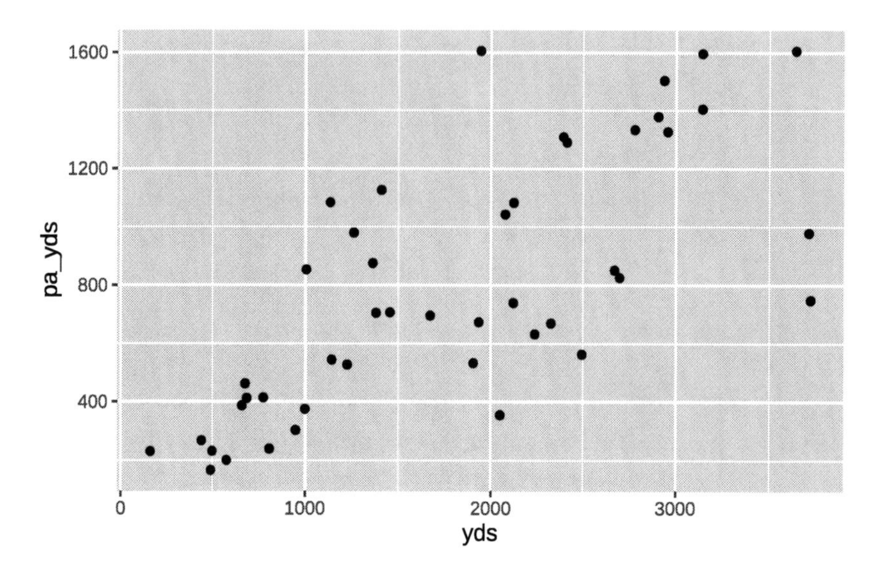

Figure 4.2: A simplistic plot using `geom_point()`

4.4 Building a Data Viz: A Step-by-Step Process

Our newly created scatter plot is an excellent starting point for a more finely detailed visualization. While we are able to see the relationship between non-play action passing yards and those attempts that included play action, we are unable to discern which specific point is which quarterback – among other issues. To provide more detail and to "prettify" the plot, let's discuss doing the following:

> **i Note**
>
> Our data visualization "to do list":
>
> 1. Adding team colors to each point.
> 2. Increasing the size of each point.
> 3. Adding player names to each point.
> 4. Increasing the number of ticks on each axis.
> 5. Provide the numeric values in correct format (that is, including a `,` to correctly show thousands).
> 6. Rename the title of each axis.

7. Provide a title, subtitle, and caption for the plot.
8. Add mean lines to both the x-axis and y-axis
9. Change `theme` elements to make data viz more appealing.

4.4.1 Adding Team Colors to Each Point

> **Note**
>
> Much like anything in the R language, there are multiple ways to go about adding team colors (and logos, player headshots, etc.) to visualizations.
>
> First, we can merge team color information into our `play_action_data` and then manually set the colors in our `geom_point` call.
>
> Second, we can use the `nflplotR` package (which is part of the `nflverse`) to bring the colors in.
>
> Both examples will be included in the below example.

To start, we will load team information using the `load_teams` function within `nflreadr`. In this case, we are requesting that the package provide only the 32 current NFL teams by including the `current = TRUE` argument. Conversely, setting the argument to `current = FALSE` will result in historical NFL teams being included in the data (the Oakland Raiders and St. Louis Rams, for example). We will also use the `select()` verb from `dplyr` to gather just the variables we know we will need (`team_abbr`, `team_nick`, `team_color`, and `team_color2`.

> **Important**
>
> We are only including the `team_abbr` variable in this example because we are going to create the plot both with and without the use of `nflplotR`. As of the writing of this book, the newest development version of the package is 1.1.0.9004 and does not yet (if ever) provide support to use team nicknames. Because of this, we must include `team_abbr` in our merge since it is the team name version that is standardized for use in `nflplotR`.

After collecting the team information needed, we can conduct a `left_join()` to match the information by `team` in `play_action_data` and `team_nick` in the team information from `load_teams()` and then confirm the merge was successful by viewing the columns names in `play_action_data` with `colnames()`.

```
teams <- nflreadr::load_teams(current = TRUE) %>%
  select(team_abbr, team_nick, team_color, team_color2)

play_action_data <- play_action_data %>%
  left_join(teams, by = c("team" = "team_nick"))

colnames(play_action_data)
```

With the team color information now built into our `play_action_data`, we can include the correct team color for each point by including the `color` argument within our `geom_point`.

```
ggplot(data = play_action_data, aes(x = yds, y = pa_yds)) +
  geom_point(color = play_action_data$team_color)
```

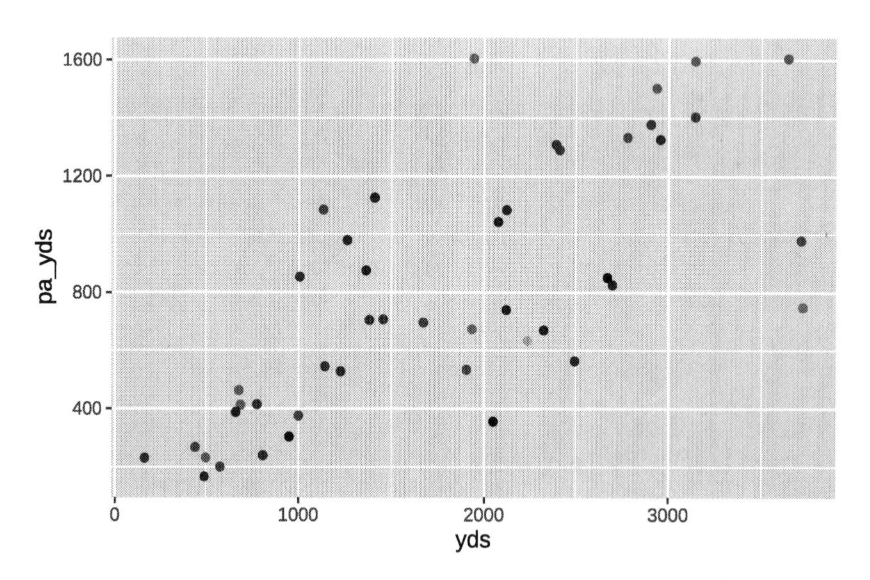

Figure 4.3: Adding team colors to the plot

! Important

You may notice that we used the $ special operator to extract the `team_color` information within our `play_action_data` data frame. This is an extremely important distinction, as using the `aes()` argument from `ggplot` and not using the $ operator will result in a custom scale color being applied to each team, without the colors being correctly associated to a team.

To see this for yourself, you can run the example following code. **Remember, this is an incorrect approach and serves to only highlight why the $ special operator was used in the above code.**

```
ggplot(data = play_action_data, aes(x = yds, y = pa_yds)) +
  geom_point(aes(color = team_color))
```

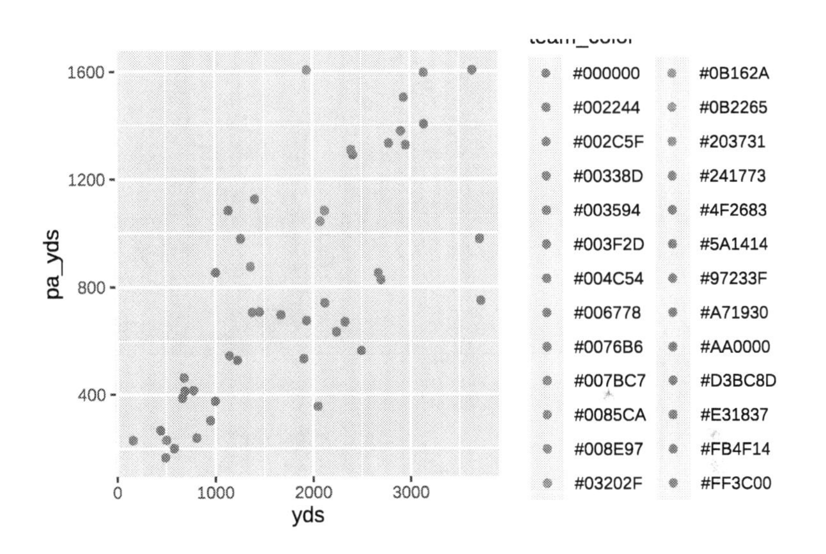

Figure 4.4: In order to bring team colors into the plot, we must use the $ special operator and not the `aes()` argument

As mentioned, the same result can be achieved using the `nflplotR` package. The following code will do so:

```
ggplot(data = play_action_data, aes(x = yds, y = pa_yds)) +
  geom_point(aes(color = team_abbr)) +
  nflplotR::scale_color_nfl(type = "primary")
```

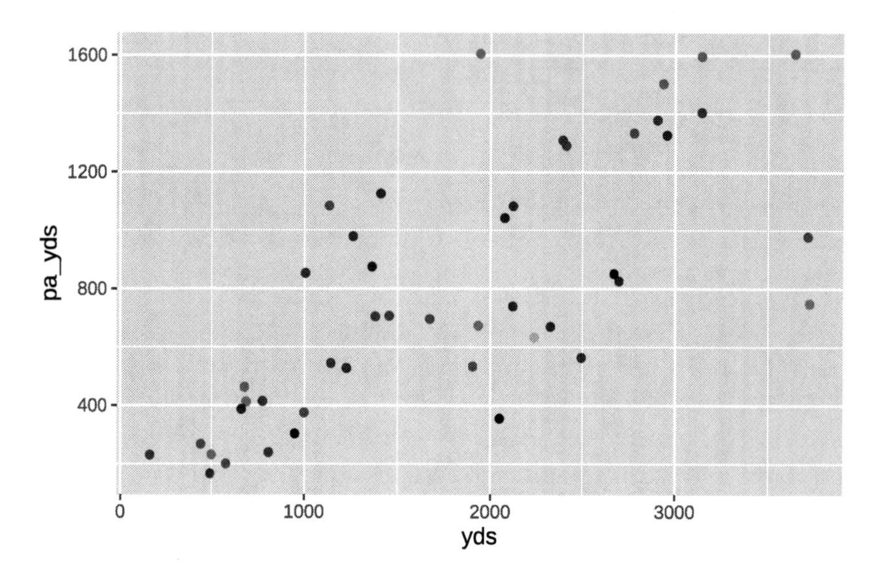

Figure 4.5: Adding team colors to the plot with the `nflplotR` package

In the above example, you will notice that we are including the team color information in an `aes()` call within the `geom_point()` function. This is because we ultimately control the specifics of the custom scale through the use of `scale_color_nfl` in the `nflplotR` package, which also allows us to select whether we want to display the primary or secondary team color.

Given the two examples above, a couple items regarding the use of `nflplotR` should become apparent.

1. If your data already includes team names in `team_abbr` format (that is: BAL, CIN, DET, DAL, etc.), then using `nflplotR` is likely a more efficient option as you do not need to merge in team color information. In other words, our `play_action_data` information could contain just the variables for `player`, `team_abbr`, `yds`, and `pa_yds` and `nflplotR` would still work as the package, "behind the scenes," automatically correlates the `team_abbr` with the correct color for each team.
2. However, if your data does not include teams in `team_abbr` format and you must merge in information manually, it is likely more efficient to use the `$` special operator to bring the team colors in without using the `aes()` call within `geom_point()`.

Finally, because we have both `team_color` and `team_color2` – the primary and secondary colors for each team – in the data, we can get fancy and create points that are filled with the primary team color and outlined by the secondary team color. Doing so simply requires changing the type of our `geom_point`. In the below example, we are specifying that we want a specific type of `geom_point` by using `shape = 21` and then providing the `fill` color and the outline color with `color`. In each case, we are again using the `$` special operator to select the primary and secondary color associated with each team.

```
ggplot(data = play_action_data, aes(x = yds, y = pa_yds)) +
  geom_point(shape = 21,
             fill = play_action_data$team_color,
             color = play_action_data$team_color2)
```

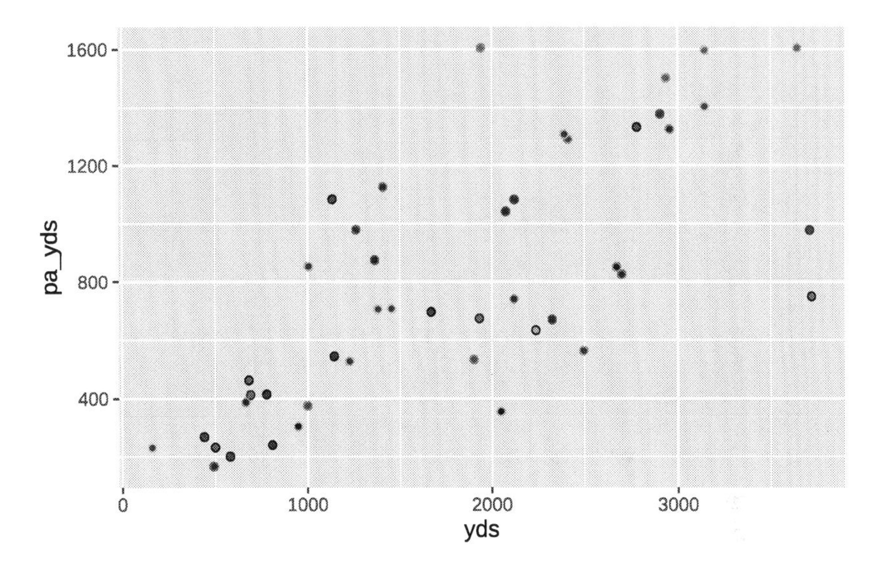

Figure 4.6: Using a different `geom_poitn()` shape in order to utilize both team colors

With team colors correctly associated with each point, we can turn back to our "to do list" to see what part of the job is next.

Note

Our data visualization "to do list":

1. ~~Adding team colors to each point.~~
2. Increasing the size of each point.
3. Adding player names to each point.
4. Increasing the number of ticks on each axis.
5. Provide the numeric values in correct format (that is, including a , to correctly show thousands).
6. Rename the title of each axis.
7. Provide a title, subtitle, and caption for the plot.
8. Add mean lines to both the x-axis and y-axis.
9. Change `theme` elements to make data viz more appealing.

4.4.2 Increasing the Size of Each Point

Determining when and how to resize the individual points in a scatterplot is a multifaceted decision. To that end, there is no hard and fast rule for doing so as it depends on both the specific goals and context of the visualization. There are, however, some general guidelines to keep in mind:

1. **Data density:** if you have a lot of data points within your plot, making the points smaller may help to reduce issues of overlapping/overplotting. Not only is this more aesthetically pleasing, but it can also help in making it easier to see patterns.
2. **Importance of individual points:** if certain points within the scatterplot are important, we may want to increase the size of those specific points to make them standout.
3. **Visual aesthetics:** the size of the points can be adjusted simply for visual appeal.
4. **Contextual factors:** can the size of the points be used to highlight even more uniqueness in the data? For example, given the right data structure, we can size individual dots to show the spread in total attempts across the quarterbacks.

Given the above guidelines, the resizing of the points in our `play_action_data` data frame is going to be a strictly aesthetic decision. We cannot, as mentioned above, alter the size of each specific points based on each quarterback's number of attempts as the data provides attempts for *both* play action and non-play action passes. Moreover, we *could* create a new column that add both attempt numbers to get a QB's cumulative total but that does not have a distinct correlation to the data on either axis.

Caution

For the sake of educational purposes, we can alter the size of each specific point to correlate to the total number of play action attempts for each quarterback (and then divide this by 25 in order to decrease the size of the points to fit them all onto the plot).

Again: it is important to point out that this not a good approach to data visualization, as the size of the points correlate to just one of the variables being explored in the plot.

```
ggplot(data = play_action_data, aes(x = yds, y = pa_yds)) +
  geom_point(shape = 21,
             fill = play_action_data$team_color,
             color = play_action_data$team_color2,
             size = play_action_data$pa_att / 25)
```

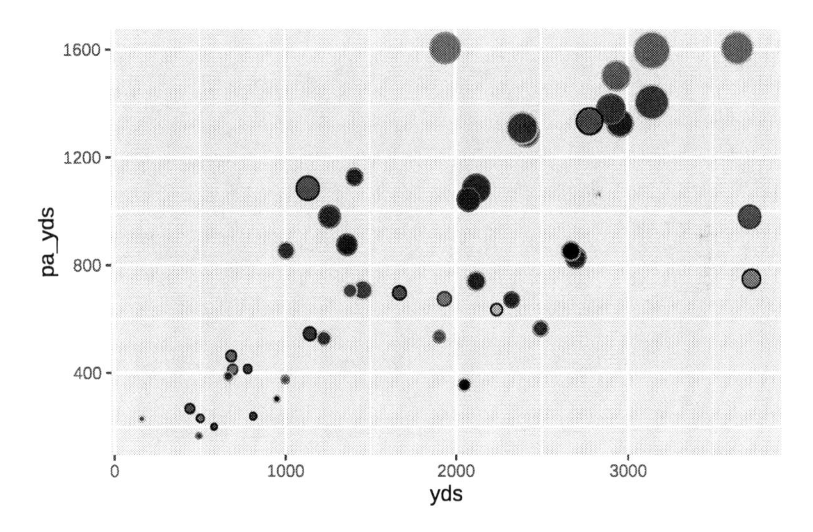

Figure 4.7: An incorrect approach to sizing the points, as the size correltes to just one of the variables

In order to maintain correct visualization standards, we can resize the points for nothing more than aesthetic purposes (that is: making them bigger so they are easier to see). To do so, we still add the `size` argument to our `geom_point` but providing a numeric value to apply uniformly across all the points. To process of selecting the numeric value is a case of trial and error – inputting and running, changing and running, and changing and running again until you find the size that provides easier to see points without adding overlap into the visualization.

```
ggplot(data = play_action_data, aes(x = yds, y = pa_yds)) +
  geom_point(shape = 21,
             fill = play_action_data$team_color,
             color = play_action_data$team_color2,
             size = 4.5)
```

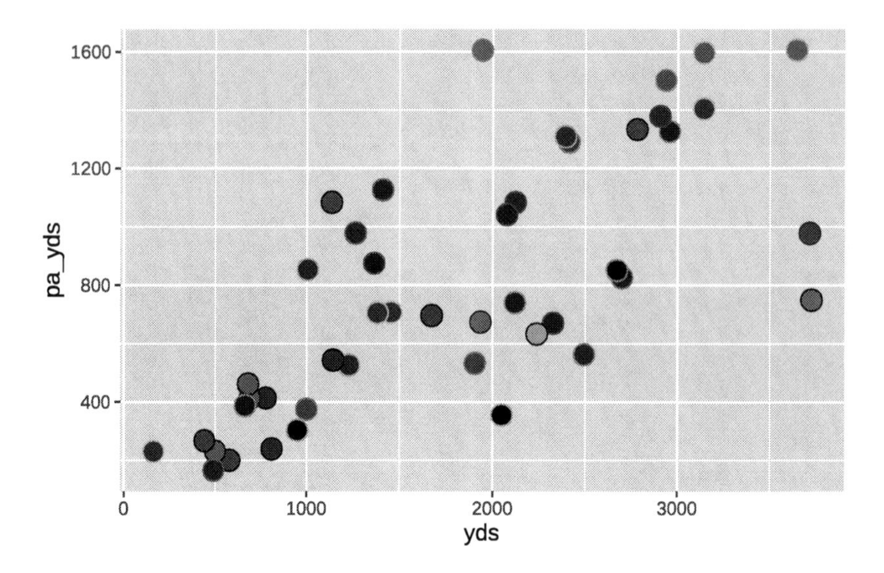

Figure 4.8: Resizing the points for aesthetic purposes

With the size of each point adequately adjusted, we can move on to the next part of our data visualization "to do list."

Note

Our data visualization "to do list":

1. ~~Adding team colors to each point.~~
2. ~~Increasing the size of each point.~~
3. Adding player names to each point.
4. Increasing the number of ticks on each axis.
5. Provide the numeric values in correct format (that is, including a , to correctly show thousands).
6. Rename the title of each axis.
7. Provide a title, subtitle, and caption for the plot.
8. Add mean lines to both the x-axis and y-axis.
9. Change `theme` elements to make data viz more appealing.

4.4.3 Adding Player Names to Each Point

While the our current plot includes team-specific colors for the points, we are still not able to discern – for the most part – which player belongs to which point. To rectify this, we will turn to using the `ggrepel` package, which is designed to improve the readability of text labels on plots by automatically repelling overlapping labels, if any. `ggrepel` operates with the use of two main functions: `geom_text_repel` and `geom_label_repel`. Both provide the same end result, with the core difference being `geom_label_repel` adding a customized label under each player's name.

We can do a bare minimum addition of the player names by adding one additional line of code using `geom_text_repel`, wrapping it in an `aes()` call, and specifying which variable in the `play_action_data` is the `label` we would like to display.

```
ggplot(data = play_action_data, aes(x = yds, y = pa_yds)) +
  geom_point(shape = 21,
             fill = play_action_data$team_color,
             color = play_action_data$team_color2,
             size = 4.5) +
  geom_text_repel(aes(label = player))
```

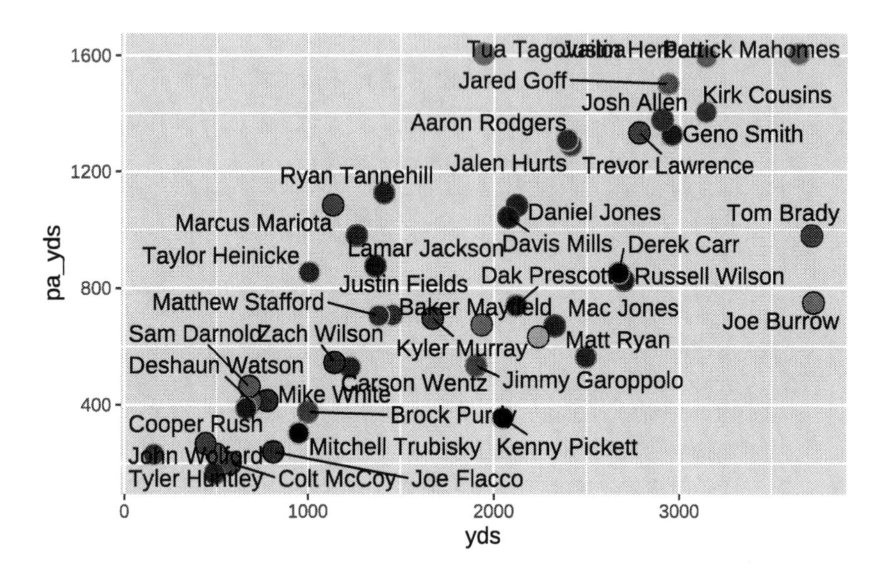

Figure 4.9: Adding player names to the plot

While it is a good first attempt at adding the names, many of them are awkwardly close to the respective point. Luckily, the `ggrepel` package provides plenty of built-in customization options:

`ggrepel` Option[1]	Description of Option
seed	a random numeric seed number for the purposes of recreating the same layout
force	force of repulsion between overlapping text labels
force_pull	force of attraction between each text label and its data point
direction	move the text labels in either "x" or "y" directions
nudge_x	adjust the starting x-axis starting position of the label
nudge_y	adjust the starting y-axis starting position of the label
box.padding	padding around the text label
point.padding	padding around the labeled data point
arrow	renders the line segment as an arrow

[1] As listed on the `ggrepel` website: https://ggrepel.slowkow.com/articles/examples.html

Of the above options, the our current issue with name and point spacing can be resolved by including a numeric value to the `box.padding`. Moreover, we can control the look and style of the text (such as size, font family, font face, etc.) in much the same way. To make these changes, we can set the `box.padding` to 0.45, set the size of the text to three using `size` as well as switch the font to "Roboto" using `family`, and – finally – make it bold using `fontface`.

```
ggplot(data = play_action_data, aes(x = yds, y = pa_yds)) +
  geom_point(shape = 21,
             fill = play_action_data$team_color,
             color = play_action_data$team_color2,
             size = 4.5) +
  geom_text_repel(aes(label = player),
                  box.padding = 0.45,
                  size = 3,
                  family = "Roboto",
                  fontface = "bold")
```

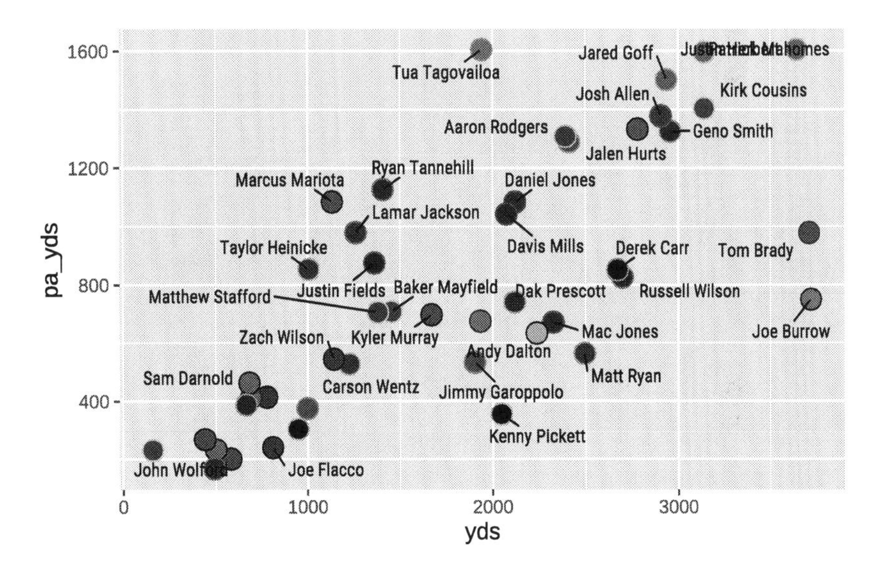

Figure 4.10: Adding options from `ggrepel` to the player names

The plot, as is, is understandable in that we are able to associate each point with a specific quarterback to examine how a quarterback's total passing yardage is split between play action and non-play action passes. While the graph could hypothetically "standalone" as is – minus a need for a title – we can still do work on it to make it more presentable. Let's return to our data visualization "to do list."

> **i Note**
>
> Our data visualization "to do list":
>
> 1. ~~Adding team colors to each point.~~
> 2. ~~Increasing the size of each point.~~
> 3. ~~Adding player names to each point.~~
> 4. Increasing the number of ticks on each axis.
> 5. Provide the numeric values in correct format (that is, including a , to correctly show thousands).
> 6. Rename the title of each axis.
> 7. Provide a title, subtitle, and caption for the plot.
> 8. Add mean lines to both x-axis and y-axis.
> 9. Change `theme` elements to make data viz more appealing.

4.4.4 Editing Axis Ticks and Values

Because steps four and five from the above "to do list" can be accomplished with the same package, we will lump both together and complete them at once.

Let's first examine the idea of increasing the number of ticks on each axis. The "axis ticks" refer to the specific spots on each axis, wherein a numeric data point resides. With our current visualization, we currently have "1000, 2000, 3000" on the x-axis and "400, 800, 1200, 1600" on the y-axis.

We may want to increase the number of axis ticks in this visualization as it can provide a more detailed view of the data being presented. Typically, increasing the number of tickets will show more granularity in the data and make it easier to interpret the values represented by each point. In this specific case, we can look at the cluster of points represented by Andy Dalton, Mac Jones, and Matt Ryan. Given the current structure of the axis ticks, we can guess that Matt Ryan has roughly 2,500 yards on non-play action passing attempts. Given we know Matt Ryan's amount, we can make guesses that Andy Dalton may be around 2,300 and Mac Jones somewhere between the two.

By increasing the number of values on each axis, we have the ability to see more specific results. Conversely, we must be careful to not add too many so that the data becomes overwhelming to interpret. Much like the size of `geom_point` was a case of trial and error, so is selecting an appropriate amount of ticks.

However, before implementing these changes, we need to segue into a discussion on continuous and discrete data.

! Important

When implementing changes to either the x- or y-axis in `ggplot`, you will be working with either continuous or discrete data, and using the `scale_x_continuous` or `scale_x_discrete` functions (replacing `x` with `y` when working with the opposite axis). In either case, both functions allow you to customize the axis of a plot but are used for different types of data.

`scale_x_continuous` is used for continuous (or numeric) data, where the axis is represented by a continuous range of numeric values. The values within a continuous axis can take on any number within the given range.

`scale_x_discrete` is used for discrete data (or often character-based data). You will see this function used when working with variables such as player names, teams, college names, etc. In any case, discrete data is limited to a specific set of categories.

Please know that `ggplot` will throw an error if you try to apply a continuous scale to discrete data, or the opposite, that reads: `Error: Discrete value supplied to continuous scale`.

In the case of our current plot, we now know we will be using the `scales_x_continuous` and `scale_y_continuous` functions as both contain continuous (numeric) data. To make the changes, we can turn to the `scales` package and its `pretty_breaks` function to change the number of "breaks" (or ticks) on each axis. By placing `n = 6` within the `pretty_breaks` argument, we are requesting a total of six axis ticks on both the x- and y-axis.

```
ggplot(data = play_action_data, aes(x = yds, y = pa_yds)) +
  geom_point(shape = 21,
             fill = play_action_data$team_color,
             color = play_action_data$team_color2,
             size = 4.5) +
  geom_text_repel(aes(label = player),
                  box.padding = 0.45,
                  size = 3,
                  family = "Roboto",
                  fontface = "bold") +
  scale_x_continuous(breaks = scales::pretty_breaks(n = 6)) +
  scale_y_continuous(breaks = scales::pretty_breaks(n = 6))
```

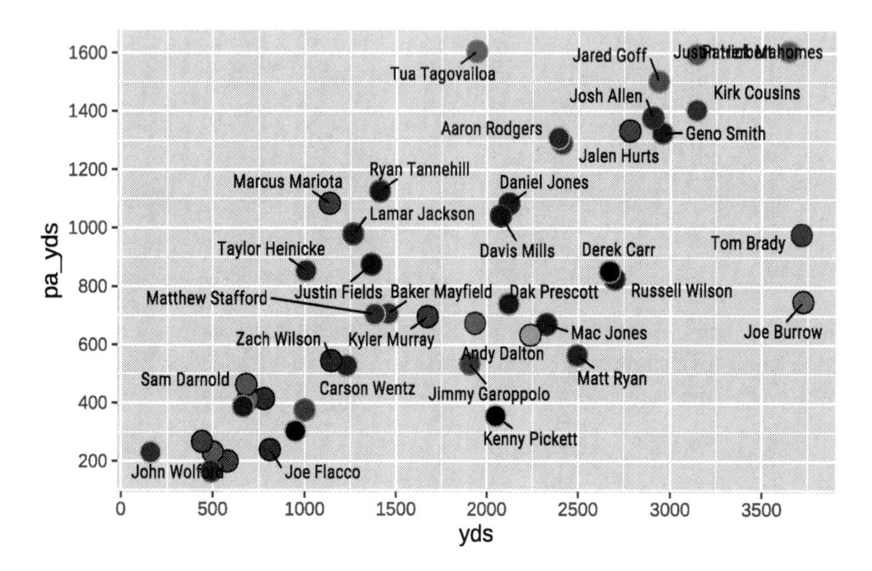

Figure 4.11: Changing numbers of breaks on each axis with `pretty_breaks()`

Despite our request to build the plot with six ticks on each axis, you will see the generated visualization includes seven on the x-axis and eight on the y-axis. *This does not mean your code with `pretty_breaks` did not work.* Instead, the `pretty_breaks` function is designed to internally determine the best axis tick optimization based on your requested number. To that end, the function determined that seven and eight ticks, respectively, was the most optimized way to display the data given our desire to have at least six on each.

With the number of axis ticks corrected, we can turn our attention to getting the labels of the axis ticks into correct numeric format. Within the same `scale_x_continuous` or `scale_y_continuous` arguments, we will use the `labels` function, combined with another tool from the `scales` package to make the adjustments.

```
ggplot(data = play_action_data, aes(x = yds, y = pa_yds)) +
  geom_point(shape = 21,
             fill = play_action_data$team_color,
             color = play_action_data$team_color2,
             size = 4.5) +
  geom_text_repel(aes(label = player),
                  box.padding = 0.45,
                  size = 3,
                  family = "Roboto",
                  fontface = "bold") +
```

```
  scale_x_continuous(breaks = scales::pretty_breaks(n = 6),
                     labels = scales::label_comma()) +
  scale_y_continuous(breaks = scales::pretty_breaks(n = 6),
                     labels = scales::label_comma())
```

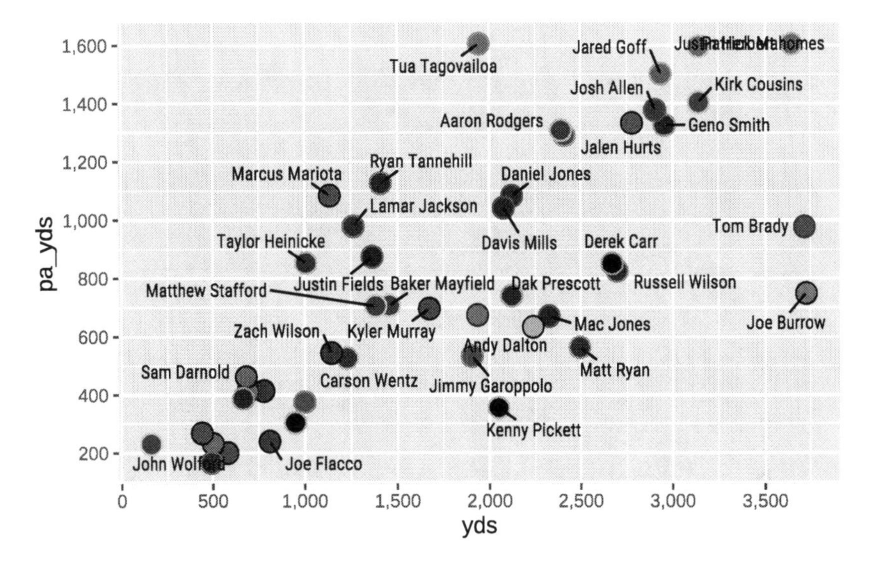

Figure 4.12: Adding commas to the axis values using `label_comma()`

By adding a `labels` option to both `scale_x_continuous` and `scale_y_continuous`, we can use the `label_comma()` option from within the `scales` package to easily add a comma into numbers that are in the thousands.

With much of the heavy lifting for our visualization now complete, we can move on to the final steps in our "to do list."

> **i** Note
>
> **Our data visualization "to do list":**
>
> 1. ~~Adding team colors to each point.~~
> 2. ~~Increasing the size of each point.~~
> 3. ~~Adding player names to each point.~~
> 4. ~~Increasing the number of ticks on each axis.~~
> 5. ~~Provide the numeric values in correct format (that is, including a , to correctly show thousands).~~

6. Rename the title of each axis.
7. Provide a title, subtitle, and caption for the plot.
8. Add mean lines to both x-axis and y-axis.
9. Change `theme` elements to make data viz more appealing.

4.4.5 Changing Axis Titles and Adding Title, Subtitle, and Caption

Much like our last section, we can work on changing the title of each axis and adding a title, subtitle, and caption for the plot within one section, as all this is added and/or changed by using `labs()`.

```
ggplot(data = play_action_data, aes(x = yds, y = pa_yds)) +
  geom_point(shape = 21,
             fill = play_action_data$team_color,
             color = play_action_data$team_color2,
             size = 4.5) +
  geom_text_repel(aes(label = player),
                  box.padding = 0.45,
                  size = 3,
                  family = "Roboto",
                  fontface = "bold") +
  scale_x_continuous(breaks = scales::pretty_breaks(n = 6),
                     labels = scales::label_comma()) +
  scale_y_continuous(breaks = scales::pretty_breaks(n = 6),
                     labels = scales::label_comma()) +
  labs(x = "Non-Play Action Yards",
       y = "Play Action Yards",
       title = "Cumulative Passing Yards",
       subtitle = "Non-Play Action vs. Play Action",
       caption = "*Introduction to NFL Analytics with R*<br>
               **Bradley J. Congelio**")
```

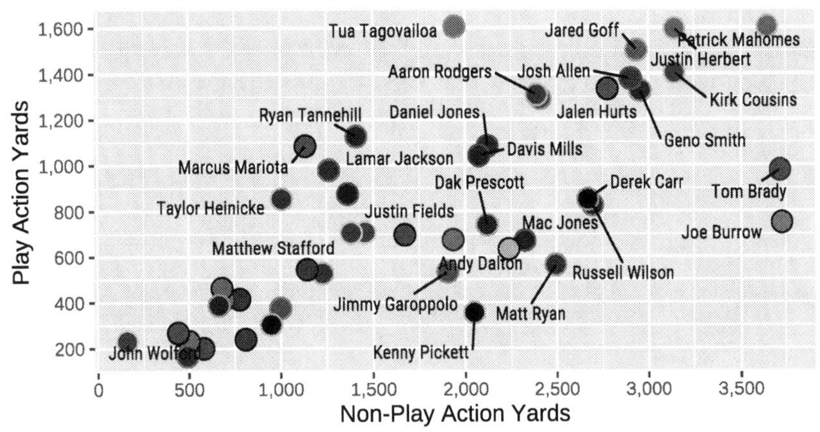

Figure 4.13: Adding the plot's title and subtitle

Within the new `labs()`, we are placing a total of five items: `x` (allowing us to name the x-axis outside the confines of what it is called in the beginning `aes()` call, `y` (allowing us to name the y-axis), `title` (allowing us to add a title to the top of the plot), `subtitle` (allowing us to add a subtitle below the title and provide more contextual information), and `caption` (allowing us to provide information about, where the graph come from and who designed it). We will explore ways to change the font, size, color, and more of these items when we move on to the last item of our "to do list."

> **i Note**
>
> **Our data visualization "to do list":**
>
> 1. ~~Adding team colors to each point.~~
> 2. ~~Increasing the size of each point.~~
> 3. ~~Adding player names to each point.~~
> 4. ~~Increasing the number of ticks on each axis.~~
> 5. ~~Provide the numeric values in correct format (that is, including a , to correctly show thousands).~~
> 6. ~~Rename the title of each axis.~~
> 7. ~~Provide a title, subtitle, and caption for the plot.~~
> 8. Add mean lines to both x-axis and y-axis.
> 9. Change theme elements to make data viz more appealing.

4.4.6 Adding Mean Lines to Both x-axis and y-axis

Adding mean (or average) lines to both the x-axis and y-axis allows us to visualize, where each quarterback falls within one of four sections (according to the amount of passing yards in both situations). Adding the lines is done with the inclusion of two additional geoms to the existing plot (in this case `geom_hline` and `geom_vline`).

```
ggplot(data = play_action_data, aes(x = yds, y = pa_yds)) +
   geom_point(shape = 21,
              fill = play_action_data$team_color,
              color = play_action_data$team_color2,
              size = 4.5) +
  geom_text_repel(aes(label = player),
                  box.padding = 0.45,
                  size = 3,
                  family = "Roboto",
                  fontface = "bold") +
  scale_x_continuous(breaks = scales::pretty_breaks(n = 6),
                     labels = scales::label_comma()) +
  scale_y_continuous(breaks = scales::pretty_breaks(n = 6),
                     labels = scales::label_comma()) +
  labs(x = "Non-Play Action Yards",
       y = "Play Action Yards",
       title = "Cumulative Passing Yards",
       subtitle = "Non-Play Action vs. Play Action") +
  geom_hline(yintercept = mean(play_action_data$pa_yds),
             linewidth = .8, color = "black",
             linetype = "dashed") +
  geom_vline(xintercept = mean(play_action_data$yds),
             linewidth = .8, color = "black",
             linetype = "dashed")
```

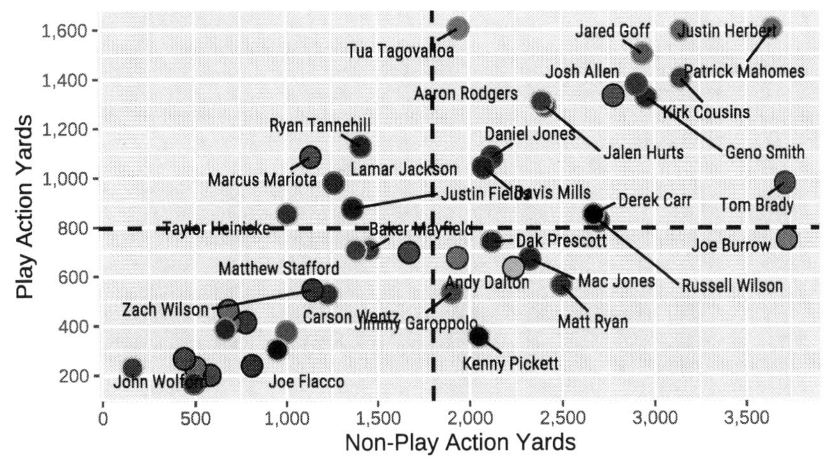

Figure 4.14: Adding mean lines incorrectly, as they are top of every layer

In the above output, we used `geom_hline` and `geom_vline` to draw a dashed line at the average for both `pa_yds` at the `yintercept` and `yds` at the `xintercept`. Because of this, we can see that – for example – Marcus Mariota is above average for play action yards, but below average for non-play action yards. Additionally, Matthew Stafford, Baker Mayfield, Kyler Murray, and many others are all below average in both metrics while Jared Goff, Justin Herbert, Patrick Mahomes, and others are well above average in both.

However, adding the `geom_hline` and `geom_vline` at the very end of the `ggplot` code creates an issues (and one I've intentionally created for the purposes of education). As you look at the plot, you will see that the dashed line runs **on top** of the names and dots in the plot. This is because `ggplot` follows a very specific ordering of layering.

> **!** Important
>
> **In `ggplot`, it is important to remember that items in a plot are layered in the order in which they are added to the plot.** This process of layering is important because it ultimately determines which items end up on top of others, which can have significant implications on the visual appearance of the plot.

> As we've seen so far in the process, each layer of a plot is added by including a `geom_`. The first layer added will *always* be at the very bottom of the plot, with each additional layer building on top of the previous layers.

Because of the important layering issue highlighted above, it is visually necessary for us to move the `geom_hline` and `geom_vline` to the beginning of the `ggplot` code so both are layered underneath everything else in the plot (`geom_point` and `geom_text_repel` in this case). As well, we can apply the `alpha` option to each to slightly decrease each line's transparency.

```
ggplot(data = play_action_data, aes(x = yds, y = pa_yds)) +
  geom_hline(yintercept = mean(play_action_data$pa_yds),
             linewidth = .8,
             color = "black",
             linetype = "dashed",
             alpha = 0.5) +
  geom_vline(xintercept = mean(play_action_data$yds),
             linewidth = .8,
             color = "black",
             linetype = "dashed",
             alpha = 0.5) +
   geom_point(shape = 21,
             fill = play_action_data$team_color,
             color = play_action_data$team_color2,
             size = 4.5) +
  geom_text_repel(aes(label = player),
                  box.padding = 0.45,
                  size = 3,
                  family = "Roboto",
                  fontface = "bold") +
  scale_x_continuous(breaks = scales::pretty_breaks(n = 6),
                     labels = scales::label_comma()) +
  scale_y_continuous(breaks = scales::pretty_breaks(n = 6),
                     labels = scales::label_comma()) +
  labs(x = "Non-Play Action Yards",
       y = "Play Action Yards",
       title = "Cumulative Passing Yards",
       subtitle = "Non-Play Action vs. Play Action")
```

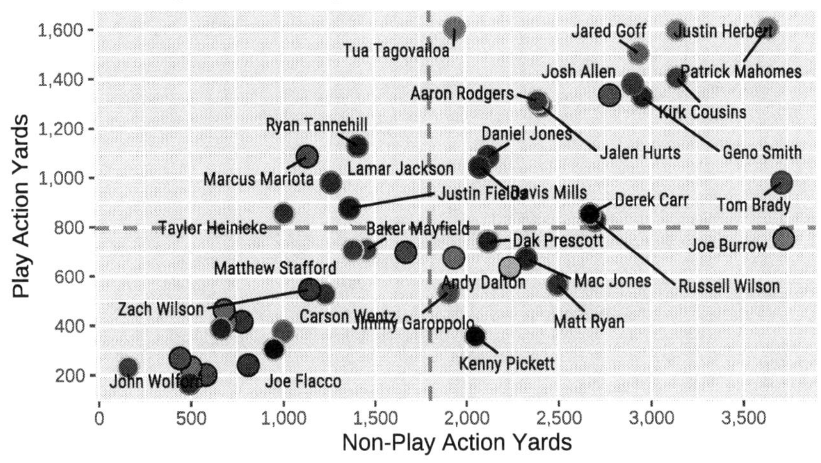

Figure 4.15: Adding mean lines at beginning of code so they are a bottom layer of the plot

By moving the average lines to the top of our `ggplot` code, both are now layered under the other two `geom_` and are not visually impacting the final plot.

4.4.6.1 Adding Mean Lines with `nflplotR`

> **Tip**
>
> Even though we were not able to use `nflplotR` to handle the colors in this plot because the data lacked a corresponding `team_abbr` variable, we can still use `nflplotR` to add our mean lines – and I actually recommend doing so, as it requires less lines of code (thus less typing). See below for an example.

```
ggplot(data = play_action_data, aes(x = yds, y = pa_yds)) +
  geom_mean_lines(aes(v_var = yds, h_var = pa_yds),
                  size = 0.8,
                  color = "black",
                  linetype = "dashed",
                  alpha = 0.5) +
```

```
geom_point(shape = 21,
           fill = play_action_data$team_color,
           color = play_action_data$team_color2,
           size = 4.5) +
geom_text_repel(aes(label = player),
                box.padding = 0.45,
                size = 3,
                family = "Roboto",
                fontface = "bold") +
scale_x_continuous(breaks = scales::pretty_breaks(n = 6),
                   labels = scales::label_comma()) +
scale_y_continuous(breaks = scales::pretty_breaks(n = 6),
                   labels = scales::label_comma()) +
labs(x = "Non-Play Action Yards",
     y = "Play Action Yards",
     title = "Cumulative Passing Yards",
     subtitle = "Non-Play Action vs. Play Action")
```

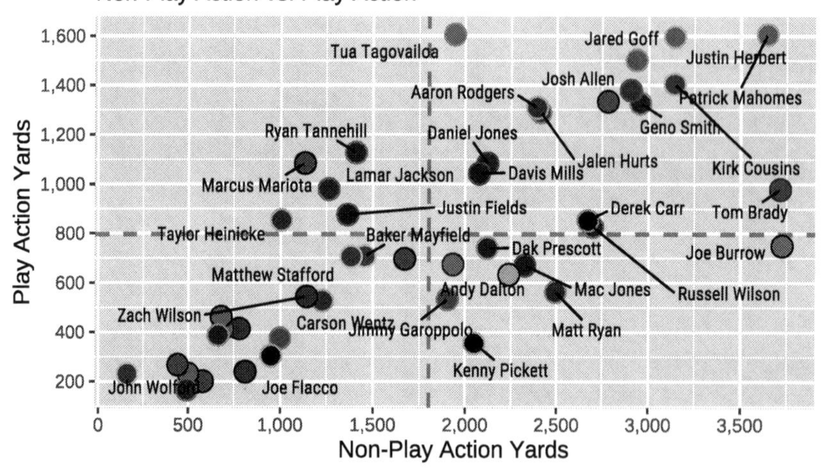

Figure 4.16: Adding mean lines with nflplotR

By using the `geom_mean_lines()` function within `nflplotR`, we can construct both of the lines together rather than needing to provide a `geom_hline()` and a `geom_vline()` argument. Because of this, we can also provide the size, width, and type of our line just once (rather than repeating it again like we had to in the former method).

We can now move onto the final item on our data visualization "to do list."

> **i** Note
>
> **Our data visualization "to do list":**
>
> 1. ~~Adding team colors to each point.~~
> 2. ~~Increasing the size of each point.~~
> 3. ~~Adding player names to each point.~~
> 4. ~~Increasing the number of ticks on each axis.~~
> 5. ~~Provide the numeric values in correct format (that is, including a , to correctly show thousands).~~
> 6. ~~Rename the title of each axis.~~
> 7. ~~Provide a title, subtitle, and caption for the plot.~~
> 8. ~~Add mean lines to both x-axis and y-axis.~~
> 9. Change theme elements to make data viz more appealing.

4.4.7 Making Changes to Theme Elements

There is a laundry list of options to be explored when it comes to editing your plot's theme elements to make it look exactly as you want. Currently, according to the `ggplot2` website, the following is a comprehensive list of elements that you can tinker with.

```
line,
rect,
text,
title,
aspect.ratio,
axis.title,
axis.title.x,
axis.title.x.top,
axis.title.x.bottom,
axis.title.y,
axis.title.y.left,
axis.title.y.right,
axis.text,
axis.text.x,
axis.text.x.top,
axis.text.x.bottom,
axis.text.y,
axis.text.y.left,
```

```
    axis.text.y.right,
    axis.ticks,
    axis.ticks.x,
    axis.ticks.x.top,
    axis.ticks.x.bottom,
    axis.ticks.y,
    axis.ticks.y.left,
    axis.ticks.y.right,
    axis.ticks.length,
    axis.ticks.length.x,
    axis.ticks.length.x.top,
    axis.ticks.length.x.bottom,
    axis.ticks.length.y,
    axis.ticks.length.y.left,
    axis.ticks.length.y.right,
    axis.line,
    axis.line.x,
    axis.line.x.top,
    axis.line.x.bottom,
    axis.line.y,
    axis.line.y.left,
    axis.line.y.right,
    legend.background,
    legend.margin,
    legend.spacing,
    legend.spacing.x,
    legend.spacing.y,
    legend.key,
    legend.key.size,
    legend.key.height,
    legend.key.width,
    legend.text,
    legend.text.align,
    legend.title,
    legend.title.align,
    legend.position,
    legend.direction,
    legend.justification,
    legend.box,
    legend.box.just,
    legend.box.margin,
    legend.box.background,
    legend.box.spacing,
    panel.background,
```

```
    panel.border,
    panel.spacing,
    panel.spacing.x,
    panel.spacing.y,
    panel.grid,
    panel.grid.major,
    panel.grid.minor,
    panel.grid.major.x,
    panel.grid.major.y,
    panel.grid.minor.x,
    panel.grid.minor.y,
    panel.ontop,
    plot.background,
    plot.title,
    plot.title.position,
    plot.subtitle,
    plot.caption,
    plot.caption.position,
    plot.tag,
    plot.tag.position,
    plot.margin,
    strip.background,
    strip.background.x,
    strip.background.y,
    strip.clip,
    strip.placement,
    strip.text,
    strip.text.x,
    strip.text.x.bottom,
    strip.text.x.top,
    strip.text.y,
    strip.text.y.left,
    strip.text.y.right,
    strip.switch.pad.grid,
    strip.switch.pad.wrap
```

It's not likely that we will encounter all these theme elements in this book. But, the ones we do use, we will use heavily. For example, I prefer to design all of my data visualizations without the "axis ticks" (those small lines sticking out from the plot just above, or beside, each yardage number).

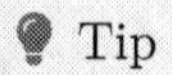
Tip

Please note the keywords in the above paragraph: "**I prefer.**"

Nearly all the work you conduct within the `theme()` of your data visualizations are just that – your preference. I very much have a "personal preference" that unites all the data viz work that I do and share for public consumption.

You can feel free to follow along with my preferences, including the use of the upcoming `nfl_analytics_theme()` I will provide, or to make slight (or major!) adjustments to everything we cover in the coming section to make it fit your artistic vision.

Be creative and do not be afraid to experiment with all the options available to you in `theme()`.

Let's start by removing the ticks on both the x- and y-axis. To do so, we will add the `theme()` argument at the end of your prior `ggplot` code and then start building out each and every change we want to make from the above list of options.

```
ggplot(data = play_action_data, aes(x = yds, y = pa_yds)) +
  geom_mean_lines(aes(v_var = yds, h_var = pa_yds),
                  size = 0.8,
                  color = "black",
                  linetype = "dashed",
                  alpha = 0.5) +
  geom_point(shape = 21,
             fill = play_action_data$team_color,
             color = play_action_data$team_color2,
             size = 4.5) +
  geom_text_repel(aes(label = player),
                  box.padding = 0.45,
                  size = 3,
                  family = "Roboto",
                  fontface = "bold") +
  scale_x_continuous(breaks = scales::pretty_breaks(n = 6),
                     labels = scales::label_comma()) +
  scale_y_continuous(breaks = scales::pretty_breaks(n = 6),
                     labels = scales::label_comma()) +
  labs(x = "Non-Play Action Yards",
       y = "Play Action Yards",
```

```
      title = "Cumulative Passing Yards",
      subtitle = "Non-Play Action vs. Play Action") +
  theme(
    axis.ticks = element_blank())
```

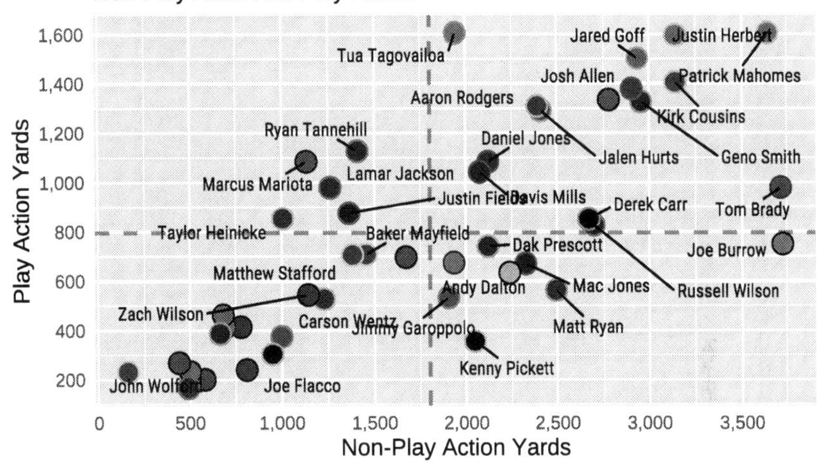

Figure 4.17: Editing theme to remove axis ticks

Within the `theme()` argument, we take the specific element we wish to change (from the above list of possibilities from the `ggplot2` website) and then provide the instruction on what to do. In this case, since we wish to completely remove the `axis.ticks` from the entire plot, we can provide `element_blank()` which removes them. We can continue making changes to the plot's elements by adding all of these preferences to out `theme()`.

```
ggplot(data = play_action_data, aes(x = yds, y = pa_yds)) +
  geom_mean_lines(aes(v_var = yds, h_var = pa_yds),
                  size = 0.8,
                  color = "black",
                  linetype = "dashed",
                  alpha = 0.5) +
  geom_point(shape = 21,
             fill = play_action_data$team_color,
             color = play_action_data$team_color2,
             size = 4.5) +
```

```
geom_text_repel(aes(label = player),
                box.padding = 0.45,
                size = 3,
                family = "Roboto",
                fontface = "bold") +
scale_x_continuous(breaks = scales::pretty_breaks(n = 6),
                   labels = scales::label_comma()) +
scale_y_continuous(breaks = scales::pretty_breaks(n = 6),
                   labels = scales::label_comma()) +
labs(x = "Non-Play Action Yards",
     y = "Play Action Yards",
     title = "Cumulative Passing Yards",
     subtitle = "Non-Play Action vs. Play Action") +
theme(
  axis.ticks = element_blank(),
  axis.title = element_text(family = "Roboto",
                            size = 10,
                            color = "black"),
  axis.text = element_text(family = "Roboto",
                           face = "bold",
                           size = 10,
                           color = "black"),
  plot.title.position = "plot",
  plot.title = element_text(family = "Roboto",
                            size = 16,
                            face = "bold",
                            color = "#E31837",
                            vjust = .02,
                            hjust = 0.5),
  plot.subtitle = element_text(family = "Roboto",
                               size = 12,
                               color = "black",
                               hjust = 0.5),
  plot.caption = element_text(family = "Roboto",
                              size = 8,
                              face = "italic",
                              color = "black"),
  panel.grid.minor = element_blank(),
  panel.grid.major =  element_line(color = "#d0d0d0"),
  panel.background = element_rect(fill = "#f7f7f7"),
  plot.background = element_rect(fill = "#f7f7f7"),
  panel.border = element_blank())
```

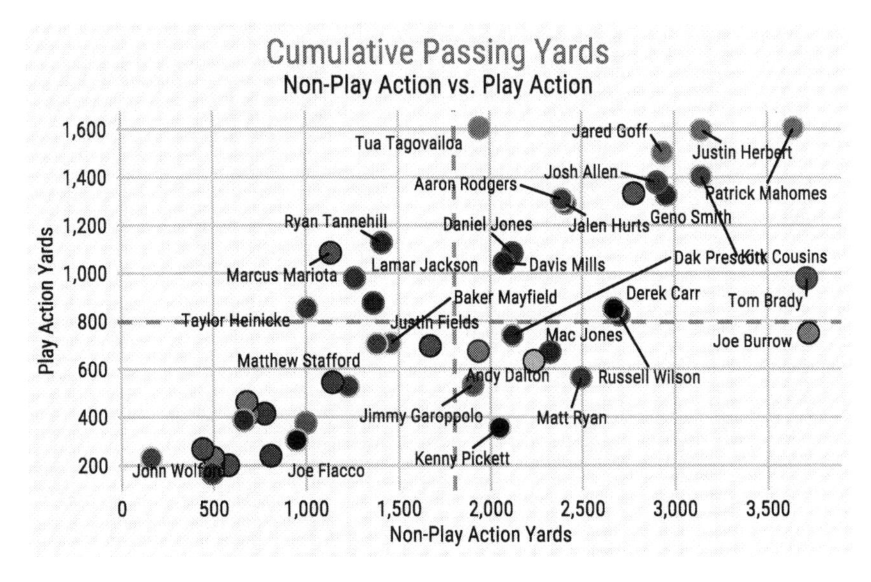

Figure 4.18: Adding a multitude of theme elements to the plot

There is a lot going on now in our `theme()` argument. Importantly, you may notice that we used `element_blank()` to remove the axis tick marks, but then switched to using `element_text()` and `element_rect()` for the reminder of the edits with our `theme()`. Both are one of four theme elements that can be modified in the above fashion.

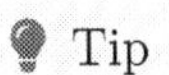 Tip

The Four Theme Elements You Can Edit

When working on editing your plot to your liking, you can make changes using one of four theme elements:

1. `element_blank()` – this used to entirely remove an element from the plot like we did with `axis.ticks`.
2. `element_rect()` – this is used to make changes to the borders and backgrounds of a plot.
3. `element_lines()` – this is used to make changes to any element in the plot that is a line.
4. `element_text()` – this is used to make change to any element in the plot that is text.

We first made changes to the text of the title associated with the x- and y-axis and made edits to the text of the numeric values for each yardage distance.

This process is started by using `axis.title` and `axis.text` in conjunction with `element_text()`, since that is the specific element type we are wishing to edit. Within our `element_text()`, we provided the numerous argument on how we wished to edit the text by providing the `family` (or the font), the `face`, the `size`, and the `color`.

After, we got a little fancy in our edits to our plots title and subtitle. I knew that I wanted to center both directly in the middle of the plot. Rather than figuring out the specific horizontal adjustment needed, I used the `plot.title.position()` argument and set it to `"plot"`, which used the entire width of our plot as the reference point for where to center the plot title and subtitle.

To take advantage of this, we followed by using the `plot.title()` argument to set the title's horizontal adjustment to 0.5 (`hjust = 0.5`). As you may guess, the inclusion of 0.5 instructs the output to center the title (and the subtitle in the ensuing edit) directly over the middle of the plot (as calculated through our prior use of `plot.title.position()`.

Our next significant change occurred by changing the aesthetics of the plot's grid lines, background, and border. Because we are working with line and background elements, we switch from `element_text()` and begin to use either `element_line()` or `element_rect()` (as well as again using `element_blank()` to completely remove the panel's minor grid lines).

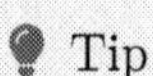

In a plot, which are the minor grid lines and which are the major?

In a ggplot2 plot, **minor grid lines** are those lines that hit either the x- or y-axis between the continuous or discrete values. Conversely, **major grid lines** are the lines that hit the axis at the same spot as the data values.

In the case of the current plot, our major grid lines are those that hit the x-axis at 500, 1,000, 1,500, and so on and hit the y-axis at 400, 600, 800, etc. The minor grid lines met the axis between the major grid lines.

After removing the panel's minor grid lines (again, a personal preference of mine), we also change the color of the panel's major grid lines, then change the color of the plot's background (both using `element_rect()`). The end result is a aesthetically pleasing data visualization.

4.5 Creating Your Own ggplot2 Theme

As mentioned, I have a distinctive "brand and look" for the data visualizations I create that make use of the same design elements and choices. Rather than copy and paste those into each and every `ggplot` piece of code I write, I've opted to consolidate all the `theme()` element changes into my own `nfl_analytics_theme()` function. In much the same way, I've created a "quick and easy" theme for use in this book. To get started, running the following chunk of code will create a function titled `nfl_analytics_theme` and place it into your RStudio environment.

```
nfl_analytics_theme <- function(..., base_size = 12) {

    theme(
      text = element_text(family = "Roboto",
             size = base_size),
      axis.ticks = element_blank(),
      axis.title = element_text(color = "black",
                                 face = "bold"),
      axis.text = element_text(color = "black",
                                face = "bold"),
      plot.title.position = "plot",
      plot.title = element_text(size = 16,
                                 face = "bold",
                                 color = "black",
                                 vjust = .02,
                                 hjust = 0.5),
      plot.subtitle = element_text(color = "black",
                                    hjust = 0.5),
      plot.caption = element_text(size = 8,
                                   face = "italic",
                                   color = "black"),
      panel.grid.minor = element_blank(),
      panel.grid.major =  element_line(color = "#d0d0d0"),
      panel.background = element_rect(fill = "#f7f7f7"),
      plot.background = element_rect(fill = "#f7f7f7"),
      panel.border = element_blank())
}
```

You may notice that the elements in the above theme creation are quite similar to the ones we passed into our previous plot. And that is true, and the results will be nearly 99.9% identical. After wrapping our `theme()` inside a function, indicated by the opening and closing curly brackets `{ }`, we can provide our desired theme element appearances as we would within a regular `ggplot2` code block.

However, in the above theme function, we have streamlined the basis a bit by indicating a `base_size` of all the text, when means all text output will be in size 12 font unless specifically indicated in the element (for example, we have the `plot.title` set to have a `size` of 16). As well, the same process was done for font (Roboto) so there was not a need to type it repeatedly into every element.

Based on the above example, you are free to create as detailed a theme function as you desire. The beauty of creating your own theme like above is you will no longer need to edit each portion of the every plot element.

```
ggplot(data = play_action_data, aes(x = yds, y = pa_yds)) +
  geom_mean_lines(aes(v_var = yds, h_var = pa_yds),
                  size = 0.8,
                  color = "black",
                  linetype = "dashed",
                  alpha = 0.5) +
  geom_point(shape = 21,
             fill = play_action_data$team_color,
             color = play_action_data$team_color2,
             size = 4.5) +
  geom_text_repel(aes(label = player),
                  box.padding = 0.45,
                  size = 3,
                  family = "Roboto",
                  fontface = "bold") +
  scale_x_continuous(breaks = scales::pretty_breaks(n = 6),
                     labels = scales::label_comma()) +
  scale_y_continuous(breaks = scales::pretty_breaks(n = 6),
                     labels = scales::label_comma()) +
  labs(x = "Non-Play Action Yards",
     y = "Play Action Yards",
     title = "**Cumulative Passing Yards**",
     subtitle = "*Non-Play Action vs. Play Action*",
     caption = "*An Introduction to NFL Analytics with R*<br>
     **Brad J. Congelio**") +
  nfl_analytics_theme()
```

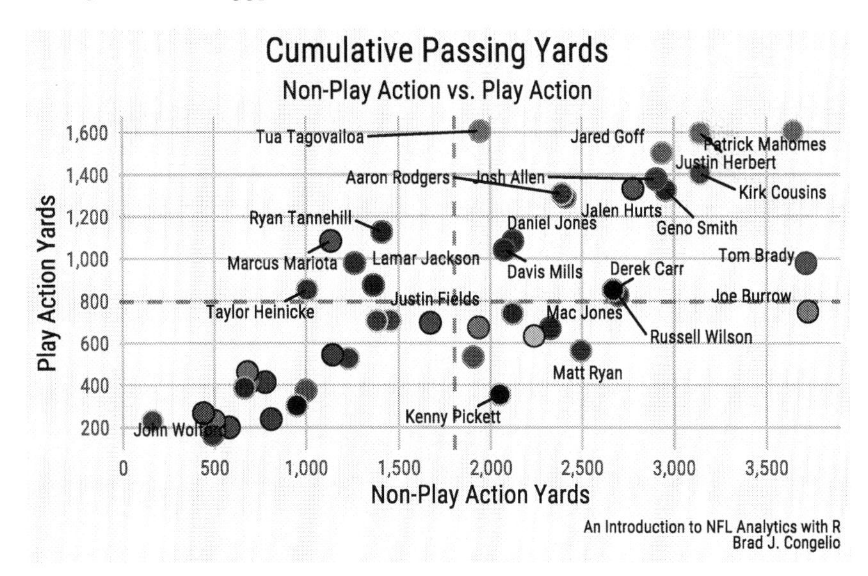

Figure 4.19: Consolidating theme element changes into `nfl_analytics_theme()`

As you can see, we just consolidated 28 lines of code into a **single line of code by wrapping all our theme elements into an easy to construct function.**

Unfortunately, you will notice that the resulting plot does not output with the title ("Cumulative Passing Yards") in Kansas City red like in our original. This is because, in our `nfl_analytics_theme()` function, the `color` for `axis.title()` is set to `"black"`. Thankfully, we can make this small edit within our `ggplot` code to switch the title back to Kansas City red, highlighting the idea that – despite the theme being built into a function – we still have the ability to make necessary edits on the fly without including all 28 lines of code. With the `nfl_analytics_theme()` active, we can still add additional `theme` elements as needed to make modifications, as seen below.

```
ggplot(data = play_action_data, aes(x = yds, y = pa_yds)) +
  geom_mean_lines(aes(v_var = yds, h_var = pa_yds),
                  size = 0.8,
                  color = "black",
                  linetype = "dashed",
                  alpha = 0.5) +
  geom_point(shape = 21,
             fill = play_action_data$team_color,
             color = play_action_data$team_color2,
```

```
              size = 4.5) +
    geom_text_repel(aes(label = player),
                    box.padding = 0.45,
                    size = 3,
                    family = "Roboto",
                    fontface = "bold") +
    scale_x_continuous(breaks = scales::pretty_breaks(n = 6),
                       labels = scales::label_comma()) +
    scale_y_continuous(breaks = scales::pretty_breaks(n = 6),
                       labels = scales::label_comma()) +
    labs(x = "Non-Play Action Yards",
       y = "Play Action Yards",
       title = "**Cumulative Passing Yards**",
       subtitle = "*Non-Play Action vs. Play Action*",
       caption = "*An Introduction to NFL Analytics with R*<br>
       **Brad J. Congelio**") +
    nfl_analytics_theme() +
    theme(plot.title = element_markdown(color = "#E31837"))
```

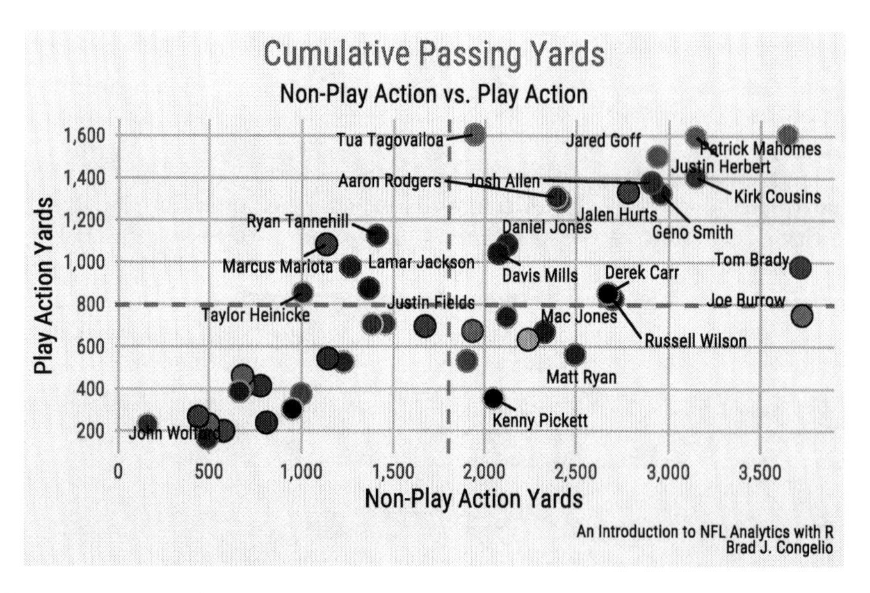

Figure 4.20: Editing the color of the plot's title after including `nfl_analytics_theme()`

4.6 Using Team Logos in Plots

To explore the process of placing team logos into a plot, let's stick with our previous example of working with play action passing data but explore it at the team level rather than individual quarterbacks. To gather the data, we can use `vroom` to read it in from the book's GitHub.

```
team_playaction
        <- vroom("http://nfl-book.bradcongelio.com/team-pa")
```

The resulting `team_playaction` data contains nearly identical information to the previous QB-level data. However, there is a slight change in how Sports Info Solutions charts passing yards on the team level. You will notice a column for both `net_yds` and `gross_yds`. When charting passing attempts in football, a player's individual passing yards are aggregated under **gross yards** with all lost yardage resulting from a sack being included. On the other hand, team passing yards are always presented in **net yards** and any lost yardage from sacks are not included. Case in point, the `gross_yds` number in our `team_playaction` data is greater than the `net_yds` for all 32 NFL teams. In any case, we will build our data visualization using `net_yds`.

In order to include team logos in the plot, we must first merge in team logo information (again with the understanding that we could use `nflplotR` if team abbreviations were included in the data). We can collect the team logo information using the `load_teams()` function within `nflreadR`. There are three different variations of each team's logo available in the resulting data: (1) the logo from ESPN, (2) the logo from Wikipedia, and (3) a pre-edited version of the logo that is cropped into a square.

> ℹ Note
>
> **There are slight differences in disk space, pixels, and utility in the provided ESPN, Wikipedia, and squared versions of the logos.**
>
> The Wikipedia versions are, generally, smaller in size. The Arizona Cardinals logo, for example, is just 9.11 KB in size from Wikipedia while the ESPN version of the logo is 20.6 KB. This difference in disk size is the result of each image's dimensions and pixels. The ESPN version, with the larger disk size, is also a higher quality image that is scaled in 500x500 dimensions (and 500 pixels). The

Wikipedia version is scaled in 179x161 dimensions at 179 pixels and 161 pixels, respectively. The squared version of the logo is a 200x200 image at 200 pixels with a background matching the team's primary color.

What does this mean? The ESPN version of the logo is better for those applications, where the logo will be large and you do not want any loss of quality. The Wikipedia version, conversely, is better suited for applications, like ours: for use in a small-scale data visualization. We do not plan on "blowing up" the image, thus losing quality and the smaller disk space size of the images allows for a slightly quicker rendering time when we use the `ggimage` package. While sparingly used, the squared version of the logo can be used in certain data visualization applications that requires the team logo to quickly and easily "merge" into the background of the plot (more on this later in this chapter, though).

Because of the above explanation, we can collect just the team nicknames and the Wikipedia version of the logo to merge into our existing `team_playaction` data.

```
teams <- nflreadr::load_teams(current = TRUE) %>%
  select(team_nick, team_logo_wikipedia)

team_playaction <- team_playaction %>%
  left_join(teams, by = c("team" = "team_nick"))

team_playaction
```

With the data now containing the correct information, we can build a basic version of our data visualization using the same `geom_point` as above, and continue to configure the information on both the x- and y-axis, to verify that everything is working correctly before switching out `geom_point` for `geom_image` in order to bring the team logos into the plot.

```
ggplot(data = team_playaction, aes(x = net_yds,
        y = pa_net_yds)) +
  geom_hline(yintercept = mean(team_playaction$pa_net_yds),
             linewidth = 0.8,
             color = "black",
             linetype = "dashed") +
```

```
  geom_vline(xintercept = mean(team_playaction$net_yds),
             linewidth = 0.8,
             color = "black",
             linetype = "dashed") +
  geom_point() +
  scale_x_continuous(breaks = scales::pretty_breaks(n = 6),
                     labels = scales::label_comma()) +
  scale_y_continuous(breaks = scales::pretty_breaks(n = 6),
                     labels = scales::label_comma()) +
  labs(title = "**Net Passing Yards Without Play Action vs.
          With Play Action**",
     subtitle = "*2022 NFL Regular Season*",
     caption = "*An Introduction to NFL Analytics with R*<br>
     **Brad J. Congelio**",
     x = "Net Yards Without Play Action",
     y = "Net Yards With Play Action") +
  nfl_analytics_theme()
```

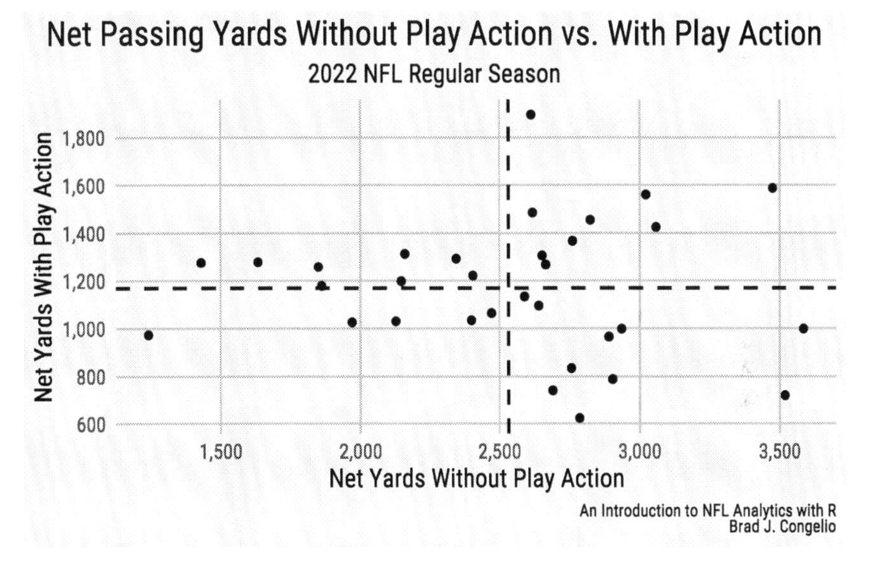

Figure 4.21: Building the plot before adding team logos instead of points

The resulting plot, constructed in nearly an identical manner to our above example, looks correct and we are ready to swap out `geom_point` for team logos.

! Important

Before proceeding with this next step, be sure that you have the **ggimage** package installed and loaded.

```
ggplot(data = team_playaction, aes(x = net_yds,
        y = pa_net_yds)) +
  geom_hline(yintercept = mean(team_playaction$pa_net_yds),
             linewidth = 0.8,
             color = "black",
             linetype = "dashed") +
  geom_vline(xintercept = mean(team_playaction$net_yds),
             linewidth = 0.8,
             color = "black",
             linetype = "dashed") +
  geom_image(aes(image = team_logo_wikipedia)) +
  scale_x_continuous(breaks = scales::pretty_breaks(n = 6),
                     labels = scales::label_comma()) +
  scale_y_continuous(breaks = scales::pretty_breaks(n = 6),
                     labels = scales::label_comma()) +
  labs(title = "**Net Passing Yards Without Play Action vs.
  StringTokWith Play Action**",
     subtitle = "*2022 NFL Regular Season*",
     caption = "*An Introduction to NFL Analytics with R*<br>
     **Brad J. Congelio**",
       x = "Net Yards Without Play Action",
       y = "Net Yards With Play Action") +
  nfl_analytics_theme()
```

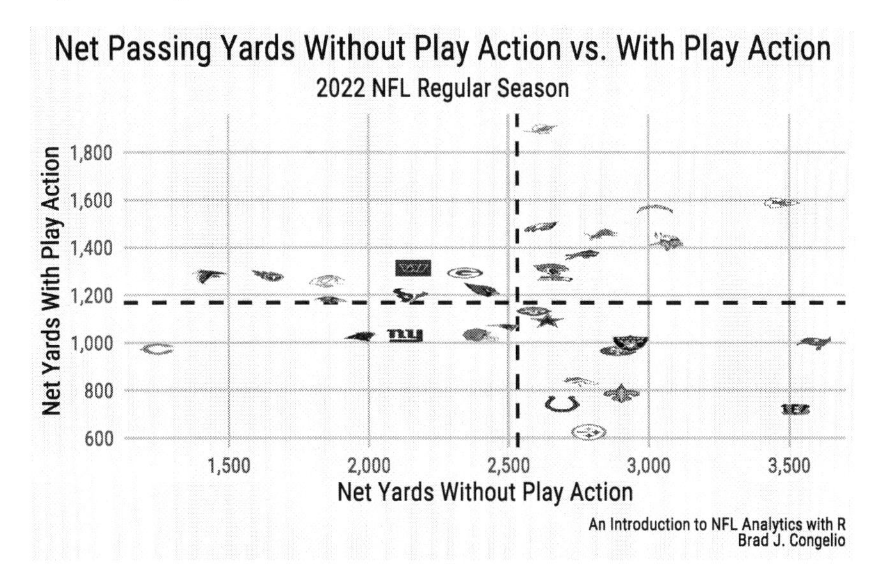

Figure 4.22: Using team logos without including an aspect ratio

By using the `geom_image` function, we are able to wrap the `image` argument within an aesthetics call (that is, `aes()`) and then stipulate that the `team_logo_wikipedia` variable is to serve as the image associated with each data point.

But, wait: **the image looks horrible, right?** Indeed, the logos are pixelated, are skewed in shape, and are just generally unpleasant to look at.

That is because we failed to provide an aspect ratio for the team logos. In this case, we need to add `asp = 16/9`.

> **i Note**
>
> **Why are we including a specific aspect ratio of 16/9 for each team logo? Good question.**
>
> An aspect ratio of 16/9 refers to the proportional relationship between the width and height of a rectangular display or image. In this case, the width of the image is 16 units, and the height is 9 units. Importantly, this aspect ratio is commonly used for widescreen displays, including most (if not all) modern televisions, computer monitors, and smartphones.

> As well, the 16/9 aspect ratio is sometimes referred to as 1.78:1, which means that the width is 1.78 times the height of the image. This aspect ratio is wider than the traditional 4:3 aspect ratio that was common used in older television and CRT-based computer monitors.

We can make the quick adjustment in our prior code to provide the correct aspect ratio for each of our team logos:

```
ggplot(data = team_playaction, aes(x = net_yds,
       y = pa_net_yds)) +
  geom_hline(yintercept = mean(team_playaction$pa_net_yds),
             linewidth = 0.8,
             color = "black",
             linetype = "dashed") +
  geom_vline(xintercept = mean(team_playaction$net_yds),
             linewidth = 0.8,
             color = "black",
             linetype = "dashed") +
  geom_image(aes(image = team_logo_wikipedia), asp = 16/9) +
  scale_x_continuous(breaks = scales::pretty_breaks(n = 6),
                     labels = scales::label_comma()) +
  scale_y_continuous(breaks = scales::pretty_breaks(n = 6),
                     labels = scales::label_comma()) +
  labs(title = "**Net Passing Yards Without Play Action vs.
        With Play Action**",
     subtitle = "*2022 NFL Regular Season*",
     caption = "*An Introduction to NFL Analytics with R*<br>
     **Brad J. Congelio**",
     x = "Net Yards Without Play Action",
     y = "Net Yards With Play Action") +
  nfl_analytics_theme()
```

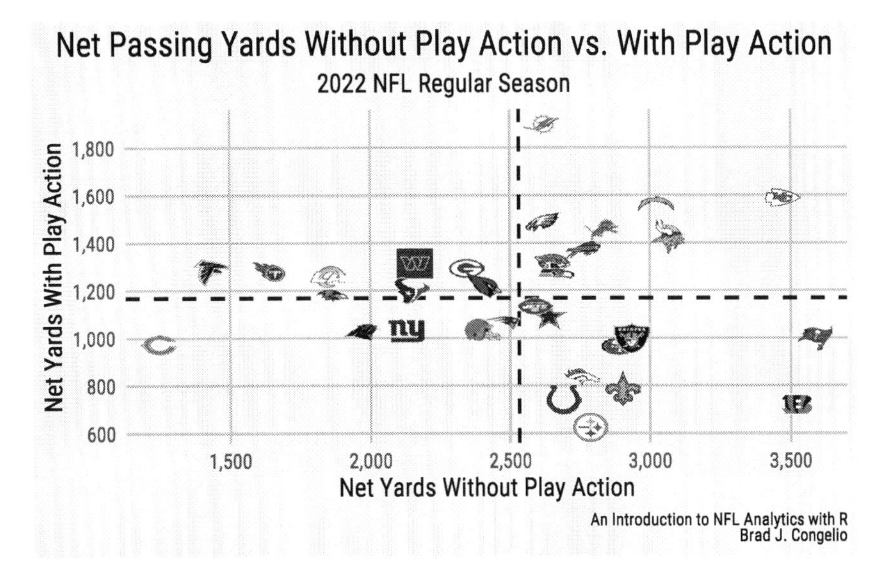

Figure 4.23: Using team logos with the correct 16/9 aspect ratio

4.7 Creating a geom_line Plot

A `geom_line()` plot is useful when you want to display the trends and/or relationships in data over a continuous variable (such as seasons). In that context a `geom_line()` plot can be used to explore player statistics over time (such as passing yards and rushing yards), or to do the same but at the team level, or even comparing one team against another by including two (or more!) lines on one plot.

Like all other `geom_` types, the basic foundation of a line graph can be created by, first, calling the `ggplot()` function and then adding `geom_line()` after. To begin building our first line graph, let's read in the below data that contains information regarding the fourth-down attempt percentages by the Philadelphia Eagles into a data frame titled `eagles_fourth_downs`.

```
eagles_fourth_downs
<- vroom("http://nfl-book.bradcongelio.com/phi-4th-downs")
```

The data frames includes information regarding the `season`, the `total` fourth downs in each season, the number of fourth-down attempts in `total_go`, and the conversion percentage in `go_pct`. As well, the data has already been joined with information from `nflreadr::load_teams()` to include colors and logos.

Let's build the foundation of the plot by using `ggplot()` and `geom_line()` and placing the `season` on the x-axis and the `go_pct` on the y-axis.

```
ggplot(eagles_fourth_downs, aes(x = season, y = go_pct)) +
  geom_line()
```

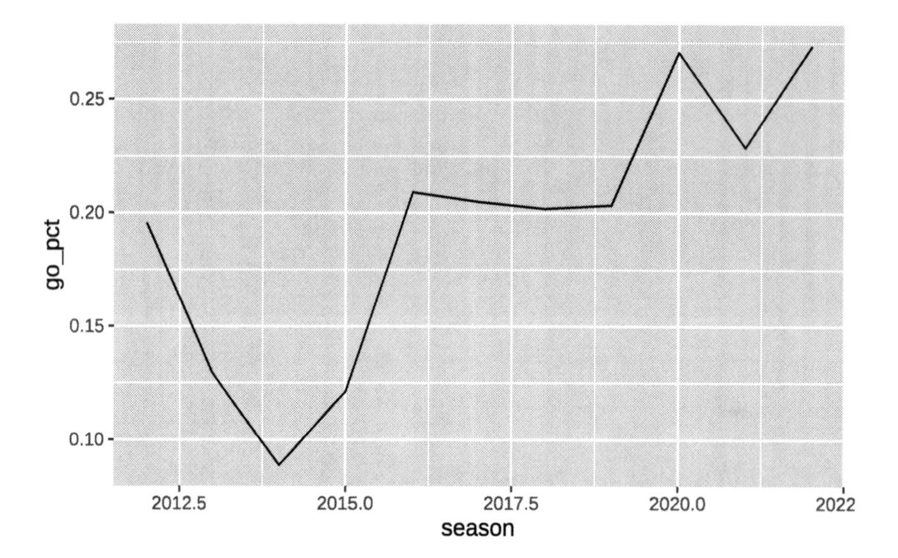

Figure 4.24: Building the foundation of a line plot

The above output is a very basic line graph. Based on this output, we can start making modifications to multiple items to reach the end result in a step-by-step fashion. First, let's explore making changes to scale of both the x and y-axis. Given that this is an examination of a yearly statistic, the x-axis should include every season in the data frame. The y-axis can be changed to be whole numbers, and to include the % after each number.

```
ggplot(eagles_fourth_downs, aes(x = season, y = go_pct)) +
  geom_line() +
  scale_x_continuous(breaks = seq(2012, 2022, 1)) +
  scale_y_continuous(breaks = scales::pretty_breaks(),
                     labels = scales::percent_format())
```

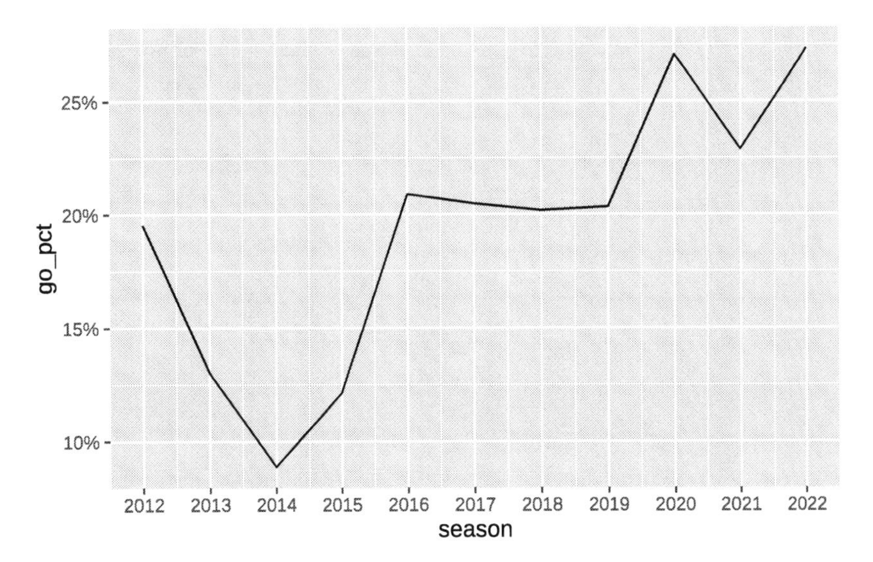

Figure 4.25: Editing axis details

Please note that difference in how the **breaks** on each scale was handled. Because we want each year on the x-axis, we manually controlled the output by using the **seq()** function to start the x-axis at 2012 and to increase by 1 until the final number of 2022 is reached. On the other hand, we used the **pretty_breaks()** function from **scales** to format the number of breaks on the y-axis and then used **percent_format()** to edit the labels to be whole numbers with the percentage sign included.

Next, we can add more **geom_** types to highlight the general percentage for each season on the x-axis. We will add one **geom_point()** to to indicate the percentage for each season. And then, for aesthetic purposes, we will add a *second* **geom_point()** that is larger than the original, with a color that matches the eventual background, to give the visual impression that the line doesn't quite "reach" the point. Lastly, because we are dealing with aesthetic issues like size and color, we will increase the size of the **geom_line()** and also add the Eagles' secondary team color to it.

```
ggplot(eagles_fourth_downs, aes(x = season, y = go_pct)) +
  geom_line(size = 2,
            color = eagles_fourth_downs$team_color2) +
  geom_point(size = 5, color = "#f7f7f7") +
  geom_point(size = 3,
            color = eagles_fourth_downs$team_color) +
  scale_x_continuous(breaks = seq(2012, 2022, 1)) +
```

```
  scale_y_continuous(breaks = scales::pretty_breaks(),
                     labels = scales::percent_format())
```

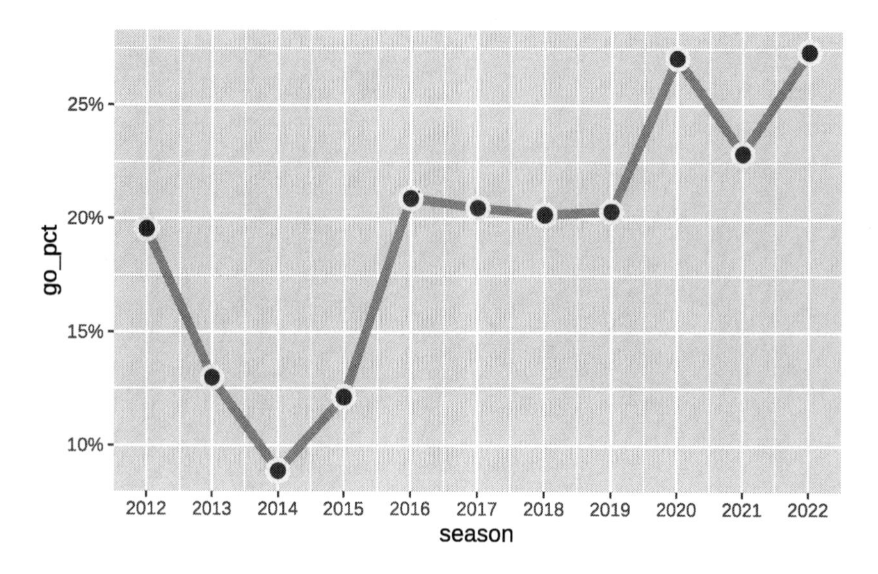

Figure 4.26: Adding points to the line

The order in which we apply all three `geom_` types is important because, as previously discussed, a `ggplot()` is layered with the first items being placed under the items that come next. In this case, `geom_line()` is under the first `geom_point()` which is then placed under the second `geom_point()`.

To complete the visualization, we can attached our custom `nfl_analytics_theme()` and then use `xlab`, `ylab`, and `labs` to edit the titles of the each axis and to include a title, subtitle, and caption.

```
ggplot(eagles_fourth_downs, aes(x = season, y = go_pct)) +
  geom_line(size = 2,
            color = eagles_fourth_downs$team_color2) +
  geom_point(size = 5, color = "#f7f7f7") +
  geom_point(size = 3,
            color = eagles_fourth_downs$team_color) +
  scale_x_continuous(breaks = seq(2012, 2022, 1)) +
  scale_y_continuous(breaks = scales::pretty_breaks(),
                     labels = scales::percent_format()) +
  nfl_analytics_theme() +
  xlab("Season") +
```

```
    ylab("Go For It Percentage") +
    labs(title = "**Philadelphia Eagles: Fourth-Down Attempt
             Percentage**",
       subtitle = "*2012 - 2022: Regular Season*",
       caption = "*An Introduction to NFL Analytics with R*<br>
       **Brad J. Congelio**")
```

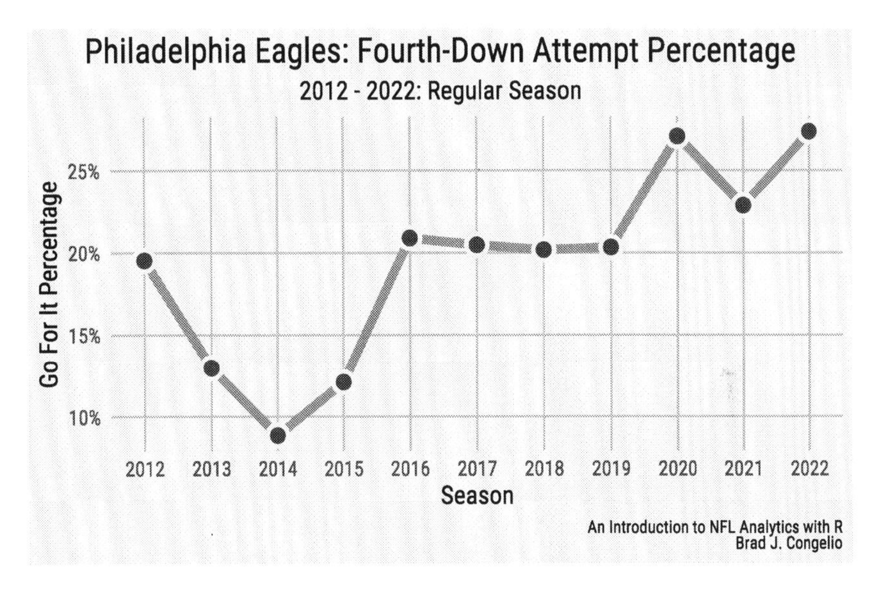

Figure 4.27: Adding the finishing touches to the 'geom_line() plot

The resulting visualization shows cases how often the Philadelphia Eagles went for it on fourth down between 2012 and 2022. The upward trend, especially starting in 2019, generally follows an established trend among NFL head coaches. Given that, how does the Eagles' trend compare to the rest of the NFL?

We can visualize this by adding a second `geom_line` that shows the averaged `go_pct` for the other 31 NFL teams.

4.7.1 Adding Secondary `geom_line` for NFL Averages

The `combined_fourth_data` data frame created by running the below code includes the same information as before for the Philadelphia Eagles, but also includes the `NFL` as a `posteam` for the 2012–2022 seasons along with the total fourth down attempt and conversion rate.

```
combined_fourth_data
  <- vroom("http://nfl-book.bradcongelio.com/4th-data")
```

The addition of a second `geom_line()` to the plot – one that belongs to the Eagles and one that belongs to averaged number for all other NFL teams – requires a few new additions to our previous code. First, in the `ggplot()` call, we've added `group = posteam`. Because the data is structured so that `PHI` and `NFL` are both observations in the `posteam` column, the use of `group = posteam` instructs `ggplot()` to "split" that information into two and to create a `geom_line()` for each. Second, we've added `scale_color_manual` to the code and manually assigned the appropriate hex code colors to each like (`#004C54` for the Eagles and `#013369` for the rest of the NFL).

```
ggplot(combined_fourth_data, aes(x = season,
                                 y = go_pct,
                                 group = posteam)) +
  geom_line(size = 2, aes(color = posteam)) +
  scale_color_manual(values = c("#013369", "#004C54")) +
  geom_point(size = 5, color = "#f7f7f7") +
  geom_point(size = 3, aes(color = posteam)) +
  nfl_analytics_theme() +
  scale_x_continuous(breaks = seq(2012, 2022, 1)) +
  scale_y_continuous(breaks = scales::pretty_breaks(),
                     labels = scales::percent_format()) +
  nfl_analytics_theme() +
  theme(legend.position = "none") +
  xlab("Season") +
  ylab("Go For It Percentage") +
  labs(title = "**Philadelphia Eagles vs. NFL: Fourth-Down
        Attempt Percentage**",
     subtitle = "*2012 - 2022: Regular Season*",
     caption = "*An Introduction to NFL Analytics with R*<br>
     **Brad J. Congelio**")
```

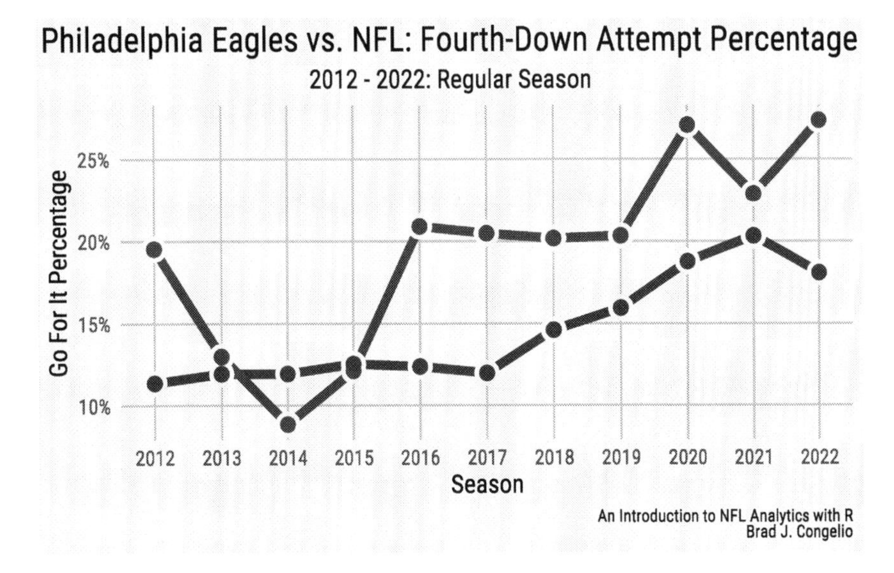

Figure 4.28: Adding a secondary line to compare to the NFL average

The resulting visualization does a very good job at showing the upward trend in head coaches going for it on fourth down and that the Eagles are consistently more aggressive in such situations than the rest of the NFL.

4.8 Creating a geom_area Plot

A `geom_area()` plot is similar to a `geom_line()` plot in that it is also often used to display quantitative data over a continuous variable. The core difference, however, is that the `geom_area()` plot provides the ability to fill in the surface area between the x-axis and the line. To showcase this, the `mahomes_epa` data frame created by running the below code contains Patrick Mahomes' average EPA for each week of the 2022 regular season, as well as the opponent and the opponent's team logo.

```
mahomes_epa
    <- vroom("http://nfl-book.bradcongelio.com/mahomes_epa")
```

The basics of the completed plot can be output by placing `week` on the x-axis and `mean_qb_epa` on the y-axis and then using `geom_area()`. Also included in the `geom_area()` are two arguments (`fill` and `alpha`). The hex color for the Kansas City Chiefs' secondary color is provided to "fill" in the area and then the `alpha` is set to 0.4 to make it slightly transparent.

```
ggplot(data = mahomes_epa, aes(x = week, y = mean_qb_epa)) +
  geom_area(fill = "#FFB612", alpha = 0.4)
```

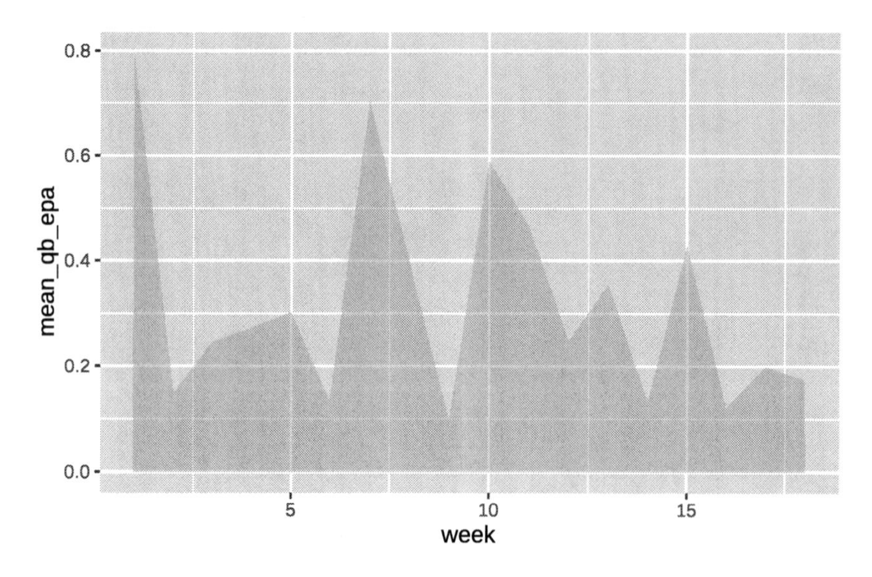

Figure 4.29: Building the foundation of a `geom_area()` plot

To help the `geom_area()` "pop" a bit more off the background, we will now included a `geom_line()` that "rides" across the top of the `geom_area()` and also add `geom_smooth()` to show the week-by-week trend of Mahomes' average EPA.

```
ggplot(data = mahomes_epa, aes(x = week, y = mean_qb_epa)) +
  geom_smooth(se = FALSE, color = "black",
  linetype = "dashed") +
  geom_area(fill = "#FFB612", alpha = 0.4) +
  geom_line(color = "#E31837", size = 1.5)
```

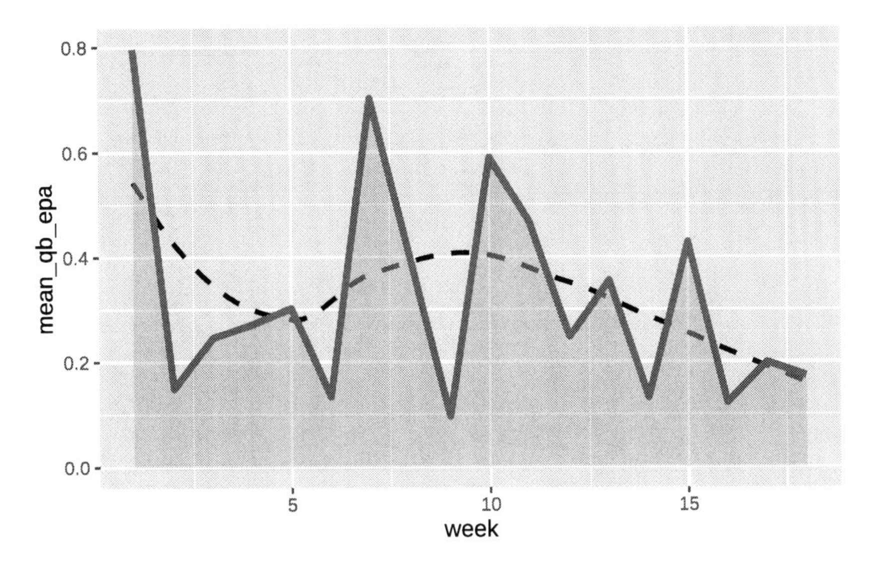

Figure 4.30: Adding `geom_smooth()` and 'geom_line() with team color

While the plot does provide information regarding Mahomes' average EPA, we still do not know who he was playing on any given week. To add this information, we can use the `geom_image()` function from `ggimage` to add the logo of each opponent to the points of the `geom_area()`.

```
ggplot(data = mahomes_epa, aes(x = week, y = mean_qb_epa)) +
  geom_smooth(se = FALSE, color = "black",
  linetype = "dashed") +
  geom_area(fill = "#FFB612", alpha = 0.4) +
  geom_line(color = "#E31837", size = 1.5) +
  geom_image(aes(image = team_logo_wikipedia),
  size = 0.045, asp = 16/9)
```

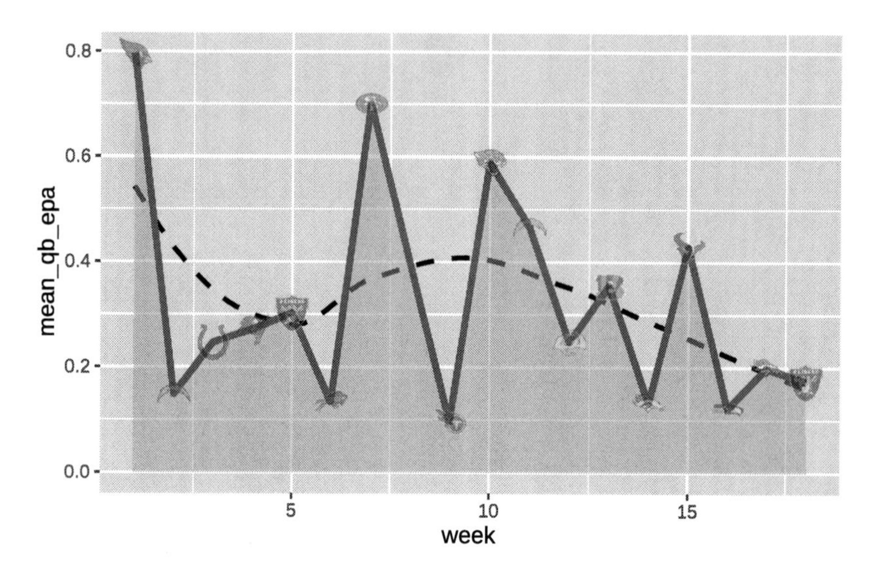

Figure 4.31: Adding logos of the opponent to the line

With the logos of each opponent added, we can turn to the finishing touches including adding our custom theme, changing the title of each axis, and adding a title. We also need to make edits to the scale of each axis, again using `seq()` to include all 18 weeks on the x-axis.

```
ggplot(data = mahomes_epa, aes(x = week, y = mean_qb_epa)) +
  geom_smooth(se = FALSE, color = "black",
  linetype = "dashed") +
  geom_area(fill = "#FFB612", alpha = 0.4) +
  geom_line(color = "#E31837", size = 1.5) +
  geom_image(aes(image = team_logo_wikipedia), size = 0.045,
  asp = 16/9) +
  scale_x_continuous(breaks = seq(1,18,1)) +
  scale_y_continuous(breaks = scales::pretty_breaks()) +
  nfl_analytics_theme() +
  xlab("Week") +
  ylab("Average QB EPA") +
  labs(title = "**Patrick Mahomes: Mean QB EPA per Week**",
       subtitle = "*2022 Regular Season*",
       caption = "*An Introduction to NFL Analytics with R*<br>
       **Brad J. Congelio**")
```

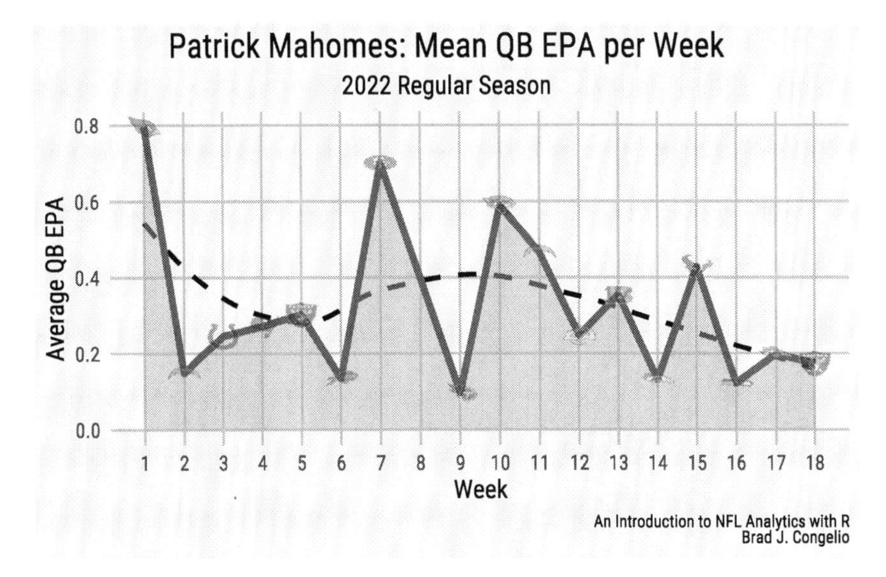

Figure 4.32: Adding finishing touches to the `geom_area()` plot

The finished visualization indicates that Mahomes' had his highest average EPA in week 1 against the Arizona Cardinals (at roughly 0.8) before dropping considerably for week two through six. However, after a spike in week 7 against the 49ers, the `geom_smooth()` shows a continued downward trend for the remainder of the season, including four poor performances (by Mahomes' standard, anyways) against the Broncos, Seahawks, and Raiders to end the season.

4.9 Creating a `geom_col` Plot

To begin creating our `geom_col` plot, we will use `vroom` to gather the necessary data into a data frame titled `qb_thirddown_data`. The resulting data frame includes the top ten quarterback in "third down aggressiveness." The metric is calculated by first gathering the total number of 3rd down passing attempts by each quarterback with 10 or less yards to go. A pass attempt is considered "aggressive" if the total air yards is equal to – or greater than – the needed yards to go. Based on this calculation, Tua Tagovailoa was the most aggressive quarterback on 3rd downs during the 2022 regular season.

Because this is *not* a metric that is constructed against a continuous variable (such as `season` in the `geom_line()` examples), we turn to using `geom_col()`.

```
qb_thirddown_data
<- vroom("http://nfl-book.bradcongelio.com/qb-3rd-data")
```

To start, we will place `passer_id` on the x-axis and `qb_agg_pct` on the y-axis. However, to make sure the columns are plotted in descending order, we use the `reorder()` function to arrange `passer_id` in descending order by the `qb_agg_pct` variable.

```
ggplot(data = qb_thirddown_data, aes(x = reorder(passer_id,
       -qb_agg_pct), y = qb_agg_pct)) +
         geom_col()
```

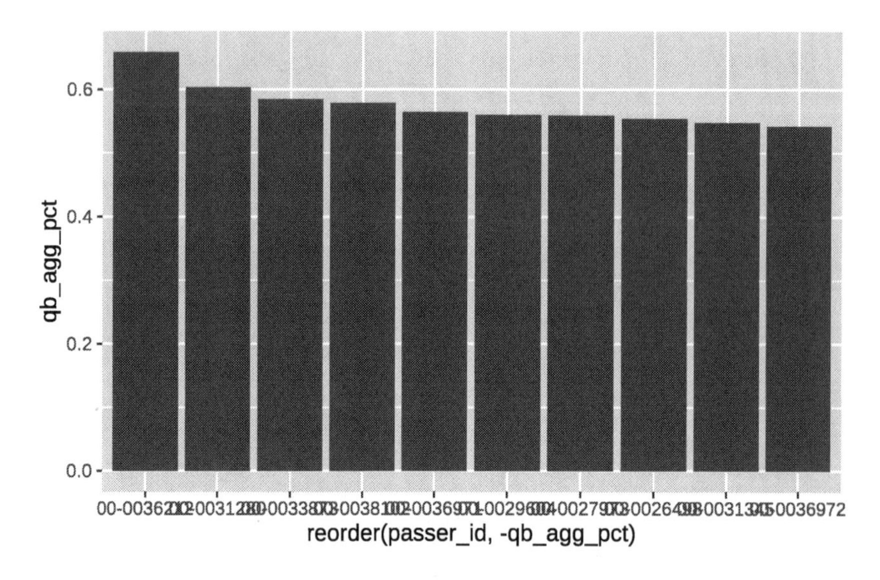

Figure 4.33: Reordering the quarterbacks in descending order by aggressiveness percentage

The resulting plot is a solid foundation and just needs aesthetic adjustments and additions to bring it to the final version. We will first use the `fill` argument to add the respective team colors to each column and then the `color` argument to add the secondary team color as an outline to each column.

```
ggplot(data = qb_thirddown_data,
       aes(x = reorder(passer_id, -qb_agg_pct),
           y = qb_agg_pct)) +
  geom_col(fill = qb_thirddown_data$team_color,
           color = qb_thirddown_data$team_color2)
```

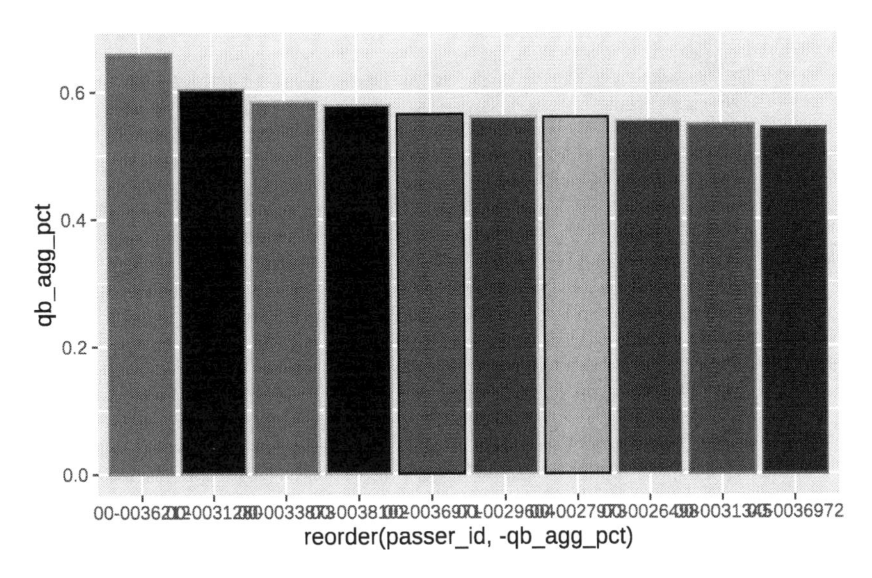

Figure 4.34: Add team colors to each column

Next we can make adjustments to each axis. Using `scale_y_continuous`, we will set the number of breaks with `pretty_breaks()` and then use `percent_format()` to adjust the scale into whole numbers with the accompanying percentage sign. The `expand()` argument is used is set to `c(0,0)` to get rid of the "dead space" between the bottom of each column and the y-axis.

We can also provide the `xlab()`, `ylab()`, and include our custom `nfl_analytics_theme()` to the plot.

```
ggplot(data = qb_thirddown_data,
       aes(x = reorder(passer_id, -qb_agg_pct),
           y = qb_agg_pct)) +
  geom_col(fill = qb_thirddown_data$team_color,
           color = qb_thirddown_data$team_color2) +
  scale_y_continuous(breaks = scales::pretty_breaks(),
                     labels = scales::percent_format(),
```

```
                             expand = c(0,0)) +
    xlab("") +
    ylab("QB Aggressiveness on 3rd Down") +
    nfl_analytics_theme()
```

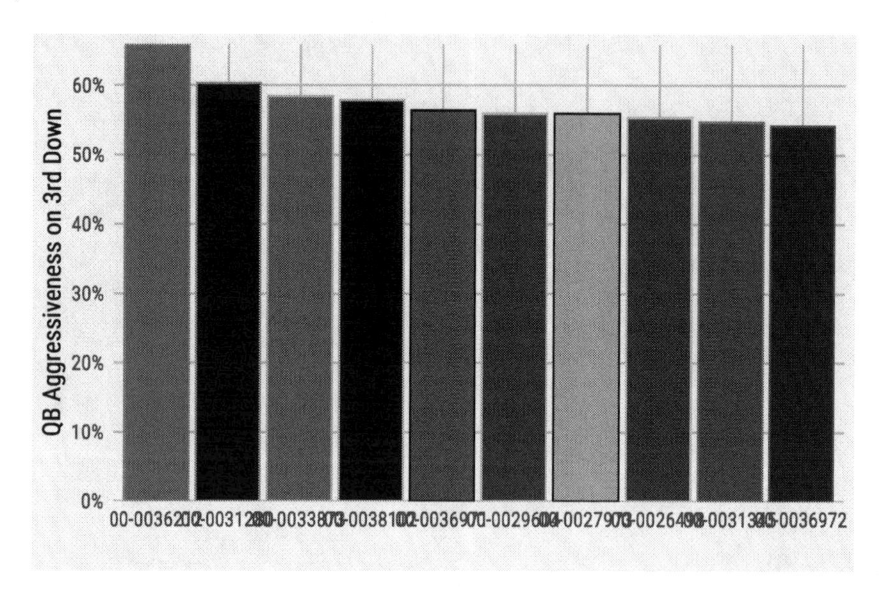

Figure 4.35: Use the `scales` package to edit each axis

We will provide two ways to identify which column belong to which quarterback: by adding the player name to the column and by including each player's headshot at the bottom of each.

To place the player name on each column, we will use `geom_text()` wrapped inside the `with_outer_glow()` function (which is part of the `ggfx` package). The `with_outer_glow()` function allows us to apply a drop shadow effect to the text, making sure that it is easy to read on the colored background of each column. The `geom_text` function accepts the arguments relating to the `angle`, horizontal adjustment `hjust`, `color`, font family, font face, and size of the text. The `with_outer_glow` function is used to apply the `sigma` (how dark or light to make the shadow/glow), `expand` (how far to spread out the glow/shadow), and `color`.

```
ggplot(data = qb_thirddown_data,
        aes(x = reorder(passer_id, -qb_agg_pct),
                y = qb_agg_pct)) +
  geom_col(fill = qb_thirddown_data$team_color,
```

```
          color = qb_thirddown_data$team_color2) +
  with_outer_glow(geom_text(aes(label = passer),
                            position =
                            position_stack(vjust = .98),
                            angle = 90, hjust = .98,
                            color = "white",
                            family = "Roboto",
                            fontface = "bold", size = 8),
                  sigma = 6, expand = 1, color = "black") +
  scale_y_continuous(breaks = scales::pretty_breaks(),
                     labels = scales::percent_format(),
                     expand = c(0,0)) +
  xlab("") +
  ylab("QB Aggressiveness on 3rd Down") +
  nfl_analytics_theme()
```

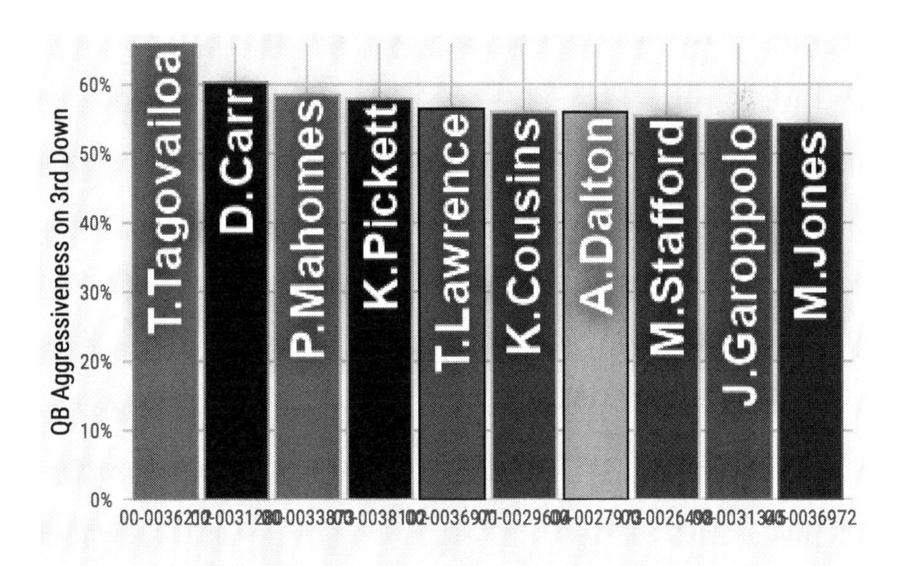

Figure 4.36: Adding player names with text effects using the `ggfx` package

To add player headshot to the bottom of each column, we will use the `element_nfl_headshot` function from the `nflplotR` package. Inside the `theme()` option, we use `element_nfl_headshot()` to replace the `axis_text.x()`. It is important to remember to use `passer_id` on the x-axis (rather than `passer_name` or something similar). The `nflplotR` package uses the `passer_id` to automatically pull the correct headshot for each player. Without including `passer_id` in the x-axis, the resulting headshots will be the "blank" NFL picture.

```
ggplot(data = qb_thirddown_data,
       aes(x = reorder(passer_id, -qb_agg_pct),
           y = qb_agg_pct)) +
  geom_col(fill = qb_thirddown_data$team_color,
           color = qb_thirddown_data$team_color2) +
  with_outer_glow(geom_text(aes(label = passer),
                  position = position_stack(vjust = .98),
                  angle = 90,
                  hjust = .98,
                  color = "white",
                  family = "Roboto",
                  fontface = "bold",
                  size = 8),
                  sigma = 6, expand = 1, color = "black") +
  scale_y_continuous(breaks = scales::pretty_breaks(),
                     labels = scales::percent_format(),
                     expand = c(0,0)) +
  theme(axis.text.x = nflplotR::element_nfl_headshot
        (size = 1)) +
  xlab("") +
  ylab("QB Aggressiveness on 3rd Down") +
  nfl_analytics_theme()
```

Figure 4.37: Using nflplotR to include headshots with the `element_nfl_headshots()` function

To complete the plot, we can add the `title`, `subtitle`, and `caption` using the `labs` function.

```
ggplot(data = qb_thirddown_data,
        aes(x = reorder(passer_id, -qb_agg_pct),
                y = qb_agg_pct)) +
  geom_col(fill = qb_thirddown_data$team_color,
           color = qb_thirddown_data$team_color2) +
  with_outer_glow(geom_text(aes(label = passer),
                  position = position_stack(vjust = .98),
                  angle = 90,
                  hjust = .98,
                  color = "white",
                  family = "Roboto",
                  fontface = "bold",
                  size = 8),
                  sigma = 6, expand = 1, color = "black") +
  scale_y_continuous(breaks = scales::pretty_breaks(),
                     labels = scales::percent_format(),
                     expand = c(0,0)) +
  theme(axis.text.x = nflplotR::element_nfl_headshot
        (size = 1)) +
  xlab("") +
  ylab("QB Aggressiveness on 3rd Down") +
  nfl_analytics_theme() +
  labs(title = "**QB Aggressiveness on Third Down**",
       subtitle =
               "*Numbers of Times Air Yards >= Yards To Go*",
       caption =
               "*An Introduction to NFL Analytics with R*<br>
       **Brad J. Congelio**")
```

Figure 4.38: Adding fishing touches to the `geom_col()` plot

As highlighted in the resulting plot, Tagovailoa was the only QB in the league to go above 60% in aggressiveness (Derek Carr was an even 60%). In this case, our domain knowledge of the NFL is important as we will recall that Tua was in concussion protocol for portions of the 2022 season. His aggressive percentage could be slightly inflated because of the fewer games compared to the other quarterbacks.

4.9.1 An Alternative `geom_col` Plot

The `geom_col()` plot is incredibly versatile, allowing you to create multiple types of visualizations. For example, we can use the data frame below, `epa_per_play`, to graph each teams average EPA per play during the 2022 regular season. However, rather than plot it in a standard column format like above, we will plot it in a "left-to-right" design with each team diverging from the 0.00 EPA per Play.

```
epa_per_play
  <- vroom("http://nfl-book.bradcongelio.com/epa-per-play")
```

The `ggplot()` code required to output the visualization is quite similar to our above `geom_col()` plot for QB aggressiveness. There are several distinct differences, however.

1. We've included a `fill = if_else()` argument in `geom_col()`. If a team has a positive EPA per play, the column for that team will be blue. If the team has a negative EPA per play, the column will be red.
2. The use of `scale_fill_identity()` is used because the data has already been scaled (by EPA per play) and already represents aesthetic values that are native to `ggplot2()`.
3. `scale_y_discrete` is used because the y-axis is team names rather than a continuous variables. Moreover,`expand = c(.05, 0))` is included to make sure that the Chiefs and Texans logos are not clipped during the processing of the plot.

```
ggplot(data = epa_per_play,
       aes(x = epa_per_play,
               y = reorder(posteam, epa_per_play))) +
  geom_col(aes(fill = if_else(epa_per_play >= 0,
                               "#2c7bb6", "#d7181c")),
           width = 0.2) +
  geom_vline(xintercept = 0, color = "black") +
  scale_fill_identity() +
  geom_image(aes(image = team_logo_wikipedia),
             asp = 16/9, size = .035) +
  xlab("EPA Per Play") +
  ylab("") +
  scale_x_continuous(breaks = scales::pretty_breaks(n = 6)) +
  scale_y_discrete(expand = c(.05, 0)) +
  nfl_analytics_theme() +
  theme(axis.text.y = element_blank()) +
  labs(title = "**Expected Points Added per Play**",
     subtitle = "*2022 Regular Season*",
     caption = "*An Introduction to NFL Analytics with R*<br>
     **Brad J. Congelio**")
```

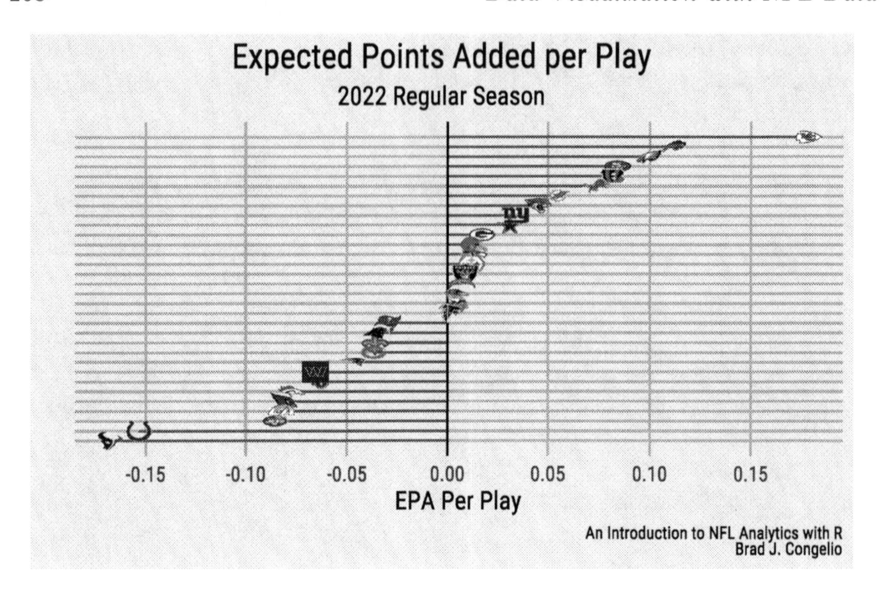

Figure 4.39: An alternate way to design a `geom_col()` plot

It should not be a surprise that the Chiefs outperform – by a large margin – the rest of the NFL. The Texans and Colts, on the other hand, averaged -0.15 expected points added per play during the 2022 regular season.

4.10 Creating a `geom_box` Plot

We can create a `geom_box()` plot to examine the distribution of air yards for each playoff team from the 2022 season. A box plot is a suitable way to view distribution but to also view the "skewness" of a metric by also including the interquartile ranges, averages, and any outliers in the data.

```
playoff_airyards_plot
<- vroom("http://nfl-book.bradcongelio.com/playoff-airyards")
```

The `geom_boxplot()` function includes one more argument – `color`. In this case, we have set the argument to `black`. We will also include the `geom_jitter()` function, which will plot each pass underneath the box plot.

This is not a requirement for a box plot, but does add an additional level of aesthetics and detail to the final plot. Last, we use the `element_nfl_logo` function to place the respective team logo under each box plot.

```
ggplot(data = playoff_airyards_plot, aes(x = posteam,
                                          y = air_yards,
                                          fill = posteam)) +
  geom_jitter(color = playoff_airyards_plot$team_color,
              alpha = 0.09) +
  geom_boxplot(color = "black") +
  nflplotR::scale_fill_nfl() +
  scale_y_continuous(breaks = scales::pretty_breaks()) +
  nfl_analytics_theme() +
  ylab("Air Yards") +
  xlab("") +
  labs(title = "**Distribution of Regular Season Air Yards**",
     subtitle = "*2022 NFL Playoff Teams*",
     caption = "*An Introduction to NFL Analytics with R*<br>
     **Brad J. Congelio**") +
  theme(axis.text.x = element_nfl_logo(size = 1.25))
```

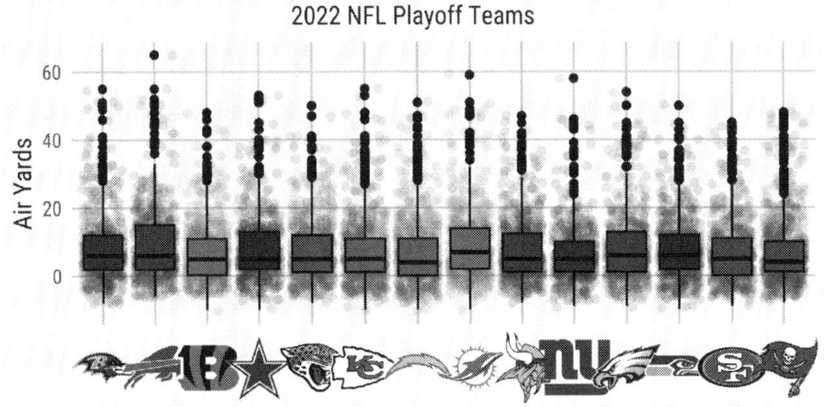

Figure 4.40: The finished `geom_boxplot` plot

In the visualization, each box plot represent the interquartile range (IQR) of air yards distribution for each team. The black line represents the median average, while the area below the black line is the lower quartile (Q1), or the 25th percentile, while the area above is the upper quartile (Q3), or the 75th percentile. The "whisker" coming from both ends of the box plot represent the minimum and maximum amounts (where the minimum is `Q1 - 1.5 * IQR` and the maximum is `Q3 + 1.5 * IQR`). The black dots above the whiskers indicate outliers.

Based on the results, the Miami Dolphins had the highest median air yards of the 2022 playoff teams while the New York Giants had the smallest spread (indicated by the 25th and 75th percentiles being closest to the median).

4.11 Creating a `geom_ridges` Plot

We can take the same idea used in the `geom_boxplot()` graph to reformat it to be used in a `geom_ridges()` plot, which is another option to represent the distribution of a continuous variable.

```
playoff_airyards_ridges_plot
<- vroom("http://nfl-book.bradcongelio.com/ay-ridges")
```

Before creating the plot, we will create a function titled `meansd` that will assist in including lines within the `geom_ridges()` to show the mean and standard deviation for each team.

The `meansd` function is included in the `geom_density_ridges()` arguments. We want to show `quantile_lines`, so we provide `meansd` to the `quantile_fun` argument and then set the `color` of the quantile lines.

```
### creating function to add SD, mean, and mean + sd
meansd <- function(x, ...) {
  mean <- mean(x)
  sd <- sd(x)
  c(mean - sd, mean, mean + sd)
}

### plotting the data
ggplot(data = playoff_airyards_ridges_plot,
```

```
      aes(x = air_yards,
              y = reorder(posteam,
                      -air_yards),
              fill = posteam)) +
  geom_density_ridges(quantile_lines = T,
                      quantile_fun = meansd,
                      color = "#d0d0d0") +
  nflplotR::scale_fill_nfl() +
  scale_x_continuous(breaks = scales::pretty_breaks()) +
  nfl_analytics_theme() +
  xlab("Air Yards") +
  ylab("") +
  labs(title = "**Distribution of Regular Season Air Yards**",
     subtitle = "*2022 NFL Playoff Teams*",
     caption = "*An Introduction to NFL Analytics with R*<br>
     **Brad J. Congelio**") +
  theme(axis.text.y = element_nfl_logo(size = 1.25))
```

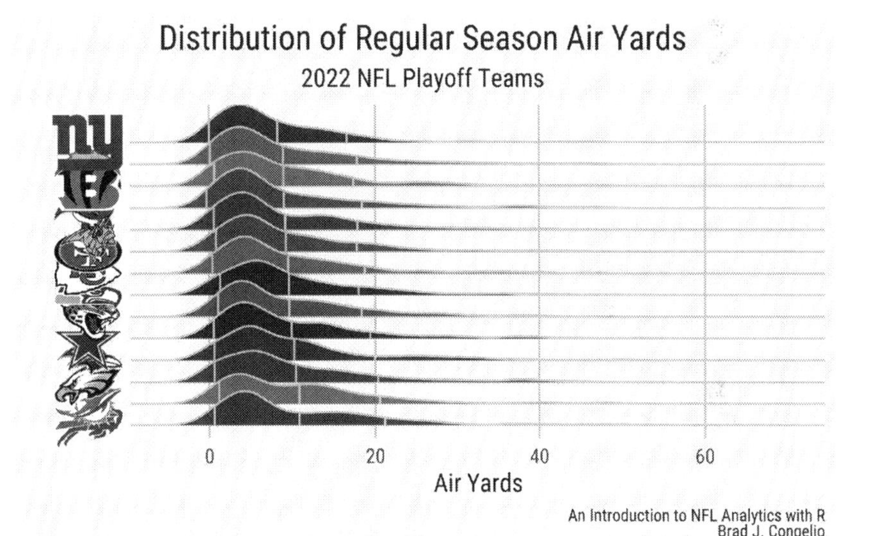

Figure 4.41: The finished `geom_ridges()` plot

4.12 Creating a geom_violin Plot

```
yards_gained_compared
<- vroom("http://nfl-book.bradcongelio.com/yards_gained")
```

```
custom_labeller <- function(variable, value){
  new_labels <- c("1st Down", "2nd Down", "3rd Down")
  return(new_labels[value])
}

ggplot(data = yards_gained_compared, aes(x = posteam,
                                         y = yards_gained,
                                         group = posteam,
                                         fill = posteam)) +
  geom_violin() +
  stat_summary(fun.y = mean, fun.ymin = mean, fun.ymax = mean,
               geom = "crossbar",
               width = 0.50,
               position = position_dodge(width = .25)) +
  facet_wrap(~down, labeller =
          labeller(down = custom_labeller)) +
  scale_fill_manual(values = c("#E31837", "#013369")) +
  ylab("Average Yards Gained") +
  xlab("") +
  labs(title = "*Averaged Yards Gained per Down*",
     subtitle = "**KC Chiefs vs. NFL: 2022 Regular Season**",
     caption = "*An Introduction to NFL Analytics with R*<br>
     **Brad J. Congelio**") +
  theme(axis.text.x = nflplotR::element_nfl_logo(size = 1.5),
      legend.position = "none",
      strip.text = element_text(family = "Roboto",
      face = "bold"),
      strip.background = element_rect(fill = "#f7f7f7")) +
  nfl_analytics_theme()
```

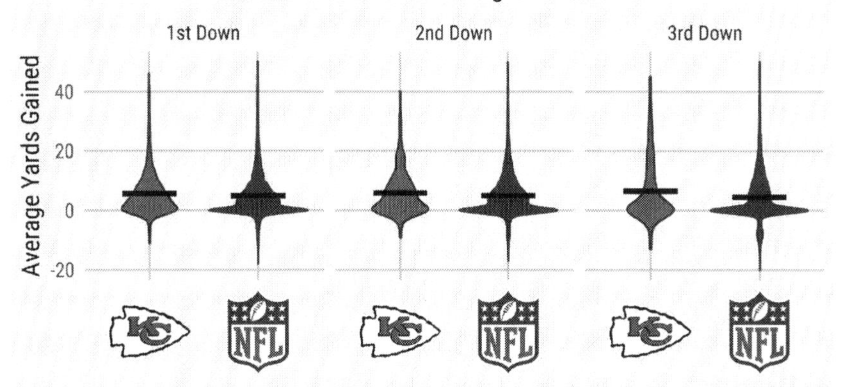

Figure 4.42: The finished `geom_violin()` plot

4.13 Creating a geom_histogram Plot

```
def_yards_allowed
<- vroom("http://nfl-book.bradcongelio.com/def-yards")
```

```
ggplot(data = def_yards_allowed, aes(x = yards_allowed,
       fill = defteam)) +
  geom_histogram(binwidth = 50) +
  nflplotR::scale_fill_nfl(type = "primary") +
  scale_x_continuous(breaks = seq(200, 500, 100)) +
  scale_y_continuous(breaks = scales::pretty_breaks()) +
  facet_wrap(vars(defteam), scales = "free_x") +
  nfl_analytics_theme() +
  theme(strip.text = element_nfl_wordmark(size = .5),
        strip.background = element_rect(fill = "#F7F7F7")) +
  xlab("Defensive Yards Allowed") +
  ylab("Total Count") +
  labs(title = "*Histogram of Defensive Yards Allowed*",
```

```
    subtitle = "**2022 NFL Regular Season**",
    caption = "*An Introduction to NFL Analytics with R*<br>
    **Brad J. Congelio**")
```

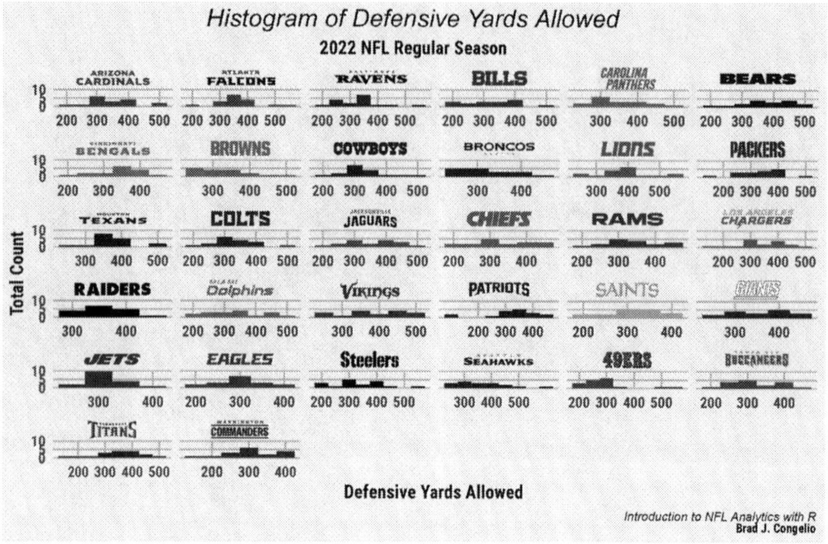

Figure 4.43: The finished `geom_histgram()` plot

4.14 Saving Your ggplot2 Creations

The most convenient way to save your finished plot is with the `ggsave()` function, which will save the last plot created. The function accepts a multitude of different arguments, including `filename`, `device`, `path`, `scale`, `width`, `height`, and `dpi`.

In most cases, you will be able to save your plot using just the `filename` and `dpi` arguments.

```
ggsave("this-is-your-filename.png", dpi = 400)
```

In the above example, the plot will be outputted in `.png` format into your current working directory (you can see where the plot will be saved by using `getwd()`). The image `dpi` – or "dots per inch" – ultimately controls the plot's resolution. A higher number will allow you to increase the size the image, but will also increase the size of the file itself. I find that using a `dpi` between 300 and 400 is suitable for nearly all plots.

4.15 Advanced Headshots Method

The ability to include player headshots onto plots is an outstanding addition to the NFL analytics universe. However, the headshots themselves quite often have "harsh" aesthetics in that the image is cut off at the shoulders and, as a result, can result in plots where the shoulders just seem to "fall off" into the abyss of the plot itself.

To correct this, we can use several different package to create circular croppings of the of the individual headshot images and then wrap the player's name around the image itself.

To showcase the process, let's create a data frame titled `complete_wr_airyards` with the below code:

```
complete_wr_airyards
<- vroom("http://nfl-book.bradcongelio.com/complete_ay")
```

To begin the process, we create a data frame called `crop_circles_data` that includes just one observation for each wide receive that has the `receiver_id` and `headshot_url` variables. After, we create a new column in `crop_circles_data` called `circle` and then use the `circle_crop` function from the `cropcircles` package to automatically crop each player's headshot into a uniform circle. The resulting image is saved locally on your device. Last, since we no longer need the `headshot_url` information, we drop that from the `crop_circles_data` data frame.

```
crop_circles_data <- complete_wr_airyards %>%
  select(receiver_id, headshot_url) %>%
  distinct(receiver_id, .keep_all = TRUE)

crop_circles_data <- crop_circles_data %>%
  mutate(circle = cropcircles::circle_crop(headshot_url))

crop_circles_data <- crop_circles_data %>%
  select(receiver_id, circle)
```

The next step, which is a modified version of an approach used by Tanya Shapiro, is multifaceted. We first use the `magick` package to create a composite of the image and place a white background behind the circular cropped headshot.

We then create a function called `plot_image_label` that will wrap each player's name around the top of the circular headshot and then proceed with producing the images with the names wrapped using `ggplot()`.

```
border <- function(im) {
  ii <- magick::image_info(im)
  ii_min <- min(ii$width, ii$height)

  img <- image_blank(width = ii_min, height = ii_min,
        color = "none")
  drawing <- image_draw(img)
  symbols(ii_min/2, ii_min/2, circles = ii_min/2.2,
          bg = "white", inches = FALSE, add = TRUE)
  dev.off()

  x = image_composite(image_scale(drawing, "x430"),
                      image_scale(im, "x400"),
                      offset = "+15+15")


  x
}

### now let's add the player's name and wrap it around
### the circle
plot_image_label<-function(image,
                           label,
                           font_color="black",
                           top_bottom="top",
                           hjust = 0.5){

  t = seq(0, 1, length.out = 100) * pi

  ### set up data
  if(top_bottom=="top"){data = data.frame(x = cos(t),
  y = sin(t))}
  else if(top_bottom=="bottom"){data=data.frame(x = cos(t),
  y = sin(t)*-1)}

  ### set up data   ### the `vjust` option here changes how
```

```
    ### close name is
    if(top_bottom=="top"){vjust=0.5}
    else if(top_bottom=="bottom"){vjust=-0.1}

    ### set up data
    if(top_bottom=="top"){ymax=1.2}
    else if(top_bottom=="bottom"){ymax=0.9}

    ### set up data
    if(top_bottom=="top"){ymin=-0.9}
    else if(top_bottom=="bottom"){ymin=-1.2}

    ### now taking the text an adding it into a ggplot with
    ###  the circle
    ggplot() +
      geom_image(aes(x=0, y=0, image = image), asp=2.4/2.1,
                 size=.7, image_fun=border) +
      scale_x_continuous(limits = c(-1.2, 1.2))+
      scale_y_continuous(limits=c(ymin, ymax))+
      geom_textpath(data = data, aes(x,y,
      label = toupper(label)),
                    linecolor=NA, color=font_color,
                    size = 16,  fontface="bold",
                    vjust = vjust, hjust=hjust,
                    family = "Roboto")+
      coord_equal()+
      theme_void()
  }
```

After the process finishes running (it may take a while depending on your computing power), we create a new column titled `new_image_path` and use `paste0` to copy the `receiver_id` for each player and add ".png" on the end.

```
crop_circles_data <- crop_circles_data %>%
  mutate(new_image_path = paste0(receiver_id, ".png"))
```

We then loop our `crop_cirlces_data` function back over the images and save them into a new images folder on our computer.

```
for(i in 1:nrow(crop_circles_data)) {

  pos = "top"
```

```
  hjust = 0.5
  path = crop_circles_data$new_image_path[i]
  plot = plot_image_label(image = crop_circles_data$circle[i],
                          label = crop_circles_data$receiver[i],
                          font_color = "black",
                          top_bottom = "top",
                          hjust = hjust)

  ggsave(filename = glue("./images/circle_headshots/{path}"),
         plot=plot, height=3.95, width=4.5)
}
```

> **! Important**
>
> If RStudio happens to stop outputting `ggplot()` visualizations at this point, please run `dev.off()`.

```
crop_circles_data <- crop_circles_data %>%
  mutate(image =
  glue("./images/circle_headshots/{new_image_path}"))
```

We can now take our new cropped headshots and bring them into a visualization created with `ggplot()`.

```
ggplot(complete_wr_ay, aes(x = air_yards, fill = posteam)) +
  geom_density(alpha = .4) +
  nflplotR::scale_fill_nfl() +
  geom_image(data = complete_wr_ay, aes(x = 40, y = 0.04,
                                        image = image),
             size = 0.5) +
  scale_x_continuous(breaks = scales::pretty_breaks()) +
  scale_y_continuous(breaks = scales::pretty_breaks()) +
  facet_wrap(~ receiver_id) +
  nfl_analytics_theme() +
  theme(strip.background = element_blank(),
        strip.text= element_blank()) +
  xlab("Air Yards") +
  ylab("Density") +
  labs(title = "**Top Ten WRs in Total Air Yards**",
     subtitle = "*Density Plot of How They Were Earned*",
     caption = "*An Introduction to NFL Analytics with R*<br>
```

```
**Brad J. Congelio**")
```

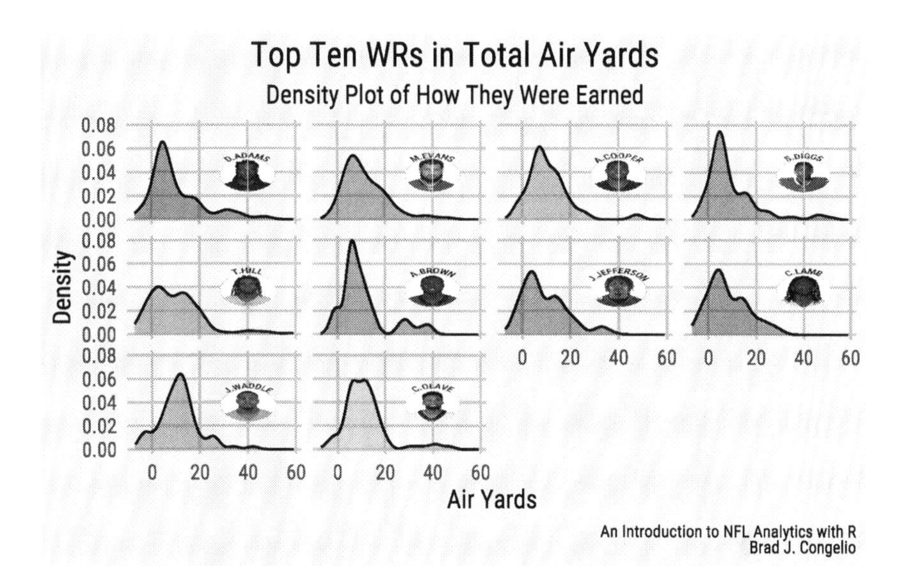

Figure 4.44: Using `magick` to create circular headshots for plots

4.16 Creating Tables with gt and gtExtras

In some instances, you may find that a plot produced by `ggplot2` is not the best medium in which to display your data. In that case, you may want to turn to use the `gt` package (and the accompanying add-on from Thomas Mock, `gtExtras`). Much like `ggplot2` stands for the "grammar of graphics," the `gt` package is short for the "grammar of tables." Taking a data frame into a `gt` table is as simple as piping into the `gt()` function, but the end result will likely not be visually pleasing. To make the necessary adjustments, you will need to begin making changes to the output by altering the core parts of a `gt` table: the table header, the stub and stub head, the column labels, the table body, and the table footer.

To provide a walk through of this process, let's read in prepared data that examines the top ten average completion percentage over expected for those attempts in a pure passing situation (that is, plays that have a predictive probability of 70% or greater of being a passing attempt).

```
pure_cpoe
<- vroom("http://nfl-book.bradcongelio.com/pure-cpoe")
```

As mentioned, we can quickly take the data and turn it into a `gt` table by piping it directly into the package's core function. **Please note that we are using the `cols_hide()` function to remove both the `headshot` and `team_logo_wikipedia` variables at first, as including them at this point in the design process results in long strings of URL data that make it difficult to grasp what the current table looks like.**

```
pure_cpoe %>%
  gt() %>%
  cols_hide(c(headshot, team_logo_wikipedia))
```

season	passer	total_attempts	mean_cpoe
2013	P.Rivers	291	10.893818
2018	P.Mahomes	261	9.293220
2020	A.Rodgers	242	8.573572
2013	R.Wilson	231	8.473162
2018	R.Wilson	277	8.448647
2011	D.Brees	311	8.285040
2018	M.Ryan	315	8.233131
2012	M.Ryan	303	8.108120
2015	R.Wilson	279	7.515386
2021	J.Burrow	343	7.117319

Using the `dplyr` to pipe the data frame directly into the `gt` package

The output is not terrible, but it does not include the headshot and team logo information yet. The output is realistically just the data frame outputted into a different format. Let's begin making incremental changes to the table by first adding a title and subtitle, using Markdown (`md()`) to provide bold and italicized styling to each.

```
pure_cpoe %>%
  gt() %>%
  cols_hide(c(headshot, team_logo_wikipedia)) %>%
  tab_header(
    title = md("**Average CPOE in Pure Passing Situations**"),
    subtitle = md("*2010-2022  |
       Pure Passing is >= 70% Chance of Pass Attempt*"))
```

Average CPOE in Pure Passing Situations

2010–2022 | Pure Passing is >= 70% Chance of Pass Attempt

season	passer	total_attempts	mean_cpoe
2013	P.Rivers	291	10.893818
2018	P.Mahomes	261	9.293220
2020	A.Rodgers	242	8.573572
2013	R.Wilson	231	8.473162
2018	R.Wilson	277	8.448647
2011	D.Brees	311	8.285040
2018	M.Ryan	315	8.233131
2012	M.Ryan	303	8.108120
2015	R.Wilson	279	7.515386
2021	J.Burrow	343	7.117319

Adding a title and subtitle to the Pure CPOE table

After adding the title and subtitle to the top of the table, we can continue to remove the `team_logo_wikipedia` information using the `cols_hide()` function and then place the player pictures into the table using the `gt_img_rows()` function from the `gtExtras` package.

```
pure_cpoe %>%
  gt() %>%
  cols_hide(team_logo_wikipedia) %>%
  tab_header(
    title = md("**Average CPOE in Pure Passing Situations**"),
    subtitle = md("*2010-2022  |  Pure Passing is >= 70%
              Chance of Pass Attempt*")) %>%
  gt_img_rows(headshot, height = 25)
```

That is considerably better. But it can be even better. For example, the column labels are still in the format from the `pure_cpoe` data frame with underscores and issues with capitalized letters. Let's make these grammatical corrections using `cols_label()` . As well, let's move the `passer` column to the far left and use the `tab_stubhead()` function, allowing us to position the names to the left of the column labels and with a vertical line separating them from the other data being presented. This also requires placing `rowname_col = "passer"` into the call to the `gt()` package.

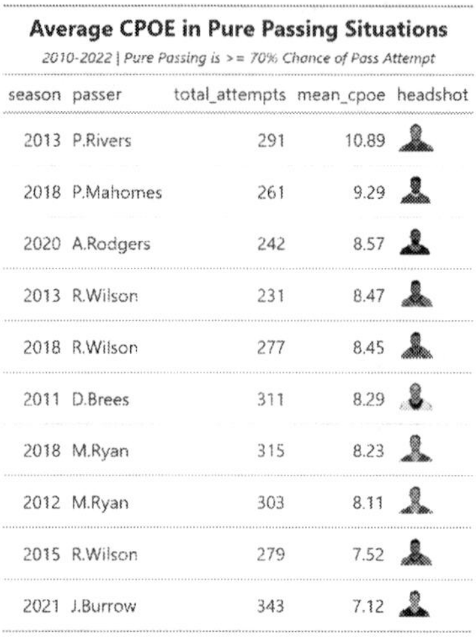

Average CPOE in Pure Passing Situations

2010-2022 | Pure Passing is >= 70% Chance of Pass Attempt

season	passer	total_attempts	mean_cpoe	headshot
2013	P.Rivers	291	10.89	
2018	P.Mahomes	261	9.29	
2020	A.Rodgers	242	8.57	
2013	R.Wilson	231	8.47	
2018	R.Wilson	277	8.45	
2011	D.Brees	311	8.29	
2018	M.Ryan	315	8.23	
2012	M.Ryan	303	8.11	
2015	R.Wilson	279	7.52	
2021	J.Burrow	343	7.12	

Figure 4.45: Adding player headshots to the table using `gt_img_rows` from `gtExtras`

```
pure_cpoe %>%
  gt(rowname_col = "passer") %>%
  cols_hide(team_logo_wikipedia) %>%
  tab_header(
    title = md("**Average CPOE in Pure Passing Situations**"),
    subtitle = md("*2010-2022  |  Pure Passing is >= 70%
                  Chance of Pass Attempt*")) %>%
  tab_stubhead(label = "Quarterback") %>%
  gt_img_rows(headshot, height = 25) %>%
  cols_label(
    passer = "Quarterback",
    headshot = "",
    season = "Season",
    total_attempts = "Attempts",
    mean_cpoe = "Mean CPOE")
```

Average CPOE in Pure Passing Situations
2010-2022 | Pure Passing is >= 70% Chance of Pass Attempt

Quarterback	Season	Attempts	Mean CPOE
P.Rivers	2013	291	10.89
P.Mahomes	2018	261	9.29
A.Rodgers	2020	242	8.57
R.Wilson	2013	231	8.47
R.Wilson	2018	277	8.45
D.Brees	2011	311	8.29
M.Ryan	2018	315	8.23
M.Ryan	2012	303	8.11
R.Wilson	2015	279	7.52
J.Burrow	2021	343	7.12

We are almost there. But, it looks strange to have the player's picture so far removed from their name. Let's place the `headshot` column as the first column and, while we are at it, use the `cols_align()` function to center align the `passer` and `total_attempts` columns.

```
pure_cpoe %>%
  gt(rowname_col = "passer") %>%
  cols_hide(team_logo_wikipedia) %>%
  tab_header(
    title = md("**Average CPOE in Pure Passing Situations**"),
    subtitle = md("*2010-2022  |  Pure Passing is >= 70%
                  Chance of Pass Attempt*")) %>%
  tab_stubhead(label = "Quarterback") %>%
  gtExtras::gt_img_rows(headshot, height = 25) %>%
  cols_label(
    passer = "Quarterback",
    headshot = "",
    season = "Season",
    total_attempts = "Attempts",
    mean_cpoe = "Mean CPOE") %>%
  cols_move_to_start(
    columns = headshot) %>%
  cols_align(align = "center",
  columns = c("passer", "total_attempts"))
```

Average CPOE in Pure Passing Situations

2010-2022 | Pure Passing is >= 70% Chance of Pass Attempt

Quarterback		Season	Attempts	Mean CPOE
P.Rivers		2013	291	10.89
P.Mahomes		2018	261	9.29
A.Rodgers		2020	242	8.57
R.Wilson		2013	231	8.47
R.Wilson		2018	277	8.45
D.Brees		2011	311	8.29
M.Ryan		2018	315	8.23
M.Ryan		2012	303	8.11
R.Wilson		2015	279	7.52
J.Burrow		2021	343	7.12

As one last visual enhancement, we will use the `gt_color_box()` function from `gtExtras` to replace the values in mean_cpoe with colorboxes that are similar to those used in data visualizations on Pro Football Focus.

```
pure_cpoe %>%
  gt(rowname_col = "passer") %>%
  cols_hide(team_logo_wikipedia) %>%
  tab_header(
    title = md("**Average CPOE in Pure Passing Situations**"),
    subtitle = md("*2010-2022  |  Pure Passing is >= 70%
                  Chance of Pass Attempt*")) %>%
  tab_stubhead(label = "Quarterback") %>%
  gt_img_rows(headshot, height = 25) %>%
  cols_label(
    passer = "Quarterback",
    headshot = "",
    season = "Season",
    total_attempts = "Attempts",
    mean_cpoe = "Mean CPOE") %>%
  cols_move_to_start(
    columns = headshot) %>%
  cols_align(align = "center",
  columns = c("passer", "total_attempts")) %>%
  gt_color_box(mean_cpoe, domain = 7:11,
               palette = "ggsci::blue_material",
               accuracy = 0.01)
```

Average CPOE in Pure Passing Situations

2010-2022 | Pure Passing is >= 70% Chance of Pass Attempt

Quarterback		Season	Attempts	Mean CPOE
P.Rivers		2013	291	10.89
P.Mahomes		2018	261	9.29
A.Rodgers		2020	242	8.57
R.Wilson		2013	231	8.47
R.Wilson		2018	277	8.45
D.Brees		2011	311	8.29
M.Ryan		2018	315	8.23
M.Ryan		2012	303	8.11
R.Wilson		2015	279	7.52
J.Burrow		2021	343	7.12

It is important to include the `accuracy = 0.01` within the `gt_color_box()`, otherwise the output of the table includes just whole numbers. At this point, the table is complete all for adding a caption at the bottom and using one of the formatted themes from `gtExtras` (I am partial to `gt_theme_538`).

```
pure_cpoe %>%
  gt(rowname_col = "passer") %>%
  cols_hide(team_logo_wikipedia) %>%
  tab_header(
    title = md("**Average CPOE in Pure Passing Situations**"),
    subtitle = md("*2010-2022  |  Pure Passing is >= 70%
                  Chance of Pass Attempt*")) %>%
  tab_stubhead(label = "Quarterback") %>%
  gt_img_rows(headshot, height = 25) %>%
  cols_label(
    passer = "Quarterback",
    headshot = "",
    season = "Season",
    total_attempts = "Attempts",
    mean_cpoe = "Mean CPOE") %>%
  cols_move_to_start(
    columns = headshot) %>%
  cols_align(align = "center",
  columns = c("passer", "total_attempts")) %>%
  gt_color_box(mean_cpoe, domain = 7:11,
               palette = "ggsci::blue_material",
               accuracy = 0.01) %>%
```

```
  tab_source_note(
    source_note =
    md("*An Introduction to NFL Analytics with R*<br>
                    **Brad J. Congelio**")) %>%
  gtExtras::gt_theme_538()
```

Average CPOE in Pure Passing Situations

2010-2022 | Pure Passing is >= 70% Chance of Pass Attempt

QUARTERBACK		SEASON	ATTEMPTS	MEAN CPOE
P.RIVERS		2013	291	10.89
P.MAHOMES		2018	261	9.29
A.RODGERS		2020	242	8.57
R.WILSON		2013	231	8.47
R.WILSON		2018	277	8.45
D.BREES		2011	311	8.29
M.RYAN		2018	315	8.23
M.RYAN		2012	303	8.11
R.WILSON		2015	279	7.52
J.BURROW		2021	343	7.12

An Introduction to NFL Analytics with R
Brad J. Congelio

4.17 Creating a Shiny App

With a solid understanding of the data visualization process, we can now turn our attention to take a completed data frame and transitioning it into a Shiny App, resulting in an interactive application for users to explore your data and findings.

To do this, we will be utilize a data frame that you will see again in Chapter 5 that is created during an XGBoost modeling process. It contains the actual rushing yards and expected rushing yards, based on predictive machine learning, for all running backs from 2016 to 2022. You can read in the data to view it with the following `vroom` code.

```
ryoe_shiny
<- vroom("http://nfl-book.bradcongelio.com/ryoe-projs")
```

The resulting `ryoe_shiny` data frame contains information regarding `actual_yards`, `season`, `down`, `ydstogo`, `red_zone`, `rusher`,

`rusher_player_id`, `posteam`, `defteam`, `exp_yards`. As you will soon see, despite being in the data frame, will will not use all of the variables while also adding some in the backend of the Shiny App. To that end, the Shiny App we will be producing will allow users to select one or more seasons to see each running back's total rushing yards, total expected yards, and the difference between the two (where a positive number indicates the running back gained more yards than the XGBoost model predicted). Users will be able to select specific downs and yards to go, as well as indicate if they want to see the information pertaining to only those rushing attempts in the red zone. We will also build in the capability for the Shiny App to automatically update and format a `ggplot` scatterplot to visualize the information.

You can see the end result and the functioning Shiny App here: NFL Analytics with R: RYOE Shiny App.

4.17.1 Requirements for a Shiny App

First, it is important to know that a Shiny App cannot be built within a typical R script. Rather, you must select "Shiny Web App" from File -> New File inside RStudio. You will then be prompted to indicate the folder in which you would like to save the files and whether or not you want a combined file or a separate `ui.R` and `server.R`. A single file containing both is acceptable for a simple Shiny App but more complex designs will require separate files to maintain organization.

1. **User Interface (`ui.R`)** – the `ui.R` file controls the layout and appearance of your Shiny App, including controlling all the elements of the user interface such as text, inputs, sliders, checkboxes, tables, and plots. The most basic requirements to construct a user interface are `fluidpage()`, `sidebarLayout()`, `sidebarPanel()`, and `mainPanel()`. The `sidebarPanel()` is what contains the options for user input, while `mainPanel()` is where the results of the input display in the web app.
2. **Server Function (`server.R`)** – the `server.R` file contains the information needed to build and modify the outputs in response to user input, as well as where you construct the inner-workings of the Shiny App (like data manipulation). Each `server.R` file will contain a `function(input, output)`, where the `input` is an object that contains the requests from the user input and `output` is the required items to be rendered.

Both the `ui.R` and `server.R` are ultimately passed into the `shinyApp()` function that outputs the final product.

4.17.2 Building Our ui.R File

Both the `ui.R` and `server.R` files will start with similar information: reading in the data, any necessary preparation, and – in our case – including the custom `nfl_analytics_theme()` to be passed into the `ggplot` output.

```
library(shiny)
library(tidyverse)
library(gt)
library(vroom)

ryoe_shiny
<- vroom("http://nfl-book.bradcongelio.com/ryoe-projs")

ryoe_shiny <- ryoe_shiny %>%
  filter(ydstogo <= 10)

teams <- nflreadr::load_teams(current = TRUE) %>%
  select(team_abbr, team_logo_wikipedia, team_color,
  team_color2)

ryoe_shiny <- ryoe_shiny %>%
  left_join(teams, by = c("posteam" = "team_abbr"))

nfl_analytics_theme <- function(..., base_size = 12) {

  theme(
    text = element_text(family = "Roboto",
                        size = base_size,
                        color = "black"),
    axis.ticks = element_blank(),
    axis.title = element_text(face = "bold"),
    axis.text = element_text(face = "bold"),
    plot.title.position = "plot",
    plot.title = element_markdown(size = 16,
                                  vjust = .02,
                                  hjust = 0.5),
    plot.subtitle = element_markdown(hjust = 0.5),
    plot.caption = element_markdown(size = 8),
    panel.grid.minor = element_blank(),
    panel.grid.major =  element_line(color = "#d0d0d0"),
    panel.background = element_rect(fill = "#f7f7f7"),
    plot.background = element_rect(fill = "#f7f7f7"),
    panel.border = element_blank(),
```

```
      legend.background = element_rect(color = "#F7F7F7"),
      legend.key = element_rect(color = "#F7F7F7"),
      legend.title = element_text(face = "bold"),
      legend.title.align = 0.5)
  }
```

After gathering the data into a data frame called `ryoe_shiny`, we use `filter()` to limited the rushes plays to just those with 10 or less yards to go, and then use the `load_teams()` function from `nflreadr` to merge in `team_color`, `team_color2`, and `team_logo_wikipedia`. After pasting in our existing code for the custom theme, we can construct the user interface.

```
ui <- fluidPage(
  titlePanel("NFL Analytics with R: RYOE Shiny App"),
  sidebarLayout(fluid = TRUE,
    sidebarPanel(
      selectInput("season", "Season:",
                  choices = unique(ryoe_shiny$season),
                  selected = NULL, multiple = TRUE),
      sliderInput("down", "Down:",
                  min = min(ryoe_shiny$down),
                  max = max(ryoe_shiny$down),
                  value = range(ryoe_shiny$down)),
      sliderInput("ydstogo", "Yards to Go:",
                  min = min(ryoe_shiny$ydstogo),
                  max = 10,
                  value = range(ryoe_shiny$ydstogo)),
      checkboxInput("red_zone", "Red Zone:", value = FALSE)),
    mainPanel(
      tableOutput("myTable"),
      plotOutput("myPlot", width = "500px"),
      width = 6)))
```

We begin constructing the user interface by calling in the `fluidPage()` function. Each "fluid page" in a Shiny application consists of rows, which are created either either with `fluidRow()`/`column()` function or, in our case, by using `sidebarLayout()`. Within `sidebarLayout()`, we can provide the arguments for both `sidebarPanel()`, which is the column that contains the user input controls, and `mainPanel()` which shows the requested outputs.

Within the `sidebarPanel()`, we begin inputting the options that users will be able to select. There are several different options in which you can allow users manipulate the data, including text inputs, selection inputs, check boxes, radio buttons, date inputs, and even file uploads. For the purposes of our application, we are providing four different options for input using four different options of input type:

1. users can select a specific season/seasons using `selectInput()`. The first `season` in the input indicates the specific column name from the data to use, while the uppercase `Season:` is the title of the input that displays on the application. Within `choices`, we use the `unique()` function to pass in the individual season values in the data (2016, 2017, 2018, 2019, 2020, 2021, and 2022). While the `selected` argument permits "pre-filling" the input with a selection, we will leave it at `NULL` and then set `multiple` to `TRUE` to allow for users to select more than one season at once.
2. Each `sliderInput()` allows the user to select from a predetermined set values: either 1–4 for down and 1–10 for yards to go. The `min` and `max` arguments are set to 1st and 4th down and 1 yard to go and 10 yards to go, respectively. For `value`, we are supplying all the numeric values in the data so that the user sees each number on the input bar.
3. Finally, the `checkboxInput` is permits users to either select/deselect the `Red Zone` argument to view the data when the offense is either in or outside of the opposing red zone.

The `mainPanel()` options takes information that is constructed within the `server.R` file (specifically, in this case, a table with results and the including scatterplot of the data) and displays it beside the user input area. We are also instructing the plot is to have a width of 500px with the totality of the `mainPanel()` having, itself, a width of 6.

4.17.3 Building Our `server.R` File

The `server.R` file begins with three different arguments.

```
server <- function(input, output) {

  selectedSeasons <- reactive({
    paste(input$season, collapse = ", ")
  })

  filteredData <- reactive({
    ryoe_shiny %>%
      filter(
```

```
            season %in% input$season,
            down >= input$down[1] & down <= input$down[2],
            ydstogo >= input$ydstogo[1] &
            ydstogo <= input$ydstogo[2],
            red_zone == input$red_zone
        )
    })
```

Each `server.R` file will begin with `function(input, output)` with the specifics of each application coming after. After, the first object created in the Shiny App is `selectedSeasons` which stores the season(s) that the user selects so that the values can be placed in the resulting `ggplot` data visualization. After, we take our `ryoe_shiny` data frame and wrap it inside the `reactive()` function and write it into a data frame called `filteredData`, thus causing `filteredData` to become what is called a "reactive expression." Reactive expressions are what give Shiny the ability to update outputs (like tables and plots, in our case) in response to user input.

The information *after* the `reactive()` function is the data that is to be updated in real time. For our Shiny App, we are using the `filter()` function to keep only the criteria from `ryoe_shiny` that the user request. That is, we are filtering for the `season` in the data frame that includes (`%in%`) the season(s) selected on the user interface. The filtering of `down` is done by instructing that the first `down` must be greater than or equal to the user input while the second `down` must be less than or equal to. The `[1]` and `[2]` including in both `down` and `ydstogo` are indices that specify the first and second elements of the value. For example, if the user selects a range of `down` from first to third, `input$down[1]` would be `1` and `input$down[2]` would be `3` (with the equivalent `tidyverse` version of it being `filter(down >= 1 & down <= 3)`. Last, if the box is checked for `red_zone`, it will filter the data to include only those plays that include the binary `1` in the `red_zone` variable.

The data, now filtered based on the user input, can be passed into the output table and plot.

```
    output$myTable <- renderTable({
      resultTable <- filteredData() %>%
        group_by(rusher) %>%
        summarise(
          total_rushes = as.integer(n()),
          total_actual_yards = as.integer(sum(actual_yards)),
          total_expected_yards = as.integer(sum(exp_yards)),
          difference = as.integer(sum(actual_yards) -
```

```
                sum(exp_yards))
    ) %>%
    rename(
      "Rusher" = rusher,
      "Total Rushes" = total_rushes,
      "Total Actual Yards" = total_actual_yards,
      "Total Expected Yards" = total_expected_yards,
      "Difference" = difference
    ) %>%
    arrange(desc(Difference)) %>%
    head(10) %>%
    as.data.frame()

      resultTable
})
```

To create the output table, we use the `renderTable()` function, another reactive element within the Shiny App. We write the information contained in `filteredData()` (which was created above during the `reactive()` filtering process) to a new data frame called `resultTable`. After, we handle the data just like we would within the `tidyverse` outside of `Shiny`, aside from ensuring the information is written out in data frame form by including `as.data.frame()` at the end. All of this is ultimately written into the `output$myTable` to be passed along to the user interface. As highlighted in the below code, outputting the `ggplot` is done in the same fashion, except for the use of `renderPlot()`. After, we create the data frame called `top10` that is also built off of the filtered data contained in `filteredData()`. After, the process includes the basic tenants of `group_by()`, `summarize()`, and building out the visualization in `ggplot()`.

```
output$myPlot <- renderPlot({
  top10 <- filteredData() %>%
    group_by(rusher) %>%
    summarize(
      team_color = last(team_color),
      total_actual_yards = as.integer(sum(actual_yards)),
      total_expected_yards = as.integer(sum(exp_yards))
    ) %>%
    top_n(10, wt = total_actual_yards)

  ggplot(data = top10, aes(x = total_actual_yards,
                           y = total_expected_yards)) +
    stat_poly_line(method = "lm", se = FALSE,
```

```
                    linetype = "dashed",
                    color = "black") +
      stat_poly_eq(mapping = use_label(c("R2", "P")),
                   p.digits = 2,
                   label.x = .20,
                   label.y = 3) +
      geom_point(color = top10$team_color, size = 3.5) +
      geom_text_repel(data = top10, aes(label = rusher)) +
      labs(
        x = "Total Actual Yards",
        y = "Total Expected Yards",
        title = paste("Actual Rushing Yards vs. Expected
                for Season(s):",
                      selectedSeasons()),
        subtitle = "Model: *LightGBM* in **tidymodels**") +
      nfl_analytics_theme()
  }, width = 500)
}
```

As you build the **Shiny** app, you can check the validity of your coding by clicking the **Run App** button in the upper-right part of the coding screen. If everything is correct, the application will run within your local system. If there are issues, the information will be provided in the "Console" area of RStudio. Once you are content with functioning and style of the application, you can publish it and host it, for free, on ShinyApps.io.

4.18 Exercises

The answers for the following answers can be found here: http://nfl-book.bradcongelio.com/ch4-answers.

4.18.1 Exercise 1

To begin, run the following code to create the data frame titled `rb_third_down`:

```
pbp <- nflreadr::load_pbp(2022) %>%
  filter(season_type == "REG")
```

```
rb_third_down <- pbp %>%
  filter(play_type == "run" & !is.na(rusher)) %>%
  filter(down == 3 & ydstogo <= 10) %>%
  group_by(rusher, rusher_player_id) %>%
  summarize(
    total = n(),
    total_first = sum(yards_gained >= ydstogo, na.rm = TRUE),
    total_pct = total_first / total) %>%
  filter(total >= 20) %>%
  arrange(-total_pct)
```

After, do the following:

1. Create a `geom_col` plot
2. Place `rusher_player_id` on the x-axis and `total_pct` on the y-axis
3. Reorder `rusher` in descending order by `total_pct`
4. On the y-axis, set the `breaks` and `labels` use the `scales::` option (note: the y-axis should be in percent format)
5. Again on the y-axis, use `expand = c(0,0)` to remove the space of the bottom of the plot
6. Use `xlab()` and `ylab()` to add a title to *only* the y-axis
7. Use the `nflplotR` package, add the primary team color as the column `fill` and secondary as `color`
8. Use the `nflplotR` package, edit the `axis.text.x` within the `theme()` option to add the player headshots
9. Add `theme_minimal()` to the plot
10. Use `labs()` to include a title and subtitle

4.18.2 Exercise 2

Run the code below to create the `offensive_epa` data frame, then follow the instructions below to complete the exercise.

```
pbp <- nflreadr::load_pbp(2022) %>%
  filter(season_type == "REG")

offensive_epa <- pbp %>%
  filter(!is.na(posteam) & !is.na(down)) %>%
  group_by(posteam) %>%
  summarize(mean_epa = mean(epa, na.rm = TRUE))
```

1. Place `mean_epa` on the x-axis and `posteam` on the y-axis, using reordering the teams by `mean_epa`
2. Use `geom_col` with the `posteam` set as both the `color` and `fill` and add a `width` of 0.75
3. Use `nflplotR` to set the `color` and `fill` to "secondary" and "primary"
4. Add a `geom_vline` with an `xintercept` of 0, add the color as "black" and a `linewidth` of 1
5. Use `nflplotR` to add team logos to the end of each bar
6. Use `theme_minimal()`
7. Remove the y-axis text with `axis.text.y` within `theme()`
8. Use `xlab`, `ylab`, and `labs` to add the x-axis label and the title/subtitle

5

Advanced Model Creation with NFL Data

5.1 Introduction to Statistical Modeling with NFL Data

The process of conducting statistical modeling with NFL data is difficult to cover in a single chapter of a book, as the topic realistically deserves an entire book as the field of modeling and machine learning is vast and complex, and continues to grow as a result of access to new data and the releases of packages (as of November 2020, over 16,000 packages are available in the R ecosystem).(Wikipedia, 2023) The list of options for conducting statistical analysis and machine learning is vast, with the more widely-used ones including `caret` (combined functions for creating predictive models), `mgcv` (for generalized additive models), `lme4` (for linear mixed effects models), `randomForest` (for creating random forests models), `torch` (for image recognition, time series forecasting, and audio processing), and `glmnet` (for lasso and elastic-net regression methods). A newer addition to the modeling and machine learning ecosystem is `tidymodels`, which is a collection of various packages that operate cohesively using `tidyverse` principles.

While it is not possible for this chapter to cover every aspect of modeling and machine learning, its goal is to provide the foundation in both to provide you with the necessary skills, understanding, and knowledge to independently explore and analyze NFL data using different methodologies and tools. The techniques covered in this chapter, from creating a simple linear regression to a predictive XGBoost model, are building blocks to all further analysis and machine learning you will conduct. In doing so, you will learn how to use built-in options such as `stats` to more technical packages such as `tidymodels`.

To that end, by mastering these introductory concepts and techniques, you not only learn the process and coding needed to create models or do machine learning, but you will begin developing an analytical mindset which will allow you to understand why one statistical approach is not correct for your data and which one is, how to interpret the results of your model, and – when

DOI: 10.1201/9781003364320-5

necessary – how to correct issues and problems in the process as they arise. The fundamentals provided in this chapter, I believe, serve as a actionable stepping stone for a further journey into the modeling and machine learning in the R programming language, with or without the use of NFL data.

5.2 Regression Models with NFL Data

Regression models are the bedrock of many more advanced machine learning techniques. For example, the "decision tree" process that is conducted in random forests models and XGBoost are created based on the split in variable relationships and neural networks are, at their core, just a intricate web of regression-like calculations. Because of this, regression models are an outstanding starting point for not only learning the basics of modeling and machine learning in R, but an important step in comprehending more advanced modeling techniques.

As you will see, a regression model is simply a mathematical determination of the relationship between two or more variables (the response and dependent variables). In the context of the NFL, a team's average points per game would be a response variable while those metrics that *cause* the average points per game are the predictor variables (average passing and rushing yards, turnovers, defensive yards allowed, etc.). The results of regression models allow us to determine which predictor variables have the largest impact on the response variable. Are passing yards a more predictive variable to average points per game than rushing yards? A regression model is capable of answering that question.

This section starts with exploring simple linear regression and then extends this knowledge to creating both multiple linear and logistic regression models which deal, respectively, with multiple predictors and categorical dependent variables. After, we will focus on creating binary and multinomial regression models.

5.2.1 Simple Linear Regression

A linear regression is a fundamental statistical technique that is used to explore the relationship between two variables – the dependent variable (also called the "response variable") and the independent variables (also called the "predictor variables"). By using a simple linear regression, we can model the relationship between the two variables as a linear equation that best fits the observed data points.

A simple linear regression aims to fit a straight line through all the observed data points in such a way that the total squared distance between the actual observations and the values predicted by the model are minimal. This line is often referred to as either the "line of best fit" or the "regression line" and it represents the interaction between the dependent and independent variables. Mathematically, the equation for a simple linear regression is as follows:

$$Y = \beta_0 + \beta_1 X + \epsilon$$

1. Y, in the above equation, is the dependent variable, where the X represents the independent variable.
2. β_o is the intercept of the regression model.
3. β_1 is the slope of the model's "line of best fit."
4. ϵ represents the error term.

To better illustrate this, let's use basic football terms using the above regression equation to compare a team's offensive points scored in a season based on how many offensive yards it accumulated. The intercept (β_o) represents the value when a team's points scored and offensive yards are both zero. The slope (β_1) represents the rate of change in Y as the unit of X changes. The error term (ϵ) is represents the difference between the actual observed values of the regression's dependent variable and the value as predicted by the model.

Using our points scored/total yards example, a team's total yards gained is the **independent variable** and total points scored is the **dependent variable**, as a team's total yardage **is what drives the change in total points** (in other words, a team's total points *is dependent* on its total yardage). A team will not score points in the NFL if it is not also gaining yardage. We can view this relationship by building a simple linear regression model in R using the `lm()` function.

> **Note**
>
> The `lm()` function is a built-in RStudio tool as part of the `stats` package and stand for "linear model." It is used, as described above, to fit a linear regression to the data that you provide. The completed regression estimates the coefficients of the data and also includes both the intercept and slope, which are the main factors in explaining the relationship between the response and predictor variables.

The `lm()` function requires just two argument in order to provide results: the formula and the data frame to use:

```
model_results <- lm(formula, data)
```

The **`formula`** argument requires that you specify both the response and predictor variables, as named in your data frame, in the structure of `y ~ x`, wherein `y` is the response variable and `x` is the predictor. In the case that you have more than one predictor variable, the `+` is used to add to the formula:

```
model_results <- lm(y ~ x1 + x2 + x3, data)
```

The returned coefficients, residuals, and other statistical results of the model are returned into your RStudio environment in a `lm` data object. There are several ways to access this data and are discussed below in further detail.

To put theory into action, **let's build a simple linear regression model that explores the relationship between the total yardage earned by a team over the course of a season and the number of points scored**. To begin, gather the prepared information into a data frame titled `simple_regression_data` by running the code below.

```
simple_regression_data
<- vroom("http://nfl-book.bradcongelio.com/simple-reg")
```

The data contains the total yardage and points scored for each NFL team between the 2012 and 2022 seasons (not including the playoffs). Before running our first linear regression, let's first begin by selecting just the 2022 data and create a basic visualization to examine the baseline relationship between the two variables.

```
regression_2022 <- simple_regression_data %>%
  filter(season == 2022)

teams <- nflreadr::load_teams(current = TRUE)

regression_2022 <- regression_2022 %>%
  left_join(teams, by = c("team" = "team_abbr"))
```

```
ggplot(regression_2022, aes(x = total_yards,
        y = total_points)) +
  geom_smooth(method = "lm", se = FALSE,
              color = "black",
              linetype = "dashed",
              size = .8) +
  geom_image(aes(image = team_logo_wikipedia), asp = 16/9) +
  scale_x_continuous(breaks = scales::pretty_breaks(),
                     labels = scales::comma_format()) +
  scale_y_continuous(breaks = scales::pretty_breaks()) +
  labs(title = "**Line of Best Fit: 2022 Season**",
       subtitle = "*Y = total_yards ~ total_points*") +
  xlab("Total Yards") +
  ylab("Total Points") +
  nfl_analytics_theme()
```

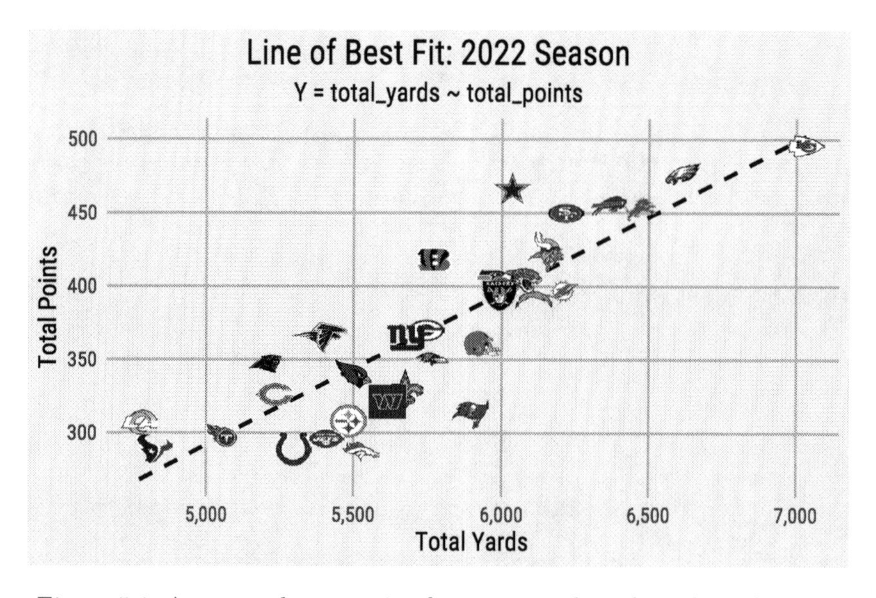

Figure 5.1: An example regression between total yards and total points

The plot shows that a regression between `total_yards` and `total_points` results in several teams – the Titans, Giants, Packers, Raiders, Jaguars, and Chiefs – being fitted nearly perfectly with the line of best fit. These teams scored points based on total yards in a *linear fashion*. The Cowboys, however, are well above the regression line. This indicates that Dallas scored more total points than what the relationship between `total_yards` and `total_points` found as "normal" for a team that earned just a hair over 6,000 total yards. The opposite is true for the Colts, Jets, and Denver. In each case, the `total_points` scored is below what is expected for teams that gained approximately 5,500 total yards.

The line of best fit can explain this relationship in slightly more detail. For example, the `total_yards` value of 5,500 cross the regression line just below the `total_points` value of 350. This means that a team that gains a total of 5,500 yards should – based on this fit – score just under 350 points during the season. Viewing it the other way, if you want your team to score 450 points during the upcoming season, you will need the offensive unit to gain roughly 6,500 total yards.

To further examine this relationship, we can pass the data into a simple linear regression model to start exploring the summary statistics.

```
results_2022 <- lm(total_points ~ total_yards,
                   data = regression_2022)
```

Using the `lm()` function, the Y variable (the dependent) is `total_yards` and the X variable (the predictor) is entered as `total_yards` with the argument that the `data` is coming from the `regression_2022` data frame. We can view the results of the regression model by using the `summary()` function.

```
summary(results_2022)
```

```
Call:
lm(formula = total_points ~ total_yards, data = regression_2022)

Residuals:
   Min     1Q Median     3Q    Max 
-71.44 -22.33   1.16  19.15  68.08 

Coefficients:
             Estimate Std. Error t value Pr(>|t|)    
(Intercept) -225.0352    65.4493   -3.44   0.0017 ** 
total_yards    0.1034     0.0113    9.14  3.6e-10 ***
---
Signif. codes:  0 '***' 0.001 '**' 0.01 '*' 0.05 '.' 0.1 ' ' 1

Residual standard error: 32 on 30 degrees of freedom
Multiple R-squared:  0.736,	Adjusted R-squared:  0.727 
F-statistic: 83.5 on 1 and 30 DF,  p-value: 3.6e-10
```

! Reading & Understanding Regression Results

You have now run and output the summary statistics for your first linear regression model that explore the relationship between an NFL team's total yards and total points over the course of the 2022 season.

But what do they mean?

The `summary()` output of any regression models contains two core components: **the residuals and the coefficients.**

Residuals

A model's residuals are the calculated difference between the actual values of the predictor variables as found in the data and the values *predicted* by the regression model. In a perfect uniform relationship, all of the values in the data frame would sit perfectly on the line of best fit. Take the below graph, for example.

```
example_fit <- tibble(
  x = 1:10,
  y = 2 * x + 3)

example_perfect_fit <- lm(y ~ x, data = example_fit)

ggplot(example_perfect_fit, aes(x = x, y = y)) +
  geom_smooth(method = "lm", se = FALSE, color = "black",
              size = .8, linetype = "dashed") +
  geom_image(image = "./images/football-tip.png",
             asp = 16/9) +
  scale_x_continuous(breaks = scales::pretty_breaks()) +
  scale_y_continuous(breaks = scales::pretty_breaks()) +
  labs(title = "A Perfect Regression Model",
       subtitle = "Every Point Falls on the Line",
       caption =
       "*An Introduction to NFL Analytics with R*<br>
       **Brad J. Congelio**") +
  xlab("Our Predictor Variable") +
  ylab("Our Response Variable") +
  nfl_analytics_theme()
```

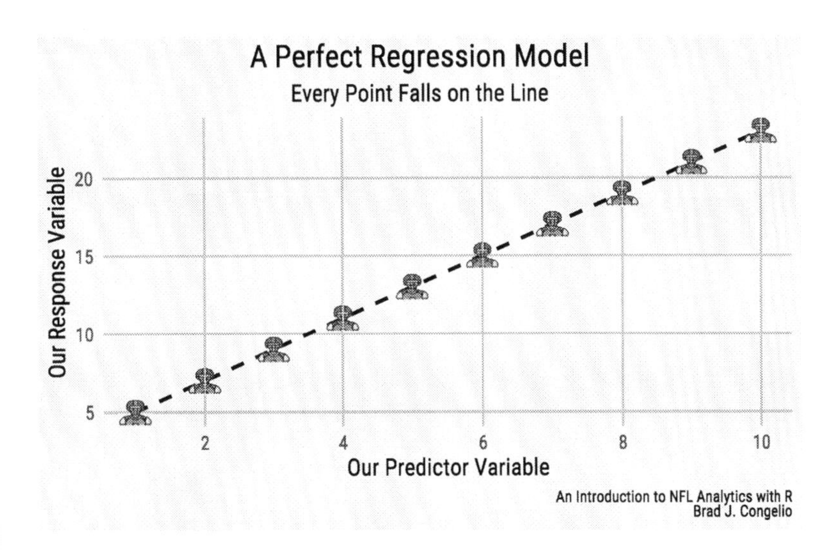

Figure 5.2: An example of a perfect line of best fit

In this example, the regression model was able to successfully "capture" the entirety of the relationship between the predictor variable on the x-axis and the response variable on the y-axis. This means that the model leaves no unexplained or undetermined variance between the variables. As a result, we could provide the model new, unseen data and the result would predict – with 100% accuracy – the resulting values of the response variable.

However, it is rare to have real-world data be perfectly situated on the line of best fit. In fact, it is more often than not a sign of "overfitting," which occurs when the model successfully discovers the "random noise" in the data. In such cases, a model with a "perfect line of fit" will perform incredibly poorly when introduced to unseen data.

A regression model that is not overfitted will have data points that do not fall on the line of best fit, but fall over and under it. The results of the regression model uses a simple formula to help us interpret the difference between those actual and estimated values:

```
residuals = observed_value - predicted_value
```

Information about these residual values are included first in our `summary(results_2022)` output and provide insight into the distribution of the model's residual errors.

Summary Distribution	Meaning
The `Min` Distribution	The `Min` distribution provides the smallest difference between the actual values of the model's predictor variable (total points) and the predicted. In the example summary, **the minimum residual is −71.443 which means that the `lm()` model predicted that one specific team scored 71.443 more points than it actually did.**
The `1Q` Distribution	The `1Q` distribution is based on the first quartile of the data (or where the first 25% of the model's residual fall on the line of best fit). **The `1Q` residual is −22.334, which means the `lm()` model predicted that 25% of the teams scored 22.334 more points than the actual values.**
The `Median` Distribution	The `Median` distribution is the residuals from the 50th percentile of the data. **The `Median` residual in the above summary is 1.157, which means that the `lm()` model – for 50% of teams – either overestimated or underestimated a teams total points by less than 1.157 points.**
The `3Q` Distribution	Covering the third quartile of the residuals, **the `3Q` Distribution is 19.145 which means that 75% of the NFL teams in the data had a total points prediction either overestimated or underestimated by less than 19.145 points.**
The `Max` Distribution	The opposite of the `Min` distribution, the `Max` distribution is the largest difference between the model's observed and predicted values for a team's total points. In this case, for one specific team, **the model predicted the team scored 68.080 points less than the actual value.**

A model's residuals allow you to quickly get a broad view of accurately it is predicting the data. Ideally, a well-performing model will return residuals that are small and distributed evenly around zero. In such cases, it is good first sign that the predictions are close to the actual value in the data and the model is not producing over or under estimates.

But that is not always the case.

For example, we can compare our above example residuals to the residuals produced by manually created data.

```
   Min. 1st Qu.  Median    Mean 3rd Qu.    Max.
 -13.70   -2.18   -0.60    0.00    4.16   10.54
```

Compared to the residuals from the above NFL data, the residuals from the randomly created data are small in comparison and are more evenly distributed around zero. Given that, it is likely that the linear model is doing a good job at making predictions that are close to the actual value.

But that does not mean the residuals from the NFL data indicate a flawed and unreliable model.

It needs to be noted that the "goodness" of any specific linear regression model is wholly dependent on both the context and the specific problem that the model is attempting to predict. To that end, it is also a matter of trusting your subject matter expertise on matters regarding the NFL.

There could be any number of reasons that can explain why the residuals from the regression model are large and not evenly distributed from zero. For example:

1. **Red-zone efficiency:** a team that moves the ball downfield with ease, but then struggles to score points once inside the 20-yardline, will accumulate `total_yards` but fail to produce `total_points` in the way the model predicts.
2. **Turnovers:** Similar to above, a team may rack up `total_yards` but ultimately continue to turn the ball over prior to being able to score.
3. **Defensive scoring:** a score by a team's defense, in this model, still counts toward `total_points` but does not count toward `total_yards`.

4. **Strength of Opponent:** At the end of the 2022 season, the Philadelphia Eagles and the New York Jets both allowed just 4.8 yards per play. The model's predicted values of the other three teams in each respective division (NFC East and AFC East) could be incorrect because such contextual information is not included in the model.

All that to say: residuals are a first glance at the results of the data and provide a broad generalization of how the model performed without taking outside contextual factors into consideration.

Coefficients

The coefficients of the model are the weights assigned to each predictor variable and provide insight into the relationship between the various predictor and response variables, with the provided table outlining the statistical metrics.

1. **Estimate** – representing the actual coefficient of the model, these numbers are the mathematical relationship between the predictor and response variables. In our case, the estimate for `total_yards` is 0.1034 which means that for each additional one yard gained we can expected a team's `total_points` to increase by approximately 0.1034.
2. **Std. Error** – the standard error is the numeric value for the level of uncertainty in the model's estimate. In general, a lower standard error indicates a more reliable estimate. In other words, if we were to resample the data and then run the regression model again, we could expect the `total_yards` coefficient to vary by approximately 0.0113, on average, for each re-fitting of the model.
3. **t value** – The t-statistic value is the measure of how many standard deviations the estimate is from 0. A larger t-value indicates that it is less likely that the model's coefficient is not equal to 0 by chance.
4. **Pr(>|t|)** – This value is the p-value associated with the hypothesis test for the coefficient. The level of significance for a statistical model is typically 0.05, meaning a value less than this results in a rejection of the null hypothesis and the conclusion that the predictor does have a statistically significant relationship with the response variable. In the case of our regression model, a *p-value* of 0.00000000036 indicates a highly significant relationship.

Returning to the summary statistics of our analysis of the 2022 season, the residuals have a wide spread and an inconsistent deviation from zero. While the `median` residual value is the closest to zero at 1.157, it is still a bit too high to safely conclude that the model is making predictions that adequately reflect the actual values. Moreover, both tail ends of the residual values (`Min` and `Max`) are a large negative and positive number, respectively, which is a possible indication that the regression model is both over- and underestimating a team's `total_points` by statistically significant amount.

However, as mentioned, this widespread deviation from zero is likely the result of numerous factors outside the model's purview that occur in any one NFL game. To get a better idea of what the residual values represent, we can plot the data and include NFL team logos.

```
regression_2022$residuals <- residuals(results_2022)

ggplot(regression_2022, aes(x = total_yards, y = residuals)) +
  geom_hline(yintercept = 0, color = "black", linewidth = .7) +
  stat_fit_residuals(size = 0.01) +
  stat_fit_deviations(size = 1.75,
     color = regression_2022$team_color) +
  geom_image(aes(image = team_logo_wikipedia), asp = 16/9,
     size = .0325) +
  scale_x_continuous(breaks = scales::pretty_breaks(),
                   labels = scales::comma_format()) +
  scale_y_continuous(breaks = scales::pretty_breaks(n = 5)) +
  labs(title = "**Total Yards & Residual Values**",
     subtitle = "*Y = total_points ~ total_yards*",
     caption = "*An Introduction to NFL Analytics with R*<br>
     **Brad J. Congelio**") +
  xlab("Total Yards") +
  ylab("Residual of Total Points") +
  nfl_analytics_theme() +
  theme(panel.grid.minor.y = element_line(color = "#d0d0d0"))
```

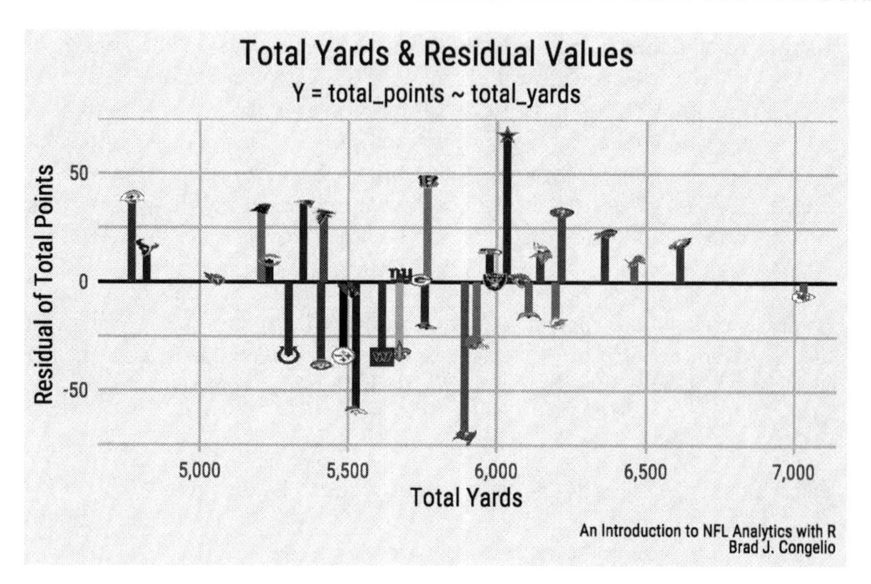

Figure 5.3: Plotting a simple linear regression's residual values

With the data visualized, it is clear that the model's `Min` distribution of -71.44 is associated with the Tampa Bay Buccaneers, while the `Max` distribution of 68.08 is the prediction for the total points earned by the Dallas Cowboys. Because a negative residual means that the model's predicted value is too high, and a positive residual means it was too low, we can conclude that the Buccaneers actually scored 71.4 points less than the results of the model, while the Cowboys scored 68.08 more than predicted.

```
`Coefficients:
             Estimate Std. Error t value      Pr(>|t|)
(Intercept) -225.0352    65.4493   -3.44        0.0017 **
total_yards    0.1034     0.0113    9.14 0.00000000036 ***
---
Signif. codes: 0 '***' 0.001 '**' 0.01 '*' 0.05 '.' 0.1 ' ' 1`
```

The `(Intercept)` of the model, or where the regression line crosses the y-axis, is -225.0350. When working with NFL data, it of course does not make sense that the `(Intercept)` is negative. Given the model is built on a team's total yards and total points, it seems intuitive that the regression line would cross the y-axis at the point of (0,0) as an NFL team not gaining any yards is highly unlike to score any points.

It is important to remember that the linear model attempts to position the regression line to come as close to all the individual points as possible. Because of this, it is not uncommon for regression line to not cross exactly where the x-axis and y-axis meet. Again, contextual factors of an NFL game are not accounted for in the model's data: strength of the opponent's defense, the quality of special teams play, defensive turnovers and/or touchdowns, field position, etc. can all impact a team's ability to score points without gaining any yardage. The lack of this information in the data ultimately impacts the positioning of the line of best fit.

The `total_yards` coefficient represents the slope of the model's regression line. It is this slope that represents how a team's total points are predicted to change with every additional gain of one yard. In this example, the `total_yards` coefficient is 0.10341 – so for every additional yard gained by a team, it is expected to add 0.10341 points to the team's cumulative amount.

The `Std. Error` summary statistic provides guidance on the accuracy of the other estimated coefficients. The `Std. Error` for the model's `(Intercept)` is quite large at 65.44927. Given the ability to resample the data from NFL terms numerous times and then allowing the linear model to predict again, this specific `Std. Error` argues that the regression line will cross the y-axis within 65.44972 of −225.03520 in either direction. Under normal circumstances, a large `Std. Error` for the `(Intercept)` would cause concern about the validity of the regression line's crossing point. However, given the nature of this data – an NFL team cannot score negative points – we should not have any significant concern about the large `Std. Error` summary statistic for the `(Intercept)`.

At 0.01132, the `Std. Error` for the `total_yards` coefficient is small and indicates that the `Estimate` of `total_yards` – that is, the increase in points per every yard gained – is quite accurate. Given repeated re-estimating of the data, the relationship between `total_yards` and `total_points` would vary by just 0.01132, either positively or negatively.

With a `t-value` of 9.135, the `total_yards` coefficient has a significant relationship with `total_points`. The value of −3.438 indicates that the `(Intercept)` is statistically different from 0 but we should still largely ignore this relationship given the nature of the data.

The model's `Pr(>|t|)` value of highly significant for `total_yards` and is still quite strong for the `(Intercept)`. The value of 0.00000000036 indicates an incredibly significant relationship between `total_yards` and `total_points`.

The linear model's `Residual Standard Error` is 32.04, which means that the average predicted values of `total_points` are 32.04 points different from the actual values in the data. The linear model was able to explain 73.56% of the variance between `total_yards` and `total_points` based on the `multiple R-squared` value of 0.7356. Additionally, the `Adjusted R-squared` value of 0.7268 is nearly identical to the multiple R2, which is a sign that the linear model is not overfitting (in this case because of the simplicity of the data).

The model's `F-Statistic` of 83.45 indicates a overall significance to the data, which is backed up by an extremely strong `p-value`.

Based on the summary statistics, the linear model did an extremely good job at capturing the relationship between a team's `total_yards` and `total_points`. However, with residuals ranging from −71.443 to 68.080, it is likely that the model can be improved upon by adding additional information and statistics. However, before providing additional metrics, we can try to improve the model's predictions by including all of the data (rather than just the 2022 season). By including 20-seasons worth of `total_yards` and `total_points`, we are increasing the sample size which, in theory, allows for a reduced impact of any outliers and an improve generalizability.

> **! Important**
>
> Working with 10+ years of play-by-play data can be problematic in that the model, using just `total_yards` and `total_points`, is not aware of changes in the overall style of play NFL. The balance between rushing and passing has shifted, there's been a philosophical shift in the coaching ranks in "going for it" on 4th down, etc. A simple linear regression cannot account for how these shifts impact the data on a season-by-season basis.

The results from including the `total_points` and `total_yards` for each NFL team from 2012 to 2022 show an improvement of the model, specifically with the residual values.

```
regression_all_seasons <- simple_regression_data %>%
  select(-season)

all_season_results <- lm(total_points ~ total_yards,
                              data = regression_all_seasons)

summary(all_season_results)
```

The residual values after including 20-seasons worth of data are a bit better. The `Median` is −1.26 which is slightly higher than just one season (`M` = 1.16). The `1Q` and `3Q` distributions are both approximately symmetric around the model's `M` value compared to just the 2022 season regression that results in a deviation between `1Q` and `3Q` (−22.33 and 19.15, respectively). The `Min` and `Max` values of the new model still indicate longtail cases on both ends of the regression line much like the 2022 model found.

Tip

To further examine the residual values, we can use a **Shapiro-Wilk Test** to test the whether results are normally distributed.

The Shapiro-Wilk Test provides two values with the output: the test statistic (provided as a `W` score) and the model's `p-value`. Scores for `W` can range between 0 and 1, where results closer to 1 mean the residuals are in a normal distribution. The `p-value` is used make a decision on the null hypothesis (that there *is* enough evidence to conclude that there is uneven distribution). In most cases, if the `p-value` is larger than the regression's level of significance (typically 0.05), than you may **reject** the null hypothesis.

We can run the Shapiro-Wilk Test on our 2012–2022 data using the `shapiro.test` function that is part of the `stats` package in R.

```
results_2012_2020 <- residuals(all_season_results)

shapiro_test_result <- shapiro.test(results_2012_2020)

shapiro_test_result
```

```

	Shapiro-Wilk normality test

data:  results_2012_2020
W = 1, p-value = 0.8
```

The `W` score for the residual is 1, meaning a very strong indication that the data in our model is part of a normal distribution. The `p-value` is 0.8, which is much larger than the regression's level of significance (0.05). As a result, we can reject the null hypothesis and again conclude that the data is in a normal distribution.

```
teams <- nflreadr::load_teams(current = TRUE)

regression_all_seasons <- left_join(regression_all_seasons,
                teams, by = c("team" = "team_abbr"))

regression_all_seasons$residuals
       <- residuals(all_season_results)
```

```
ggplot(regression_all_seasons, aes(x = total_yards,
        y = residuals)) +
  geom_hline(yintercept = 0,
             color = "black", linewidth = .7) +
  stat_fit_residuals(size = 2,
        color = regression_all_seasons$team_color) +
  stat_fit_deviations(size = 1,
        color = regression_all_seasons$team_color,
        alpha = 0.5) +
  scale_x_continuous(breaks = scales::pretty_breaks(),
                     labels = scales::comma_format()) +
  scale_y_continuous(breaks = scales::pretty_breaks(n = 5)) +
  labs(title = "**Total Yards & Residual Values: 2012-2022**",
   subtitle = "*Y = total_points ~ total_yards*",
   caption = "*An Introduction to NFL Analytics with R*<br>
   **Brad J. Congelio**") +
  xlab("Total Yards") +
  ylab("Residual of Total Points") +
  nfl_analytics_theme() +
  theme(panel.grid.minor.y = element_line(color = "#d0d0d0"))
```

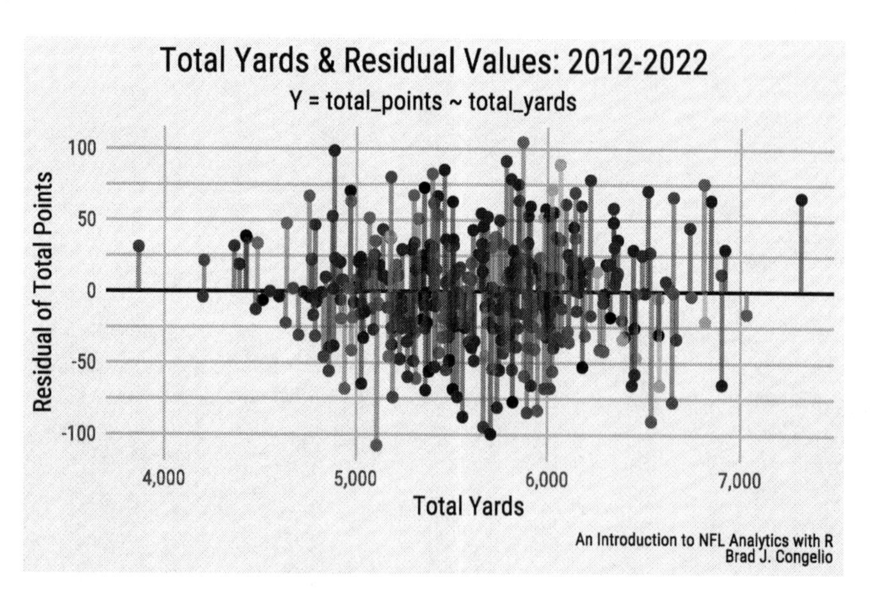

Figure 5.4: Residual values for 20 seasons of data

We can also compare the multiple R2 and adjusted R2 score between the two regression models.

```
2012 - 2022 Data:
Multiple R-squared:  0.683
Adjusted R-squared:  0.682

2022 Data
Multiple R-squared:  0.736
Adjusted R-squared:  0.727
```

The regression using just the 2022 data results in a slightly better multiple and adjusted R2 score compared to using data from the last twenty seasons of the NFL. While this does indicate that the model based on the single season is better at defining the relationship between a team's `total_yards` and `total_points` it is essential to remember that there is different underlying patterns in the data as a result of the changing culture in the NFL and, ultimately, the epp and flow of team performance as a result of high levels of parity in the league.

In order to account for this "epp and flow" in both team performance and the changing culture/rules of the NFL, we need to turn to a multiple linear regression in order to include these additional factors as it is a model that is capable of better accounting for the nuances of NFL data.

5.2.2 Multiple Linear Regression

A multiple linear regression is extremely similar to a simple linear regression (both in design and implementation in RStudio).The main difference is that a multiple linear regression allows for us to include additional predictor variables by using the `+` sign in the model's formula. The inclusion of these additional predictive variables, in theory, allows the model to compute more complex relationships in NFL data and improve on the model's final performance.

We will again create our first multiple regression linear regression with data from just the 2022 regular season that includes the same predictor (`total_yards`) and response variable (`total_points`). For additional predictors, we must consider what circumstances may lead a team to have high `total_yards` but an amount of `total_points` that would fall below the model's predicted value. We will include the following as additional predictors:

1. **Redzone Efficiency:** provided as a percentage, this is a calculation of how many times a team drove into the redzone and scored. A higher percentage is better.
2. **Redzone Touchdown Efficiency:** This is the same as redzone efficiency, but includes only the number of redzone trips divided by the total touchdowns scored from the redzone.

3. **Redzone Field Goal Efficiency:** The same as redzone touchdown efficiency, but with field goals.
4. **Cumulative Turnovers**: The total number of turnovers during the regular season.
5. **Defensive Touchdowns**: The number of touchdowns scored by each team's defensive unit.
6. **Special Teams Touchdowns**: The number of touchdowns scored by special teams (kick/punt returns).

To begin building the multiple linear regression model for the 2022 season, we can read in the data below using `vroom::vroom()`.

```
multiple_lm_data
<- vroom("http://nfl-book.bradcongelio.com/multiple-lm")

multiple_lm_data <- multiple_lm_data %>%
  filter(season == 2022)

teams <- multiple_lm_data$team
```

The data for our multiple linear regression has the same four columns as the simple linear regression (`season`, `team`, `total_points`, and `total_yards`) but also includes the new predictor variables (`rz_eff`, `rz_td_eff`, `rz_fg_eff`, `def_td`, and `spec_tds`).

> Caution
>
> Please note that, of the predictor and response variables, all of the values are in whole number format except for `rz_eff`, `rz_td_eff`, and `rz_fg_eff`. While it is not a problem to include predictors that are on differing scales (in this case, whole numbers and percentages), it may cause difficulty in interpreting the summary statistics. If this is the case, the issue can be resolved by using the `scale()` function to standardize all the predictors against one another.

The construction of the multiple linear regression model is the same process of the simple regression, with the inclusion of additional predictors to the formula using the + sign. We are also applying a `filter()` to our `multiple_lm_data` to retrieve just the 2022 season.

```
lm_multiple_2022
<- lm(total_points ~ total_yards + rz_eff + rz_td_eff
```

```
          + rz_fg_eff
          + total_to + def_td + spec_tds,
          data = multiple_lm_data)

summary(lm_multiple_2022)
```

```
Call:
lm(formula = total_points ~ total_yards + rz_eff + rz_td_eff +
    rz_fg_eff + total_to + def_td + spec_tds,
    data = multiple_lm_data)

Residuals:
   Min     1Q Median     3Q    Max
-53.03 -15.58  -0.35  14.60  43.53

Coefficients: (1 not defined because of singularities)
             Estimate Std. Error t value Pr(>|t|)
(Intercept) -459.5225   128.5557   -3.57   0.0015 **
total_yards    0.0906     0.0113    8.02  2.2e-08 ***
rz_eff       228.8853   113.5625    2.02   0.0547 .
rz_td_eff    167.2323    82.7097    2.02   0.0540 .
rz_fg_eff          NA         NA      NA       NA
total_to       0.4058     1.4934    0.27   0.7881
def_td         4.4560     4.0573    1.10   0.2826
spec_tds       5.4977     7.8674    0.70   0.4911
---
Signif. codes:  0 '***' 0.001 '**' 0.01 '*' 0.05 '.' 0.1 ' ' 1

Residual standard error: 26.9 on 25 degrees of freedom
Multiple R-squared:  0.844, Adjusted R-squared:  0.807
F-statistic: 22.6 on 6 and 25 DF,  p-value: 5.81e-09
```

The summary statistic residuals for the multiple linear regression are more evenly distributed toward the mean than our simple linear regression. Based on the residuals, we can conclude that – for 50% of the teams – the model either over or underestimated their `total_points` by just -0.35 (as listed in the `Median` residual). The interquartile range (within the `1Q` and `3Q` quartiles) are both close to the median and the `Min` and `Max` residuals both decreased significantly from our simple linear model, indicating a overall better line of fit.

We can confirm that the multiple linear regression resulted in an even distribution of the residuals by again using a Shapiro-Wilk Test.

```
    Shapiro-Wilk normality test

data:  results_lm_2022
W = 1, p-value = 0.9
```

The results of the Shapiro-Wilk test (`W = 1` and `p-value = 0.9`) confirm that residuals are indeed evenly distributed. A visualization showcases the model's even distribution of the residuals.

```
mlm_2022_fitted <- predict(lm_multiple_2022)
mlm_2022_residuals <- residuals(lm_multiple_2022)

plot_data_2022 <- data.frame(Fitted = mlm_2022_fitted,
                             Residuals = mlm_2022_residuals)

plot_data_2022 <- plot_data_2022 %>%
  cbind(teams)

nfl_teams <- nflreadr::load_teams(current = TRUE)

plot_data_2022 <- plot_data_2022 %>%
  left_join(nfl_teams, by = c("teams" = "team_abbr"))

ggplot(plot_data_2022, aes(x = Fitted, y = Residuals)) +
  geom_hline(yintercept = 0,
             color = "black", linewidth = .7) +
  stat_fit_deviations(size = 1.75,
                      color = plot_data_2022$team_color) +
  geom_image(aes(image = team_logo_espn),
             asp = 16/9, size = .0325) +
  scale_x_continuous(breaks = scales::pretty_breaks(),
                     labels = scales::comma_format()) +
  scale_y_continuous(breaks = scales::pretty_breaks()) +
  labs(title = "**Multiple Linear Regression Model: 2022**") +
  xlab("Fitted Values") +
  ylab("Residual Values") +
  nfl_analytics_theme() +
  theme(panel.grid.minor.y = element_line(color = "#d0d0d0"))
```

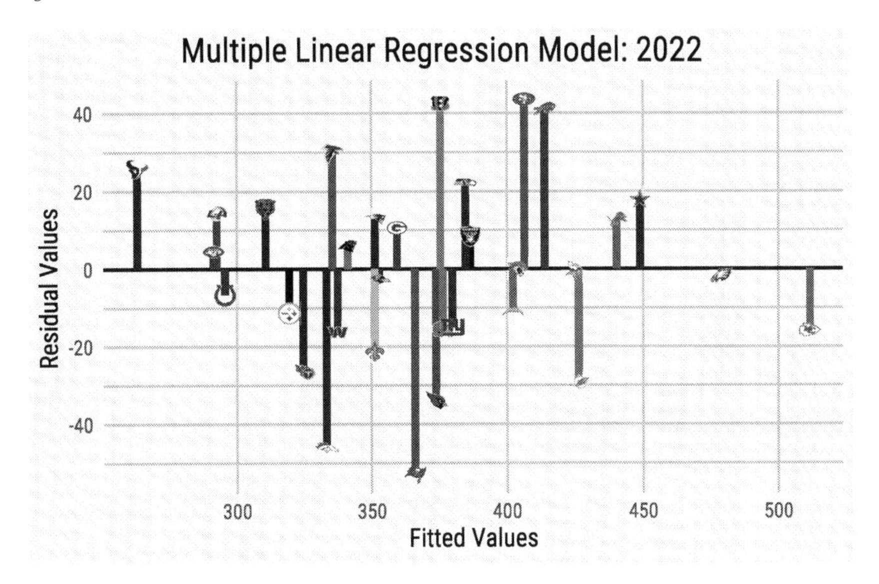

Figure 5.5: Visualizing residuals of a multiple linear regression

Just as the residual values in the summary statistics indicated, plotting the `fitted_values` against the `residual_values` shows an acceptable spread in the distribution, especially given the nature of NFL data. Despite positive results in the residual values, the summary statistics of the multiple linear regression indicates a significant issue with the data. Within the `Coefficients`, it is explained that one of the items is "not defined because of singularities."

> **! Important**
>
> "Singularities" occur in the data as a result of the dreaded multicollinearity between two or more predictors. The involved predictors were found to have a high amount of correlation between one another, meaning **that one of the variables can be predicted in a near linear fashion with one or more of the other predictive variables**. As a result, it is difficult for the regression model to correctly estimate the contribution of these dependent variables to the response variable.
>
> The model's Coefficients of our multiple linear regression shows `NA` values for the `rz_fg_eff` predictor (the percentage of times a team made a field goal in the red zone rather than a touchdown). This is because `rz_fg_eff` was one of the predictive variables strongly correlated and was the one dropped by the regression model to avoid producing flawed statistics as a result of the multicollinearity.

If you are comfortable producing the linear regression with rz_fg_eff being a dropped predictor, there are no issues with that. However, we can create a correlation plot that allows is to determine which predictors have high correlation values with others. Examining the issue allows us to determine if `rz_fg_eff` is, indeed, the predictive variable we want the regression to drop or if we'd rather, for example, drop `rz_eff` and keep just the split between touchdowns and field goals.

```
regression_corr <-
  cor(multiple_lm_data[, c("total_yards",
                           "rz_eff", "rz_td_eff",
                           "rz_fg_eff", "total_to",
                           "def_td", "spec_tds")])

melted_regression_corr <- melt(regression_corr)

ggplot(data = melted_regression_corr, aes(x = Var1,
                                          y = Var2,
                                          fill = value)) +
  geom_tile() +
  scale_fill_distiller(palette = "PuBu",
                       direction = -1,
                       limits = c(-1, +1)) +
  geom_text(aes(x = Var1, y = Var2,
  label = round(value, 2)),
            color = "black",
            fontface = "bold",
            family = "Roboto", size = 5) +
  labs(title = "Multicollinearity Correlation Matrix",
       subtitle = "Multiple Linear Regression: 2022 Data",
       caption =
       "*An Introduction to NFL Analytics with R*<br>
       **Brad J. Congelio**") +
  nfl_analytics_theme() +
  labs(fill = "Correlation \n Measure", x = "", y = "") +
  theme(legend.background = element_rect(fill = "#F7F7F7"),
        legend.key = element_rect(fill = "#F7F7F7"))
```

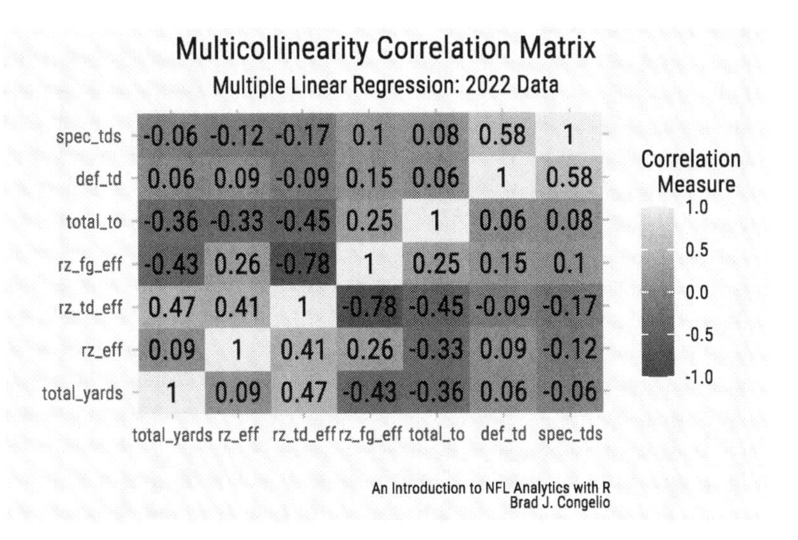

Figure 5.6: Using a correlation matrix to determine which predictor have high correlation values with others

Using a correlation plot allows for easy identification of those predictive variables that have high correlation with one another. The general rule is that two predictors become problematic in the regression model if the coefficient between the two is above 0.7 (or 0.8, given domain knowledge about the context of the data).

In our correlation plot, there are two squares (indicated by the darkest blue color) that have a value greater than 0.7 (or −0.7 in this case, as both strong and negative correlations are capable of producing multicollinearity. The two squares happen to relate to the same relationship between the `rz_fg_eff` and `rz_td_eff` predictors.

Recall that the regression model automatically removed the `rz_fg_eff` from the measured Coefficients. Given the context of the data, I am not sure that is the best decision. Because we are examining the relationship between the predictive variables and `total_points`, removing the `rz_fg_eff` variable inherently erases a core source of points in a game of football.

Because of this – and since our `rz_eff` predictor accounts for both touchdowns and field goals – I believe we can move forward on rerunning the regression without both`rz_fg_eff` and `rz_td_eff`.

To run the multiple linear regression again, without the predictors relating to red zone touchdown and field efficiency, we will drop both from our `multiple_lm_2022` data frame, rerun the regression model, and then examine the ensuing summary statistics.

```
lm_multiple_2022_edit <- multiple_lm_data %>%
  select(-rz_td_eff, -rz_fg_eff)

lm_multiple_2022_edit <- lm(total_points ~
                            total_yards + rz_eff +
                            total_to + def_td + spec_tds,
                            data = lm_multiple_2022_edit)

summary(lm_multiple_2022_edit)
```

```
Call:
lm(formula = total_points ~ total_yards + rz_eff + total_to +
    def_td + spec_tds, data = lm_multiple_2022_edit)

Residuals:
   Min     1Q Median     3Q    Max
-60.18 -14.83  -3.79  18.99  55.92

Coefficients:
             Estimate Std. Error t value Pr(>|t|)
(Intercept) -472.5877   135.8042   -3.48   0.0018 **
total_yards    0.0998     0.0109    9.13  1.4e-09 ***
rz_eff       309.1167   112.5462    2.75   0.0108 *
total_to      -0.2760     1.5388   -0.18   0.8591
def_td         3.5808     4.2670    0.84   0.4090
spec_tds       4.9584     8.3167    0.60   0.5562
---
Signif. codes:  0 '***' 0.001 '**' 0.01 '*' 0.05 '.' 0.1 ' ' 1

Residual standard error: 28.5 on 26 degrees of freedom
Multiple R-squared:  0.819, Adjusted R-squared:  0.784
F-statistic: 23.5 on 5 and 26 DF,  p-value: 7.03e-09
```

We have certainly simplified the model by removing both **`rz_td_eff`** and **`rz_fg_eff`** but the impact of this change is a fair trade off to avoid further issues with multicollinearity. Our new adjusted R is still high (0.784), only dropping a bit from the original model that included both predictors (0.807). Both models did well at explaining the amount of variance between the predictors and the response variable. While the `F-statistic` and the `p-value` are strong in both models, it is important to note that the `Residual standard error` dropped from 27 in the original model to 28 in the more simplified version. Given that this value is the average difference between the actual values and the predicted equivalents in the regression, both would ideally be smaller.

With multiple linear regression model producing acceptable results over the course of the 2022 season, we can now see if the results remain stable when produced over the course of the 2012–2022 seasons.

```
multiple_lm_data_all <- multiple_lm_data %>%
  select(-rz_td_eff, -rz_fg_eff, -season)

lm_multiple_all <- lm(total_points ~ total_yards + rz_eff +
                                    total_to + def_td + spec_tds,
                         data = multiple_lm_data_all)

summary(lm_multiple_all)
```

```
Call:
lm(formula = total_points ~ total_yards + rz_eff + total_to +
    def_td + spec_tds, data = multiple_lm_data_all)

Residuals:
   Min     1Q Median     3Q    Max
-60.18 -14.83  -3.79  18.99  55.92

Coefficients:
            Estimate Std. Error t value Pr(>|t|)
(Intercept) -472.5877   135.8042   -3.48   0.0018 **
total_yards    0.0998     0.0109    9.13  1.4e-09 ***
rz_eff       309.1167   112.5462    2.75   0.0108 *
total_to      -0.2760     1.5388   -0.18   0.8591
def_td         3.5808     4.2670    0.84   0.4090
spec_tds       4.9584     8.3167    0.60   0.5562
---
Signif. codes:  0 '***' 0.001 '**' 0.01 '*' 0.05 '.' 0.1 ' ' 1

Residual standard error: 28.5 on 26 degrees of freedom
Multiple R-squared:  0.819, Adjusted R-squared:  0.784
F-statistic: 23.5 on 5 and 26 DF,  p-value: 7.03e-09
```

The results of the multiple linear regression over data from the 2012–2022 indicates a statistically significant relationship between our predictor variables and a team's total yards. That said, two items are worth further exploration.

1. The model's Residual standard error increased closer to 30, as opposed to the values of 27 and 28 from the models built on a single season of data. This means that the model, on average, is over or underpredicting the actual values by approximately thirty points. To verify that a residual standard error of 30 is not too high, we can evaluate the value against the scale of our data based on the mean and/or median averages of the `total_points` variable. As seen below, the model's RSE as a percentage of the mean is `8.1%` and its percentage of the median is `8.2%`. Given that both values are below 10%, it is reasonable to conclude that the value of the model's residual standard error is statistically small compared to the scale of the `total_points` dependent variable.

> **Tip**
>
> ```
> total_mean_points
>         <- mean(multiple_lm_data_all$total_points)
> total_points_median
>         <- median(multiple_lm_data_all$total_points)
>
> rse_mean_percentage <- (30 / total_mean_points) * 100
> rse_median_percentage <- (30 / total_points_median) * 100
> ```

2. The `spec_tds` predictor, which is the total number of special teams touchdowns scored by a team, has a `p-value` of 0.61. This high of a `p-value` indicates that the amount of special teams touchdowns earned by a team is not a dependable predictor of the team's total points. Given the rarity of kickoff and punt returns, it is not surprising that the predictor returned a high `p-value`. If we run the regression again, without the `spec_tds` predictive variable, we get results that are nearly identical to the regression model that includes it as a predictor. The only significant difference is a decrease in the `F-statistic` from 398 to 317. Given the small decrease, we will keep `spec_tds` in the model.

Important

The final step of our multiple linear regression model is feeding it new data to make predictions on.

To begin, we need to create a new data frame that holds the new predictor variables. For nothing more than fun, let's grab the highest value from each predictive variable between the 2012–2022 seasons.

```
new_observations <- data.frame(
  total_yards = max(multiple_lm_data$total_yards),
  rz_eff = max(multiple_lm_data$rz_eff),
  total_to = max(multiple_lm_data$total_to),
  def_td = max(multiple_lm_data$def_td),
  spec_tds = max(multiple_lm_data$spec_tds))
```

This hypothetical team gained a total of 7,317 yards in one season and was incredibly efficient in the red zone, scoring 96% of the time. It also scored nine defensive touchdowns and returned a punt or a kickoff to the house four times. Unfortunately, the offense also turned the ball over a whopping total of 41 times.

We can now pass this information into our existing model using the `predict` function and it will output the predicted `total_points` earned by this hypothetical team based on the multiple linear regression model we built with 20 years of NFL data.

```
new_predictions <- predict(lm_multiple_all,
        newdata = new_observations)

new_predictions
```

The model determined, based on the new predictor variables provided, that this hypothetical team will score a total of 563 points, which is the second-highest cumulative amount scored by a team dating back to the 2012 season (the 2013 Denver Broncos scored 606 total points). In this situation, the hypothetical team has nearly double the turnovers as the 2013 Bronco (41 turnovers to 21). It is reasonable that providing this hypothetical team a lower number of turnovers would result in it becoming the highest scoring team since 2012.

5.2.3 Logistic Regression

Logistic regressions are particularity useful when modeling the relationship between a categorical dependent variable and a given set of predictor variables. While linear models, as covered in the last section, handles data with a continuous dependent variable, logistic regressions are used when the response variable is categorical (whether a team successfully converted a third down, whether a pass was completed, whether a running back fumbled on the play, etc.). In each example, there are only two possible outcomes: "yes" or "no."

Moreover, a logistic regression model does not model the relationship between the predictors and the response variables in a linear fashion. Instead, logistic regressions use seek to mutate the predictors into a values between 0 and 1. Specifically, the formula for a logistic regression is as follows:

$$P(Y = 1|X) = \frac{1}{1 + e^{-(\beta_0+\beta_1 X_1+\beta_2 X_2+...+\beta_n X_n)}}$$

1. $P(Y = 1|X)$ in the equation is the likelihood of the event $Y = 1$ taking place given the provided predictor variables (X).
2. The model's coefficients are represented by $\beta_0 + \beta_1 X_1 + \beta_2$.
3. $X1, X2, ..., Xn$ represent the model's predictor variables.
4. The 1 in the formula's numerator results in the model's probability values always being between 0 and 1.
5. The 1 in the denominator is a part of the underlying logistic function and ensures that the value is always greater than or equal to 1.
6. Finally, the e is part of the model's logarithm (or the constant of 2.71828). In short, this function allows the linear combination to be a positive value.

To better explain the formula, let's assume we want to predict the probability of any one NFL team winning a game (again, in a binary `1` for winning, or `0` for losing). These binary outcome is represented by the Y. The predictor variables, such as the team's average points scored, the average points allowed by the opposing defense, home-field advantage, and any other statistics likely to be important in making the prediction are represented by $X1, X2, ..., Xn$. The model's coefficients, or the relationship between the predictors and a team's chance of winning, are represented by $\beta_0 + \beta_1 X_1 + \beta_2$. In the event that a team's average passing yards is represented by the $B1$ coefficient, a positive number suggests that higher average passing yards results in an increase in the probability of winning. Conversely, a negative coefficient number indicates the statistic has a negative impact on winning probability.

To that end, there are three distinct types of logistic regressions: binary, multinomial, and ordinal (we will only be covering binary and multinomial in this chapter). All three allow for working with various types of dependent variables.

Binary regression models are used when the dependent variable (the outcome) has just two outcomes (typically presented in binary format – `1` or `0`). Using the logistic function, a binary regression model determines the probability of the dependent event occurring given the predictor variables. As highlighted below, a binary regression model can be used to predict the probability that a QB is named MVP at the conclusion of the season. The response variable is a binary (`1` indicating that the player was named MVP and `0` indicating that the player did not win MVP). Various passing statistics can serve as the predictor variables.

A **multinomial regression model** is an extension of the binary model in that it allows for the dependent variable to have more than two unordered categories. For example, a multinomial regression model can be used to predict the type of play a team is going to run (run or pass) based on the predictor variables (down and distance, time remaining, score differential, etc.).

An **ordinal regression model** is used when the dependent variables not only have ordered categories, but the structure and/or order of these categories contains meaningful information. Ordinal regression can be used, for example, to attempt to predict the severity of a player's injury. The ordered and meaningful categories of the dependent variable can coincide with the severity of the injury itself (mild, moderate, severe, season-ending) while the predictor variables include several other types of categories (the type of contact, the playing surface, what position the player played, how late into the season it occurred, whether the team is coming off a bye week, etc.).

5.2.3.1 Logistic Regression 1: Binary Classification

A binary regression model is used when the dependent variable is in binary format (that is, `1` for "yes" or `0` for "no). This binary represents two – and only two – possible outcomes such as a a team converting on third down or not or if a team won or lost the game. Constructing binary regression models allows use to predict the likelihood of the dependent event occurring based on the provided set of predictor variables.

We will build a binary regression model to predict the probability that an NFL QB will win the MVP at the conclusion of the season.

Let's first create a data frame, called `qb_mvp`, that contains the statistics for quarterbacks that we intuitively believe impact the likelihood of a QB being named MVP.

```
player_stats <- nflreadr::load_player_stats(2000:2022) %>%
  filter(season_type == "REG" & position == "QB") %>%
  filter(season != 2000 & season != 2005 & season != 2006 &
           season != 2012) %>%
  group_by(season, player_display_name, player_id) %>%
  summarize(
    total_cmp = sum(completions, na.rm = TRUE),
    total_attempts = sum(attempts, na.rm = TRUE),
    total_yards = sum(passing_yards + rushing_yards,
          na.rm = TRUE),
    total_tds = sum(passing_tds + rushing_tds, na.rm = TRUE),
    total_interceptions = sum(interceptions, na.rm = TRUE),
    mean_epa = mean(passing_epa, na.rm = TRUE)) %>%
  filter(total_attempts >= 150) %>%
  ungroup()
```

We are using data from over the course of 22 NFL seasons (2000 to 2022), but then removing the 2000, 2005, 2006, and 2012 seasons as the MVP for each was not a quarterback (Marshall Faulk, Shaun Alexander, LaDainian Tomlison, and Adrian Peterson, respectively). To begin building the model, we collect QB-specific metric from the `load_player_stats()` function from `nflreadr`, including: total completions, total attempts, total yards (passing + rushing), total touchdowns (passing + rushing), and the QB's average EPA (only for pass attempts).

Because the style of play in the NFL has changed between the earliest seasons in the data frame and the 2022 season, it may not be fair to compare the specific statistics to each other. Rather, we can rank the quarterbacks for each season, per statistic, in a decreasing fashion as the statistical numbers increase. For example, Patrick Mahomes led the league in passing yards in the 2022 season. As a result, he is ranked as `1` in the forthcoming `yds_rank` column while the QB with the second-most yards (Justin Herbert, with 4,739) will be ranked as `2` in the `yds_rank` column. This process allows us to normalize the data and takes into the account the change in play style in the NFL over the time span of the data frame. To add the rankings to our data, we will create a new data frame titled `qb_mvp_stats` from the existing `player_stats` and then use the `order` function from base R to provide the rankings in descending order. After, we use the `select()` function to gather the `season`, `player_display_name`, and `player_id` as well as the rankings that we created using `order()`.

```
qb_mvp_stats <- player_stats %>%
  dplyr::group_by(season) %>%
  mutate(cmp_rank = order(order(total_cmp,
```

```
             decreasing = TRUE)),
           att_rank = order(order(total_attempts,
                   decreasing = TRUE)),
           yds_rank = order(order(total_yards,
                   decreasing = TRUE)),
           tds_rank = order(order(total_tds,
                   decreasing = TRUE)),
           int_rank = order(order(total_interceptions,
                   decreasing = FALSE)),
           epa_rank = order(order(mean_epa,
                   decreasing = TRUE))) %>%
    select(season, player_display_name, player_id, cmp_rank,
             att_rank, yds_rank, tds_rank,
           int_rank, epa_rank)
```

The data, as collected from `load_player_stats()`, does not contain information pertaining to MVP winners. To include this, we can load a prepared file using data from Pro Football Reference. The data contains two variables: `season` and `player_name` wherein the name represents the player that won that season's MVP. After reading the data in, we can use the `mutate` function to create a new variable called `mvp` that is in binary (`1` representing that the player was the MVP). After, we can merge this data into our `qb_mvp_stats` data frame. After merging, you will notice that the `mvp` column has mainly `NA` values. We must set all the `NA` values to `0` to indicate that those players did not win the MVP that season.

```
pfr_mvp_data
<- vroom("http://nfl-book.bradcongelio.com/pfr-mvp")

pfr_mvp_data$mvp <- 1

qb_mvp_stats <- qb_mvp_stats %>%
  left_join(pfr_mvp_data,
  by = c("season" = "season",
          "player_display_name" = "player_name"))

qb_mvp_stats$mvp[is.na(qb_mvp_stats$mvp)] <- 0
```

The above results in a data frame with 723 observations with 6 predictive variables used to determine the probability of the QB being named MVP. We can now turn to the construction of the regression model.

> **!** Important
>
> If you have not already done so, please install `tidymodels` and load it.
>
> ```
> install.packages("tidymodels")
> library(tidymodels)
> ```

While we will not be building the binary model with the `tidymodels` package, we will be utilizing the associated `rsample` package – which is used to created various types of resamples and classes for analysis – to split our `qb_mvp_stats` data into both a training and testing set.

> **!** Important
>
> Like many of the models to follow in this chapter, it is important to have both a training and testing set of data when performing a binary regression study for three reasons:
>
> 1. **Assess model performance**. Having a trained set of data allows us to evaluate the model's performance on the testing set, allowing us to gather information regarding how we can expect the model to handle new data.
> 2. **It allowd us to avoid overfitting**. Overfitting is a process that occurs when the regression model recognizes the "statistical noise" in the training data but not necessarily the underlying patterns. When this happens, the model will perform quite well on the training data but will then fail when fed new, unseen data. By splitting the data, we can use the withheld testing data to make sure the model is not "memorizing" the information in the training set.
> 3. **Model selection**. In the case that you are evaluating several different model types to identify the best performing one, having a testing set allows you to determine which model type is likely to perform best when provided unseen data.

The process of splitting the data into a training and testing set involves three lines of code with the `rsample` package. Prior to splitting the data, we will create a new data frame titled `mvp_model_data` that is our existing `qb_mvp_stats` information after using the `ungroup()` function to overwrite any prior `group_by()` that was applied to the data frame. We first use the `initial_split` function to create a split of the data into a training and testing set. Moreover, we use the **strata** argument to conduct what is called "stratified sampling." Because there are very few MVPs in our data compared to those non-MVP players, using the **strata** argument allows us to create a training and test set with a similar amount of MVPs in each. We then use the **training** and **testing** argument to create the data for each from the initial split.

```
mvp_model_data <- qb_mvp_stats %>%
  ungroup()

mvp_model_split <- rsample::initial_split(mvp_model_data,
        strata = mvp)
mvp_model_train <- rsample::training(mvp_model_split)
mvp_model_test <- rsample::testing(mvp_model_split)
```

With the MVP data split into both a training and testing set, we use the **glm** package to conduct the regression analysis on the **mvp_model_train** data frame.

```
mvp_model <- glm(formula = mvp ~ cmp_rank + att_rank +
                   yds_rank + tds_rank + int_rank + epa_rank,
                 data = mvp_model_train, family = binomial)
```

While the model does run, and placing the **mvp_model** results in the RStudio environment, we receive the following warning message in the Console:

```
Warning message: glm.fit: fitted probabilities numerically 0 or
1 occurred
```

As the message indicates, this is not necessarily an error in the modeling process, but an indication that some of the fitted probabilities in the model are fitted numerically to 0 or 1. This most often occurs in binomial regression models when there is nearly perfect separation within the data underpinning the model, meaning there is one or more predictor variables (passing yards, passing touchdowns, etc.) that are able to perfectly classify the outcome (MVP) of the model. This type of "perfect separation" in modeling is usual when working with data that is both limited and "rare" – only one QB per season in our data is able to win the MVP award. We can create a quick histogram of the fitted probabilities, wherein multitudes of probabilities close to either 0 or 1 is a good explanation for the warning message.

```
plot(residuals(mvp_model))
```

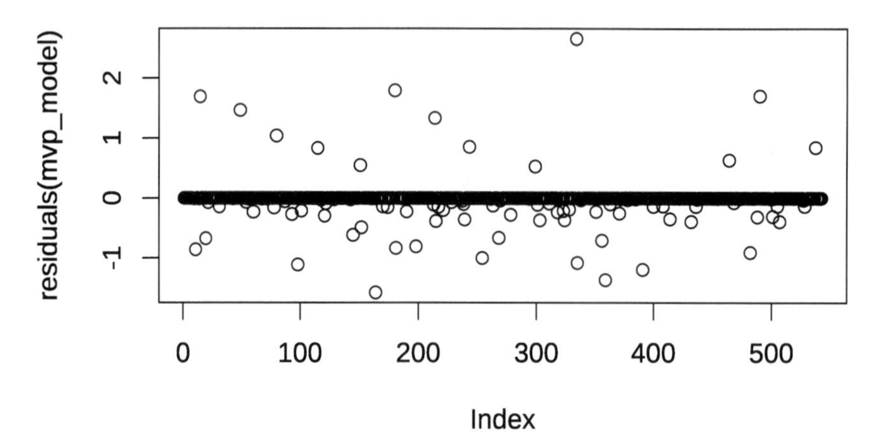

The histogram clearly shows the separation issue in the model's data. The majority of the QBs in the data have a 0% to an incredibly slim chance to win the MVP, while the bar for those with a fitted probability from 0.2 to 1.0 are barely visible in the plot. While a warning that "fitted probabilities numerically 0 or 1 occurred" is cause for further examination, we are able to diagnose the issue by using our "domain knowledge" of the data – that is, of course separation is going to occur since only one QB can be MVP in any season.

To verify that the model is predicting reasonable probabilities, we can calculate the predicted values and probabilities on the existing training data.

```
training_probs <- mvp_model_train %>%
  mutate(pred = predict(mvp_model, mvp_model_train,
                        type = "response")) %>%
  group_by(season) %>%
  mutate(pred_mvp = as.numeric(pred == max(pred,
  na.rm = TRUE)),
         mvp_prob = pred / sum(pred))
```

With the model now trained, we can take the results of the model and apply it to our withheld `mvp_model_test` data frame. Using the `predict` function, we instruct to take the modeled predictions from the trained `mvp_model` and generate further predictions on the `mvp_model_test` data frame. Because the `type = "response"` argument results in a probability between `0` and `1`, we use `ifelse` to create a "cutoff" for the probabilities, where any number above 0.5 will be `1` while any number below will be `0`.

```
test_predictions <- predict(mvp_model,
                            newdata = mvp_model_test,
                            type = "response")

test_class_predictions <- ifelse(test_predictions > 0.5, 1, 0)
```

With the model now fitted to previously unseen data (the withheld `mvp_model_test` data frame), we can calculate its accuracy.

```
accuracy <- sum(test_class_predictions == mvp_model_test$mvp)
   / nrow(mvp_model_test)

accuracy <- round(accuracy, 2)

print(paste("Accuracy:", accuracy))
```

```
[1] "Accuracy: 0.98"
```

! Important

A model that predicts with 97% accuracy seems great, right?

However, we must keep in mind the significant class imbalance in our data as visualized on the earlier histogram. Roughly 3% of the quarterbacks in the `qb_mvp_stats` data frame won the MVP award, meaning the remaining 97% of the quarterbacks did not.

Because of this, a model could always predict "not MVP" and, as a result, also have an accuracy of 97%.

While a high accuracy score like the model produced is a promising first sign, it is important to remember that accuracy – as a metric – only gives the broad picture. It does not depict how the model is performing on each individual class (that is, `0` and `1`). To get a more granular understanding of the model's performance, we can turn to metrics such as precision, recall, and the F1 score.

To begin determining the scores for each, we must conduct a bit of data preparation by ensuring that both `test_class_predictions` and `mvp_model_test$mvp` are both `as.factor()`. After, we use the `confusionMatrix` function from the `caret` package to build a matrix between the binary class predictions (from `test_class_predictions`) and the true labels (from `mvp_model_test$mvp`).

```
library(caret)

test_class_predictions <- as.factor(test_class_predictions)
mvp_model_test$mvp <- as.factor(mvp_model_test$mvp)

mvp_cm <- confusionMatrix(test_class_predictions,
mvp_model_test$mvp)
```

After constructing the confusion matrix, we can calculate the results for precision, recall, and F1.

```
precision <- mvp_cm$byClass['Pos Pred Value']

recall <- mvp_cm$byClass['Sensitivity']

f1 <- 2 * (precision * recall) / (precision + recall)

print(paste("Precision:", precision))
```

```
[1] "Precision: 0.988636363636364"
```

```
print(paste("Recall:", recall))
```

```
[1] "Recall: 0.994285714285714"
```

```
print(paste("F1 Score:", f1))
```

```
[1] "F1 Score: 0.991452991452991"
```

The resulting scores further indicate a well-trained and performing model.

1. **Precision** – the Positive Predictive Value is the fraction of true positives among all positive predictions (including false positives). The resulting score of 0.96 means that the model correctly predicted the MVP 96% of the time.
2. **Recall** – the Sensitivity or True Positive Rate is similar to precision in that it is also the fraction of true positive predictions, but only those among all true positives. Our recall scores of 0.994 means that the model correctly predicted the MVP 99.4% of the time.

3. **F1 Score** – the F1 score is the "harmonic mean" between precision and recall, providing a balanced combination of both. With another score of 0.98, the model's F1 Score indicates a healthy balance between both precision and recall in its predictions.

The model producing the same score for each metric is indicative that it contains an equal number of both false positives and false negatives, which is ultimately a sign of balance in the prediction making process.

Based on the results of accuracy, precision, recall, and the F1 score, our model is quite accurate. We can now create our own "fictional" data frame with corresponding predictor variables to determine the probabilities for each of the quarterbacks to be named MVP.

```
new_mvp_data <- data.frame(
  cmp_rank = c(1, 4, 2),
  att_rank = c(2, 1, 5),
  yds_rank = c(5, 3, 7),
  tds_rank = c(3, 1, 2),
  int_rank = c(21, 17, 14),
  epa_rank = c(3, 1, 5))

new_mvp_predictions <- predict(mvp_model,
                               newdata = new_mvp_data,
                               type = "response")

new_mvp_predictions
```

```
     1      2      3
0.2413 0.9350 0.0861
```

According to our fully trained and tested model, fictional quarterback 2 has just over a 90% probability of winning the MVP when compared to the two other opposing quarterbacks.

Finally, we can use the `broom` package to transform the model's information into a tidyverse-friendly data frame allowing us to visualize the information using `ggplot`.

```
tidy_mvp_model <- broom::tidy(mvp_model) %>%
  filter(term != "(Intercept)")

ggplot(tidy_mvp_model, aes(x = reorder(term, estimate),
                           y = estimate, fill = estimate)) +
  geom_col() +
  coord_flip() +
  labs(x = "Predictive Variables",
       y = "Estimate",
       title = "**Probability of a QB Being NFL MVP**",
       subtitle = "*GLM Model | F1-Score: 99.4%*",
       caption =
       "*An Introduction to NFL Analytics with R*<br>
       **Brad J. Congelio**") +
  theme(legend.position = "none") +
  nfl_analytics_theme()
```

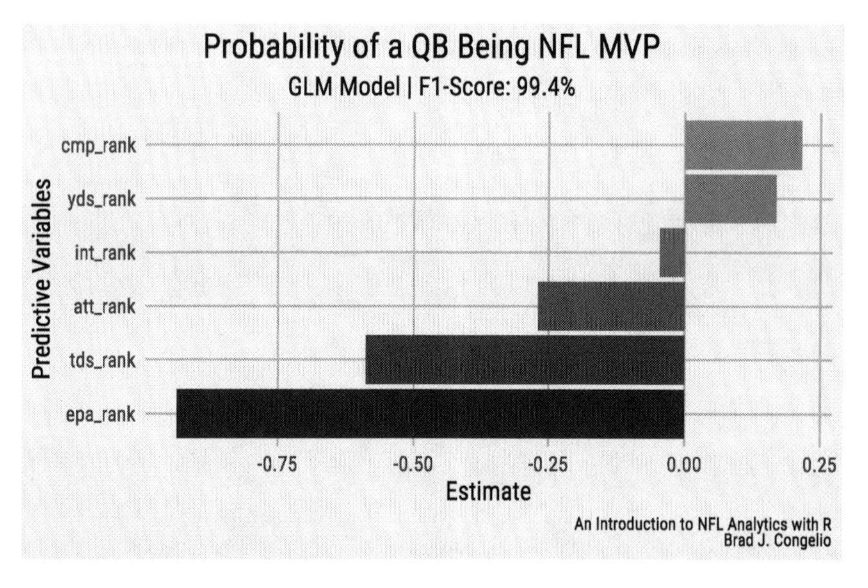

Figure 5.7: Visualizing the results of the GLM model

A negative `Estimate` for a `Predictive Variable` means that as that statistic increases (a player has a worse rank among his peers), the probable odds of that quarterback being awarded the Most Value Players decreases. For example, in our model, the `epa_rank` predictor has an `Estimate` of approximately -0.8 which means for each drop in rank, a quarterback's chance of being named MVP decreases by -0.8. Within the model, `int_rank`, `att_rank`, `yds_rank`, tds_ranks, and epa_rank align with our domain knowledge intuition: as quarterbacks perform worse in these areas, their chance of being MVP decreases.

Why, then, does a worse performance in cmp_rank provide a quarterback an increased chance in becoming MVP? Several issues could be causing this, including statistical noise, a variable that is highly correlated with cmp_rank, or a general non-linear relationship between cmp_rank and the response variable. If we were to continue working on this model, a good first step would using the filter() function to remove those quarterbacks with a limited number of attempts during the season.

5.2.3.2 Logistic Regression 2: Multinomial Regression with tidymodels

In our previous example building a binomial regression model, there were only two outcomes: 0 and 1 for whether the specific quarterback was named that season's MVP. A multinomial regression is an extension of the binomial model, but is used when there are multiple categories of outcomes rather than just two. The model itself ultimately determines the probability of each dependent variable based on the given set of predictive values. Using the softmax function to ensure that the probabilities for each dependent variable total 1, the main assumptions of the model are:

1. the dependent variable must be categorical with multiple unordered categories
2. the predictor variables can be a combination of continuous values or categorical
3. much like all the other models we've worked on, there must not be multicollinearity among the predictor values
4. the observations must be independent of each other

The specific formula for a multinomial regression is as follows:

$$P(Y_i = k|X_i) = \frac{e^{(\beta_{k0}+\beta_{k1}X_{i1}+\beta_{k2}X_{i2}+...+\beta_{kp}X_{ip})}}{\sum_{j=1}^{K} e^{(\beta_{j0}+\beta_{j1}X_{i1}+\beta_{j2}X_{i2}+...+\beta_{jp}X_{ip})}}$$

1. $P(Y_i = k|X_i)$ is the likelihood of the observation i belonging to class k given the model's predictive variables
2. In the formula, $X_{i1}, X_{i2}, ..., X_{ip}$ are the predictive variables
3. The coefficients are represented by $\beta_{k0}, \beta_{k1}, ..., \beta_{kp}$
4. Finally, K represents the number of categories the dependent variable has

As an example, let's build a multinomial regression model that predicts the type of offensive play call based on contextual factors of the game. We will build the model around five dependent categories:

1. run inside (noted as `middle` in the play-by-play data)
2. run outside (noted as `left` or `right` in the play-by-play data)
3. short pass (passes with air yards under 5 yards)
4. medium pass (passes with air yards between 6 and 10 yards)
5. long passes (passes with air yards of 11+ yards)

The construction of a multinomial regression will allow us to find the probability of each play type based on factors in the game. Such a model speaks directly to coach play-calling tendencies given the situation. To begin, we will gather data from the 2010–2022 NFL regular season. After loading the data, we will select the predictor variables and then conduct some featured engineering (which is, in this case, inspired by a presentation given by Thomas Mock on how to use `tidymodels` to predict a run or pass using binary regression)(Mock, 2020).

! Important

An Introduction to the `tidymodels` package.

As just mentioned, `tidymodels` is a collection of various packages for statistical modeling and machine learning. Importantly, the `tidymodels` package is structured much in the same way that `tidyverse` is – meaning that the packages work together to provide a unified and efficient workflow for constructing, testing, and evaluating models. A total of 43 different model types come in the base installation of `tidymodels`, ranging from simple linear regressions to more complex approaches such as neural networks and Bayesian analysis. You can view the different model types, housed within the `parsnip` package, at the `tidymodels` website: Type of models included in `tidymodels`.

Because `tidymodels` follows the `tidyverse` philosophy of "providing a uniform interface" so that "packages work together naturally," there is a general flow in which users construct a model.

1. First, data preprocessing and feature engineering is done with the `recipes` package.
2. Next, the `parsnip` package is used to create a model specification.
3. If any resampling or tuning is being conducted during the model, the `rsample` package is used to organize the necessary type of hyperparameter tuning (grid search, regular search, etc.).

4. The `workflows` package is then used to combine all of the prior steps into a singular object in your RStudio environment.
5. After training the model on the initial split of data, the `yardstick` packages allows for various forms of evaluation, including accuracy, precision, recall, F1 score, RMSE, and MAE.
6. Once you are happy with the training evaluation of the model, the `last_fit()` function conducts the process on the entire dataset to generate predictions.

Below, we will take our newly created `model_data_clean` data frame and conduct a multinomial logistic regression using the `tidymodels` package.

```
pbp <- nflreadr::load_pbp(2006:2022) %>%
  filter(season_type == "REG")

pbp_prep <- pbp %>%
  select(
    game_id, game_date, game_seconds_remaining,
    week, season, play_type, yards_gained,
    ydstogo, down, yardline_100, qtr, posteam,
    posteam_score, defteam, defteam_score,
    score_differential, shotgun, no_huddle,
    posteam_timeouts_remaining, defteam_timeouts_remaining,
    wp, penalty, half_seconds_remaining, goal_to_go,
    td_prob, fixed_drive, run_location, air_yards) %>%
  filter(play_type %in% c("run", "pass"),
         penalty == 0, !is.na(down), !is.na(yardline_100)) %>%
  mutate(in_red_zone = if_else(yardline_100 <= 20, 1, 0),
         in_fg_range = if_else(yardline_100 <= 35, 1, 0),
         two_min_drill =
         if_else(half_seconds_remaining <= 120, 1, 0)) %>%
  select(-penalty, -half_seconds_remaining)
```

There is a lot going in on the first bit of code to prepare our multinomial regression. We first collect regular season play-by-play data from 2006 to 2022. It is important to note that the 2006 season was not arbitrarily chosen. Rather, it is the first season in which the `air_yards` variable is included in the data. After gathering the data, we use the `select()` function to keep just those variables that may give an indication of what type of pay to expect next. Lastly, we conduct feature engineering by creating three new metrics in the data (`in_red_zone`, `in_fg_range`, and `two_min_drill`) and use `if_else()` to turn all three into a `1/0` binary.

Warning

Before running the next bit of code, **be sure that your installed version of dplyr is 1.1.1 or newer**.

An issue was discovered in version 1.1.0 that created major computational slowdowns when using `case_when()` and `mutate()` together within a `group_by()` variable. Prior to discovering this issue, running the below code took an absurd 22 minutes.

As a result, you will notice I resorted to using the `fcase()` function from the `data.table()` package. You can also use the `dt_case_when()` function from `tidyfast` if you prefer to use the same syntax of `case_when()` but utilize the speed of `data.table::ifelse()`.

After switching to `fcase()`, the code finished running in 17.69 seconds. After upgrading to `dplyr` 1.1.1, the `case_when()` version of the code completed in 18.94 seconds.

```
model_data <- pbp_prep %>%
  group_by(game_id, posteam) %>%
  mutate(run_inside = fcase(
    play_type == "run" & run_location == "middle", 1,
    play_type == "run", 0,
    default = 0),
    run_outside = fcase(
      play_type == "run" & (run_location == "left" |
                              run_location == "right"), 1,
      play_type == "run", 0,
      default = 0),
    short_pass = fcase(
      play_type == "pass" & air_yards <= 5, 1,
      play_type == "pass", 0,
      default = 0),
    medium_pass = fcase(
      play_type == "pass" & air_yards > 5 &
        air_yards <= 10, 1,
      play_type == "pass", 0,
      default = 0),
    long_pass = fcase(
      play_type == "pass" & air_yards > 10, 1,
      play_type == "pass", 0,
      default = 0),
```

```
    run = if_else(play_type == "run", 1, 0),
    pass = if_else(play_type == "pass", 1, 0),
    total_runs = if_else(play_type == "run",
                        cumsum(run) - 1, cumsum(run)),
    total_pass = if_else(play_type == "pass",
                        cumsum(pass) - 1, cumsum(pass)),
    previous_play = if_else(posteam == lag(posteam),
                           lag(play_type),
                           "First play of drive"),
    previous_play = if_else(is.na(previous_play),
                           replace_na("First play of drive"),
                           previous_play)) %>%
  ungroup() %>%
  mutate(across(c(play_type, season, posteam, defteam,
                  shotgun, down, qtr, no_huddle,
                  posteam_timeouts_remaining,
                  defteam_timeouts_remaining, in_red_zone,
                  in_fg_range, previous_play, goal_to_go,
                  two_min_drill), as.factor)) %>%
  select(-game_id, -game_date, -week, -play_type,
         -yards_gained, -defteam, -run, -pass,
         -air_yards, -run_location)
```

We use the `mutate()` and `case_when()` functions to create our response variables (`run_inside`, `run_outside`, `short_pass`, `medium_pass`, `long_pass`) by providing both the binary `1` and `0` arguments for the `play_type`. After, we create two more binary columns based on whether the called play was a rush or a pass and then use those two columns to calculate each team's cumulative runs and passes for each game. We conclude the feature engineering process by using the `lag()` function to provide the previous play (or, to use "First play of drive" if there was no prior play).

Before moving on to building out the model, we use `mutate(across(c())` to turn categorical variables into factors, as well as those numeric variables that are not continuous in nature. Variables **that can take on just a limited number of unique values are typically made into a factor.** For example, the `previous_play` variable is categorical and is only capable of being one of two values: `run` or `pass`. Because of this, it will be converted into a factor.

Deciding which numeric variables to convert with `as.factor()` can be a bit more tricky. We can compare the `down` and `half_seconds_remaining` variables, as you see the first in the `mutate()` function but not the second. This is because `down` is not a continuous variable as it can only take on a specific number of unique values (1, 2, 3, or 4). On the other hand, half_seconds_remaining is not continuous as there is no rhyme or reason to how it appears in the data – or, in other hands, there is no specific amount by which the half_seconds_remaining decreases for each individual play (while there *is* an ordered way in how down changes).

Because we are focusing on completing a multinomial regression, we must now convert the various play type columns (run_inside, run_outside, etc.) from binary format and then unite all five into a single response variable column titled play_call.

```
model_data <- model_data %>%
  mutate(play_call = case_when(
    run_outside == 1 ~ "run_outside",
    run_inside == 1 ~ "run_inside",
    short_pass == 1 ~ "short_pass",
    medium_pass == 1 ~ "medium_pass",
    long_pass == 1 ~ "long_pass",
    TRUE ~ NA)) %>%
  select(-run_outside, -run_inside, -short_pass,
         -medium_pass, -long_pass) %>%
  mutate(play_call = as.factor(play_call))

model_data_clean <- na.omit(model_data)
```

Before doing any feature engineering with the recipes package, we will first split the data into training and testing sets using the rsample package. The model is first trained on the training data and then evaluated on the testing data. This crucial first step ensures that the model is able to generalize well to any unseen data. As well, splitting the data helps in avoiding overfitting, which occurs when the model is able to recognize any random noise in the data. As a result, the model may perform very well on training data but then fail to reach acceptable evaluation metrics when fit on the withheld testing data.

To start the splitting process of the model_data_clean data frame, we first use set.seed() and pick a random number to run with it. The number inside set.seed() does not matter, as it is simply the initial value of what is otherwise a random-number as the rows are split – at random – into training and testing sets. However, using set.seed() allows for reproducible results in that conducting the split a second time, as the same initial value in set.seed() will result in an identical split in future splitting of the data.

We then use the initial_split function in rsample to create the multinom_play_split data frame. The initial_split process take the provided data frame and creates a single binary split of the information.

With the `initial_split()` process completed, the data is divided into a `multinom_play_train` data frame and a `multinom_play_test` data frame using the `training()` and `testing()` functions within `rsample` to conduct the process on the `multinom_play_split` data. Last, we create folds of the data using the `vfold_cv` function in `rsample` that will allow use to conduct K-fold cross validation during the training process.

```
set.seed(1984)

multinom_play_split
        <- rsample::initial_split(model_data_clean,
                strata = play_call)
multinom_play_train <- rsample::training(multinom_play_split)
multinom_play_test <- rsample::testing(multinom_play_split)

set.seed(1958)
multinom_play_folds <- rsample::vfold_cv(model_data_clean,
                                         strata = play_call)
```

The `recipes` package within `tidymodels`, first and foremost, is where the formula for the model is passed into the eventful `workflow`. However, it also provides the opportunity for further preprocessing of the data by using the `step_` function that allows for multitudes of refinement, including imputation, individual transformations, discretization, the creation of dummy variables and encodings, interactions, normalization, multivariate transformations, filters, and row operations. The complete list of possible `step_` functions can be fond on the reference section of the `recipes` website. We use several of these `step_` functions below in preparing the `recipes` for our multinomial logisitic regression model.

Tip

When building your recipe in the `tidymodels` framework, there is a general suggested order that your steps should be in to avoid issues with how the data becomes structured for the tuning/modeling process.

1. Imputation
2. Handle factor levels
3. Transformations
4. Discretize
5. Creating dummy variables and (if needed) one-hot encoding
6. Creating interactions between variables
7. Normalization steps
8. Multivariate transformations

> Not every recipe will require all of these steps, of course. But, when using more than one, it is important to follow the above order when building the recipe.

We provide the formula for our regression model (`play_call ~ .`, and then provide the name of our training data (`multinom_play_train`). We then apply the optional `update_role()` function to both the `posteam` and `season` variables from the data before adding additional steps into the recipe using the pipe operator.

1. **`update_role()`** – by applying the `update_role()` function to the `posteam` and `season` variables, we are able to retain the information without including it in the model process allowing us to investigate the prediction on a season and team basis.
2. **`step_zv()`** – the `step_zv()` function will remove any variables that contain zero variance (that is, only a single value).
3. **`step_normalize()`** – in short, the `step_normalize()` function applies both the mean and the standard deviation from the split training set to the testing set. Doing so helps prevent data leakage during the process.
4. **`step_dummy()`** – the `step_dummy()` function creates a set of binary variables from the inputted factor.

```
multinom_play_recipe <-
  recipe(formula = play_call ~ .,
  data = multinom_play_train) %>%
  update_role(posteam, new_role = "variable_id") %>%
  update_role(season, new_role = "variable_id") %>%
  step_zv(all_predictors(),
          -has_role("variable_id")) %>%
  step_normalize(all_numeric_predictors(),
                 - has_role("variable_id")) %>%
  step_dummy(down, qtr, posteam_timeouts_remaining,
             defteam_timeouts_remaining,
             in_red_zone, in_fg_range,
             two_min_drill, previous_play)
```

With the recipe for our model created, we can build the model itself and then pass both `multinom_play_recipe` and the below `multinom_play_model` into a combined `workflow()` created within the `tidymodels` framework.

```
multinom_play_model <-
  multinom_reg(penalty = tune(), mixture = tune()) %>%
  set_mode("classification") %>%
  set_engine("glmnet", family = "multinomial")

multinom_play_workflow <- workflow() %>%
  add_recipe(multinom_play_recipe) %>%
  add_model(multinom_play_model)
```

There are three important items provided to `multinom_play_model`: the **model type**, the **mode**, and the **computational engine.**

1. **Model type** – the first argument provided is the specific type of model that will be used in the analysis. As of the writing of this book, the `parsnip` package within `tidymodels` provides access to 43 various model types. You can see the complete list on the tidymodels Explore Models page. In this case, we are using the `multinom_reg()` model type. In many cases, each type of each maintains its own page on the `parsnip` website that provides further details about the various engines that be used to run the model.
2. **Mode** – Because we are utilizing a multinomial regression, the only valid `mode` for the model is "classification." The most common modes in the `tidymodels` universe are "classification" and "regression." In those circumstances where a model's mode can accept various arguments, the list of available options is provided on the engine's website (such as "censored regression," "risk regression," and "clustering").
3. **Engine** – the engine provides instructions on how the model is to be fit. In many cases, the provide engine type results in `tidymodels` calling in outside R packages (such as `randomForest` or `ranger`) to complete the process. In the case of a `multinom_reg()` model, there are six available type of engines: `nnet`, `brulee`, `glmnet`, `h20`, `keras`, and `spark`. While not wanting to over complicate the process for first-time users of `tidymodels`, it is important to know that each engine type within a model type comes with various options for tuning parameters (if that is a desired part of your model design). Our model will be using the `glmnet` engine which provides the ability to tune for `penalty` (the amount of regularization within the model) and `mixture` (the proportion of Lasso Penalty, if desired, wherein a `mixture` of `1` results in a pure Lasso model, `0` results in a ridge regression and anything between is an elastic net model). Other engines, like `brulee` provide upward of nine tuning parameters (such as `epochs`, `learn_rate`, `momentum`, and `class_weights`). The parameters you wish to tune during the training process are inputted when providing your model type. We will be conducting a tuning process for both `penalty` and `mixture`.

Note

You will notice that we've included `family = "multinomial"` in the construction of our model's engine, which makes since as the model we are using is a multinomial regression. However, the `glmnet` engine allows you to fit multiple family types, including gaussian, binomial, and Possion.

Because we provided specific tuning parameters in the model type, it is necessary to provide a grid for the process of finding the best hyperparameters – which are those parameters within the model (again, `penalty` and `mixture`) that are not provided in the original data frame. The use of these parameters ultimately control the learning process of the model and, as a result, can have a significant impact on the performance and accuracy of the final results. Because of this, we create what is called a "grid" in order to allow the model to run over and over again on a set range of each parameter, a process coined hyperparameter tuning. The end result is a model that has determined the best `penalty` and `mixture`.

We will use the `crossing` function from the `tidyr` package to manually create a grid for the hyperparameter tuning process.

```
multinom_play_grid <- tidyr::crossing(penalty = 10^seq(-6, -1,
                                      length.out = 20),
                                      mixture = c(0.05, 0.2,
                                                  0.4, 0.6,
                                                  0.8, 1))
```

1. **Penalty** – we are creating a sequence of 20 numbers that are evenly spaced on a log scale between 10^-6 and 10^-1. The training process will be conducted with each number in the `penalty` argument. A correctly tuned `penalty` allows us to avoid overfitting.
2. **Mixture** – to tune the `mixture` parameter, we create a simple sequence of 6 values between 0.05 and 1.

The model will use `multinom_play_grid` to train and evaluate against every possible combination of `penalty` and `mixture` as defined in our above arguments. With the hyperparameter tuning grid created, we now turn to actually running the model.

```
all_cores <- parallel::detectCores(logical = FALSE) - 1

registerDoParallel(all_cores)

set.seed(1988)

multinom_play_tune <- tune_grid(
  multinom_play_workflow,
  resample = multinom_play_folds,
  grid = multinom_play_grid,
  control = control_grid(save_pred = TRUE,
                         verbose = TRUE))

doParallel::stopImplicitCluster()
```

> ⚠ Warning
>
> The process of using the `paralell` package is included in the above code.
>
> The process is important if you want to speed up the training of your model as, by default, RStudio runs **one just one core of your CPU**. The processor in my computer, an Intel Core i5-10400F, has 6 cores. The `paralell` package allows us to provide more computational cores to RStudio, which in turns allows the modeling process to be conducting different tasks on different cores, then compiling the information back together when completed.
>
> When using the `parallel` package, our multinomial model took 25 minutes to complete on my computer. Out of curiosity, I spun up an Ubuntu server running R and RStudio in the Amazon cloud on an instance with 16 cores. Just this small increase in computing power dropped the model's training time to under three minutes.

As mentioned, we are using the `parallel` package to provide the model as much computational power we can. We let the package discover the exact number of cores in the computer by using the `detectCores` function and then place this number in a vector called `all_cores` after subtracting one from it (as the computer will need one free core during the modeling process to stay operational). Next, we utilize the `doParallel` package to physically begin the process of parallel processing.

Finally, we use the `tune_grid()` function, from `tune` housed within `tidymodels`, to begin the tuning process on our play-by-play data. At bare minimum, the `tune_grid()` function requires the model's `workflow` and `grid` to successfully run. In our case, we are passing in `multinom_play_workflow`, our `multinom_play_folds` resample data to allow for K-fold cross validation, and `multinom_play_grid`. Additionally, we are requesting, via the `control` argument in `grid`, to save the model's predictions and to provide verbose output in the Console regarding the current status of the tuning process. When the model completes the tuning process, be sure to run `doParallel::stopImplicitCluster()` to stop your computer from parallel processing.

> **i Note**
>
> If conducting the tuning process in a parallel processing environment, there will be not verbose output in the console regardless of how you set the argument. In many cases, even when parallel processing, I will set this option to `TRUE`, as the *lack* of verbose output is an indication itself that the model is properly tuning and still working.

When the tuning process is complete, we use the `show_best()` function to retrieve the top 5 models as determined, in this case, by AUC (area under the curve). The output of `show_best()` includes important pieces of information pertaining to specific performance metrics for each tuning of the model.

1. `penalty` and `mixture` – each respective column displays the value used from the hyperparameter tuning grid used in that specific model's configuration. In model 1, the `penalty` was set to 0.000001 while the `mixture` used was 0.2.
2. `.metric` – this column provides the type of performance metric being used to evaluate each model. As mentioned, we are used `roc_auc` to help in determining the best performing model.
3. `.estimator` – the `.estimator` indicates the specific method used to calculate the `.metric`. In the case of our model using the `glmnet` engine, the "hand and till" estimator, a method described in Hand, Till (2001), was used to determine the area `roc_auc`.
4. `mean` – this is the average of the `roc_auc` among those folds in the cross-validation with the same `penalty` and `mixture` configuration.
5. `n` – this number represents the specific number of folds used in the cross validation process.
6. `std_err` – each model's standard error metric.
7. `.config` – the unique identifier for each model.

```
show_best(multinom_play_tune, metric = "roc_auc")
```

```
`
     penalty mixture .metric .estimator  mean     n  std_err .config
       <dbl>   <dbl> <chr>   <chr>      <dbl> <int>    <dbl> <chr>
1 0.000001       0.2 roc_auc hand_till  0.668    10 0.000537 Preprocessor1_Model021
2 0.00000183     0.2 roc_auc hand_till  0.668    10 0.000537 Preprocessor1_Model022
3 0.00000336     0.2 roc_auc hand_till  0.668    10 0.000537 Preprocessor1_Model023
4 0.00000616     0.2 roc_auc hand_till  0.668    10 0.000537 Preprocessor1_Model024
5 0.0000113      0.2 roc_auc hand_till  0.668    10 0.000537 Preprocessor1_Model025`
```

There is very little difference among the five top-performing models from the tuning process. Among the five best, the `mixture` is the same (0.2), the `mean` is the same (0.668), as well as the `n` (10) and `std_err` (0.000537). The only difference between each model is found in the `penalty`, wherein a smaller number equates to a model with less regularization, allowing the model to fit more closely to the training data. On the other hand, a larger `penalty` value provides a model with more regularization, forcing the model to fit less closely to the training data. However, the amount of regularization in the model had no significant impact on any other performance metric, as they remained static across all models.

Rather than make an arbitrary choice in the model to take forward in the process, we will use the `select_by_one_std_err()` function to select the appropriate model. The function operates under the "one-standard-error-rule," which argues that we should move forward with the model whose standard error is no more than one standard error above the error of the best model. Moreover, we can sort the models in descending order by `penalty`, thus following the principles of Occam's Razor, as selecting the model with the largest `penalty` is essentially selecting simplest model we can based on the criteria of the "one-standard-error-rule."

```
final_penalty <- multinom_play_tune %>%
  select_by_one_std_err(metric = "roc_auc", desc(penalty))
```

After identifying the best model in the above step and storing it in `final_penalty`, we pass it along into the `finalize_workflow()`. This "finalized workflow" cancels out our previously constructed `workflow()` and replaces it with *only* our selected best model. We then use the `last_fit()` function to run the `finalize_workflow()` over the previously created `multinom_play_split`, which again fits the model on the training data and evaluates it on the withheld validation data.

```
multinom_final_results <- multinom_play_workflow %>%
  finalize_workflow(final_penalty) %>%
  last_fit(multinom_play_split)
```

With the `finalize_workflow()` fitted to our data, we can now make and collect predictions on both the training and test data frames and then combine both to explore the model's performance.

```
workflow_for_merging <- multinom_final_results$.workflow[[1]]

multinom_all_predictions <- bind_rows(
  augment(workflow_for_merging,
  new_data = multinom_play_train) %>%
    mutate(type = "train"),
  augment(workflow_for_merging,
  new_data = multinom_play_test) %>%
    mutate(type = "test"))
```

In the first step, we are extracting the final workflow created in the prior step and placed the information into an object called `workflow_for_merging`. Next, we create a data frame called `multinom_all_predictions` that uses the `augment` function to make predictions on both the training and testing data using the workflow that holds the parameters of the best model. After creating a column called `type` that indicates whether the play was from the training or testing data, we combine both sets of predictions using the `bind_rows()` function.

With both sets of predictions combined, we can use the `collect_predictions` function to gather the predictions into a data frame that allows us to further evaluate how the model performed. Additionally, by piping into the `roc_curve()` function, we can calculate the curve for each class in our data frame (`long_pass`, `short_pass`, etc.). The resulting `predictions_with_auc` data frame includes the true positive and false positive rates for each different play type at various classification thresholds. Including the calculated ROC curve gives us the ability to visualize the model's performance using the area under the curve (AUC).

```
predictions_with_auc
<- collect_predictions(multinom_final_results) %>%
  roc_curve(truth =
   play_call, .pred_long_pass:.pred_short_pass)
```

> **! Important**
>
> The area under the curve (AUC) is a single-value of the model's performance. In this case, we are calculating it for each play type classification. An AUC score ranges from 0 to 1, wherein a score of 1 (100%) means the model's predictions were never incorrect. A score of 0.5 (50%) indicates that the model's performance was no more predictive than flipping a coin to determine the response variable.

```
ggplot(data = predictions_with_auc, aes(1 - specificity,
                 sensitivity, color = .level)) +
  geom_abline(slope = 1, color = "black", lty = 23,
          alpha = 0.8) +
  geom_path(size = 1.5) +
  scale_x_continuous(breaks = scales::pretty_breaks(),
        labels = scales::number_format(accuracy = 0.01)) +
  scale_y_continuous(breaks = scales::pretty_breaks(),
        labels = scales::number_format(accuracy = 0.01)) +
  scale_color_brewer(palette = "Paired",
                     labels = c("Long Pass", "Medium Pass",
                              "Run Inside","Run Outside",
                              "Short Pass")) +
  coord_fixed() +
  nfl_analytics_theme() +
  theme(legend.background = element_rect("#F7F7F7"),
        legend.key = element_rect("#F7F7F7")) +
  xlab("1 - Specificity") +
  ylab("Sensitivity")
```

This visualization is a Receiver Operating Characteristic (ROC) curve, which is created by plotting the `True Positive Rate` against the `False Positive Rate` for each of the different play types in the data frame. The "arching" of the `run_outside` line toward the 0.80 mark indicates that the model is doing a commendable job at recognizing the patterns that result in the `run_outside` class against the others. The model had slightly more difficulty in distinguishing the other play types (`long_pass`, `medium_pass`, `run_inside`, and `short_pass`) from one another. But, each is still solidly above the line of no-discrimination (shown as the dashed, diagonal line in the plot), where the model's performance would be no better than random guesses.

With reasonably good results given the complexity in both NFL offenses and decision making, we can use the results of our model to determine which teams over the course of the time period were most predictable. To

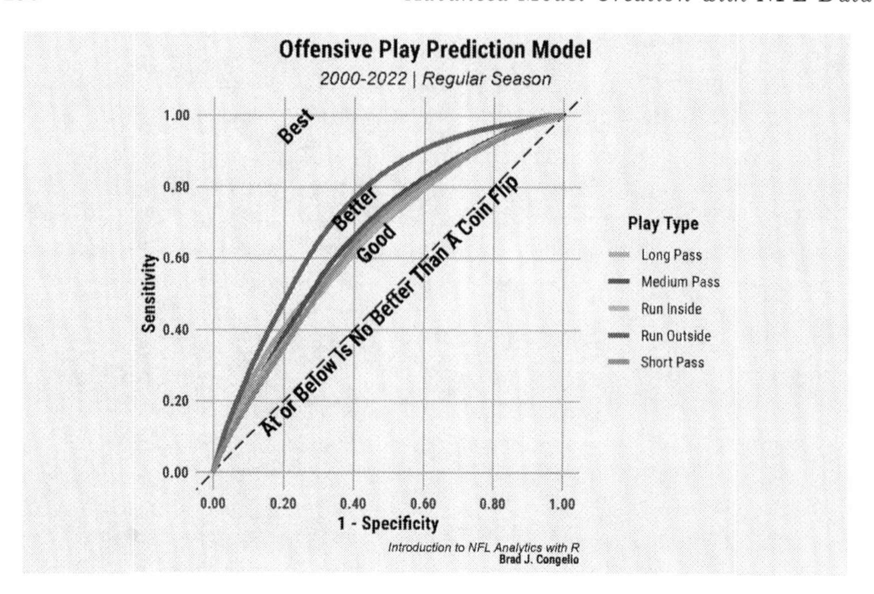

Figure 5.8: Area under curve (AOC) for the playtype predictive model

start, we will create a data frame titled `predictions_by_team` from the combined `multinom_all_predictions`. Next, we need to switch `.pred_class` to `as.character` as well as `play_call` and `posteam`. After using `summarize()` to find the `total_plays`, the `total_pred`, and `pred_pct` for each team by season, we find that the 2017 New Orleans Saints were the most predictable team in the data, with the model selecting the right play type nearly 59% of the time.

```
predictions_by_team <- multinom_all_predictions %>%
  mutate(.pred_class = as.character(.pred_class),
         play_call = as.character(play_call),
         posteam = as.character(posteam)) %>%
  group_by(season, posteam) %>%
  summarize(total_plays = n(),
            total_pred = sum(.pred_class == play_call),
            pred_pct = total_pred / total_plays * 100) %>%
  ungroup()

predictions_by_team <- as.data.frame(predictions_by_team)
```

Finally, to conclude the process of designing a multinomial regression model, we can use the data we just created in `predictions_by_team` and use knowledge we've gained in the Data Visualization chapter to place the results into a visually appealing table using `gt` and `gtExtra`.

```
teams <- nflreadr::load_teams(current = TRUE)

predictions_by_team <- predictions_by_team %>%
  left_join(teams, by = c("posteam" = "team_abbr"))

options(digits = 3)

predictions_by_team %>%
  dplyr::select(season, team_logo_wikipedia,
                total_plays, total_pred, pred_pct) %>%
  dplyr::arrange(-pred_pct) %>%
  dplyr::slice(1:10) %>%
  gt() %>%
  tab_spanner(
    label = "Most Predictable Offenses: 2006 - 2022",
    columns = c("season", "team_logo_wikipedia",
                "total_plays", "total_pred", "pred_pct")) %>%
  cols_label(
    season = "Season",
    team_logo_wikipedia = "",
    total_plays = "# of Plays",
    total_pred = "# Predicted",
    pred_pct = "Percent Predicted") %>%
  cols_align(align = "center", columns = everything()) %>%
  gt_img_rows(columns = team_logo_wikipedia, height = 25) %>%
  gt_color_rows(pred_pct, palette = "ggsci::blue_material")%>%
  gtExtras::gt_theme_espn()
```

MOST PREDICTABLE OFFENSES: 2006 - 2022

SEASON		# OF PLAYS	# PREDICTED	PERCENT PREDICTED
2017		925	543	58.7
2017		833	485	58.2
2014		929	525	56.5
2018		912	497	54.5
2015		891	485	54.4
2009		913	496	54.3
2014		998	535	53.6
2009		932	499	53.5
2014		964	513	53.2
2015		935	497	53.2

5.3 Advanced Model Creation with NFL Data

5.3.1 K-means Clustering

The K-means technique, originally titled as "bagging predictors," was formulated by Hugo Steinhaus, a Polish mathematician, in 1956 and was introduced in his paper "Sur la Division des Corps Materielsen Parties." The modern and refined version of the algorithm that is widely-used today was developed a year later by Stuart Lloyd, an electrical engineer employed at Bell Labs. The process was dubbed with the *K-means* moniker in 1967.

In short, K-means is an unsupervised machine learning algorithm that is used for the clustering process. The algorithm aims to find similarities between data points and then group, or cluster, them into the same class (wherein "K" is the defined number of clusters for the data to be grouped into). The assumption that underpins the K-means process is that a closer data point to the center of the cluster is more related than a data point that is at a further distance, with this measure of closeness typically determined through the use of Euclidean distance.

The K-means process is used in a wide range of applications, including in business to segment customers into different groups based on their purchase history or used to help develop markets for advertising and promotional efforts. In this example, we will be conducting the process on running backs from the 2022 NFL season (with at least 50 rushing attempts) over various different metrics collected from both Sports Info Solutions and Pro Football Focus, including:

1. the player's elusive rating from PFF (success of runner independent of blocking)
2. the number of broken and/or missed tackles created by the runner
3. the amount of runs that resulted in first downs
4. the number of times the runner was stuffed at the line
5. the number of times the runner used the play's designed running gap
6. the player's boom percent (rushes with an EPA of at least 1)
7. the player's bust percent (rushes with an EPA of less than −1)

Before jumping into the K-means process, a principal component analysis (PCA) is conducted on the data frame. It is common practice to do so, as the PCA process reduces the dimensionality in the data resulting in noise reduction. The PCA process takes our provided data frame and converts it into linearly correlated sets of values called principal components. The first principal component produced (PC1) accounts for the largest amount of variance in the data, while PC2 accounts for the next largest amount with the continued results being cumulative in nature. That is, if PC1 accounts for 25% of the variance in the data and PC2 accounts for 15%, the cumulative variance explained between the two principal components is 40%.

The results of conducting a PCA and a K-means on the running backs data is a division of the players into groups, where those players in the same group have similar attributes based on the variables we have collected from SIS and PFF. This clustering process allows us to produce a quantitative and data-driven way to both understand and compare running back attributes and tendencies from the 2022 season.

5.3.1.1 Principal Component Analysis

The Principal Component Analysis process is a common part of data preprocessing prior to conducting the k-means analysis, allowing us to achieve several results with the data.

1. **Variable scaling** – In many cases, the variables within a data frame will be in different units (such as in our case, as some of the statistics are provided in whole numbers, like attempts and touchdowns, while others are in "per attempt" like fumbles and designed gap.). Because of this, scaling the data ensures that each of the variables – regardless of numeric unit – contributes equally to the K-means process. Otherwise, it is possible that one variable could become dominant, thus introducing bias into the process. The scaling process computes the data frame so that each variable has a standard deviation of one and a mean of zero, with the equation to produce these results below.

$$\frac{x - 1 - \text{mean}(x)}{\text{sd}(x)}$$

2. **Dimensionality Reduction** – Our data frame has 12 variables, with each of these serving as a dimension. The PCA process seeks to find a simpler way to represent all 12 of these variables, while still capturing as much as possible of the original information. These simplified dimensions are outputted into numbered principal components (Principal Component 1, Principal Component 2, etc.) with each additional one adding to the cumulative total of variance explained in the data.
3. **Noise Reduction** – Because the principal components outputted in the PCA process are ranked in the order of variance explained, the higher-numbered ones that provide little cumulative impact are typically the ones holding the noise and other minor nuances in the data. To provide a more robust data frame to the K-means process, we can drop these components that contain unhelpful data.

To begin this process, first download the `rushing_kmeans_data`. After, we create a vector of each rusher's name and unique ID number from the `rushing_kmeans_data` data frame, as we will want to use the names as the data frame's row names during the data visualization process and have the player's ID to merge in further data, if needed (such as headshot URLs). After, we create a data frame titled `rushers_pca` that is a copy of `rushing_kmeans_data` but using `filter()` to drop both `player` and `player_id`, and then add each name back into the data frame using the **rownames** function. Now, rather than each row being associated by a number, each rusher's name will be the identification method.

```
rushing_kmeans_data
<- vroom("http://nfl-book.bradcongelio.com/kmeans-data")


rusher_names <- rushing_kmeans_data$player
rusher_ids <- rushing_kmeans_data$player_id

rushers_pca <- rushing_kmeans_data %>%
  select(-player, -player_id)

rownames(rushers_pca) <- rusher_names

rushers_pca
<- prcomp(rushers_pca, center = TRUE, scale = TRUE)
```

After that brief bit of data cleaning and preparation, we use the **prcomp** function from the **stats** package to both center and scale the variables in the data frame. The results of the PCA process includes the first look at relationship between all of our variables.

```
Standard deviations (1, .., p=7):
[1] 1.583 1.312 0.987 0.869 0.731 0.546 0.456

Rotation (n x k) = (7 x 7):
                  PC1      PC2     PC3     PC4     PC5      PC6
elusive_rating -0.557  0.04196 -0.0755  0.2275 -0.3645  0.03989
broken_missed  -0.552 -0.00659 -0.1902  0.2322 -0.3176  0.10692
fd_rush        -0.178 -0.56078 -0.0400 -0.6319 -0.0425  0.49942
stuff_percent  -0.314  0.52663  0.0670 -0.0239  0.6037  0.50434
designed_gap   -0.145 -0.08014  0.9750  0.0548 -0.1115  0.00318
boom_percent   -0.461 -0.33119 -0.0254 -0.0885  0.5526 -0.60328
bust_percent   -0.147  0.53876  0.0258 -0.6954 -0.2866 -0.34533
                    PC7
elusive_rating -0.70437
broken_missed   0.70196
fd_rush        -0.04228
stuff_percent  -0.01936
designed_gap    0.08065
boom_percent   -0.00129
bust_percent    0.04956
```

First, the results provide the standard deviations for each variable as well as the `Rotation` in which a number is provided to show how much each variable contributed to each principal component. For example, `bust_percent` and `designed_gap` both contributed very little to `PC1` with values of -0.14, while `elusive_rating` had the largest contribution with a value of -0.55. A further examination of the results show that `PC1` is perhaps a measure of a running back's ability to use his elusive ability to break tackles and break out of the backfield for big gains. Further, `PC2` has a strong contribution from `fd_rush` (-0.56) and very little from `broken_missed` (-0.006), suggesting that `PC2` is favoring those running backs that are regularly gaining short yardage on the ground and getting first downs, but not producing much EPA (based on the high contribution from `bust_percent`).

> **i Note**
>
> It is important to note that a negative sign does that indicate "negative influence," or a a lack of. Rather, a negative sign means that the specific variable is inversely related to the principal component. In `PC3`, for example, `elusive_rating` has a value of -0.399, which means that as the value of `elusive_rating` decreases, the values of `PC3` increases.

We can also visualize these results using the `factoextra` package which provides the ability to extract and visualize the output of various types of multivariate data analyzes, including a Principal Component Analysis. Using `factoextra`, we can explore the results in three different arrangements: the individual data points (in this case, the play callers) using `fviz_pca_ind()`, the variables using `fviz_pca_var()`, or a combination of both using `fviz_pca_biplot()`. As an example, let's construct a biplot that shows `PC1` on the x-axis and `PC2` on the y-axis along with the positioning of the variables and rusher names.

```
fviz_pca_biplot(rushers_pca, geom = c("point", "text"),
                ggtheme = nfl_analytics_theme()) +
  xlim(-6, 3) +
  labs(title = "**PCA Biplot: PC1 and PC2**") +
  xlab("PC1 - 35.8%") +
  ylab("PC2 - 24.6%")
```

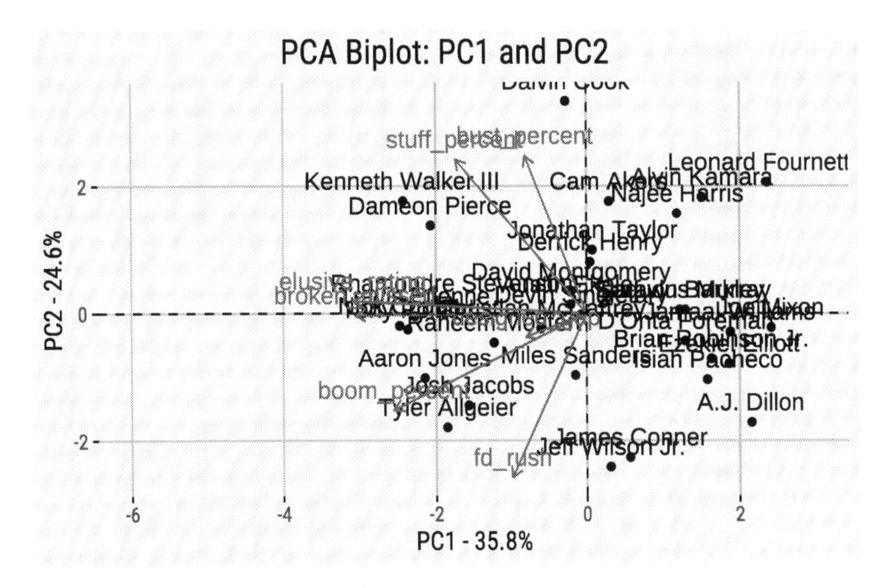

Figure 5.9: Using a biplot to view the relationship between principal components

The resulting plot contains a few unique pieces of information regarding the principal component analysis. The information on each axis provides a numeric value regarding how much variance each principal component accounts for. In this case, `PC1` captured 35.8% of the variance while `PC2` captured 24.6%. Between both, a cumulative total of 60.4% of the variance was explained.

Each point on the plot, represented by the rusher's name, is a single sample in the reduced-dimensionality space created with the PCA process. The positioning of each rusher is determined with values from the principal components, where each value is a weighted average of the original variable for each sample. In other words, the placement of each runner is dependent on their traits and attributes. The arrows and associated variable labels are also representative of the reduced-dimensionality space with the direction and length of each arrow being determined by the contribution to the variable for each principal component.

5.3.1.2 Determining the Amount of K

In our biplot, we determined that `PC1` and `PC2` accounted for 60.4% of the variance in the data. The total number of principal components we select, based on cumulative variance explained, will be the number of clusters the running backs are grouped into during the K-means process. To get a better understanding of the amount being explained, we can use the `get_eigenvalue` function from `factoextra`.

```
get_eigenvalue(rushers_pca)
```

```
       eigenvalue variance.percent cumulative.variance.percent
Dim.1       2.507            35.81                        35.8
Dim.2       1.722            24.61                        60.4
Dim.3       0.974            13.92                        74.3
Dim.4       0.755            10.79                        85.1
Dim.5       0.534             7.63                        92.8
Dim.6       0.299             4.27                        97.0
Dim.7       0.208             2.98                       100.0
```

The output provides the amount of variance explained by each principal (`Dim`) along with the cumulative value. We can see what we already discovered about `PC1` (35.8%) and `PC2` (24.6%) with the cumulative total of the two being correct at 60.4%. Going further, `PC3` explains 13.9% more of the variance in the data, bringing the cumulative total to 74.3%. Adding `PC4` brings the cumulative total to 85.1%. The fifth principal component brings the cumulative total to over 90%, with the remaining two (`PC6`, `PC7`) more than likely accounting for noise and small nuances within the data.

Unfortunately, there is no widely-accepted process to decide with certainty how many components are enough. A popular method, however, is called the "elbow method" or the Scree plot, wherein the "elbow of the curve" reflects the correct number of components to select, as the resulting cumulative amount of variance explained is not worth the noise brought in.

We can product a Scree plot using the `fviz_eig` function.

```
fviz_eig(rushers_pca, addlabels = TRUE,
         ggtheme = nfl_analytics_theme()) +
  xlab("Principal Component") +
  ylab("% of Variance Explained") +
  labs(title = "**PCA Analysis: Scree Plot**")
```

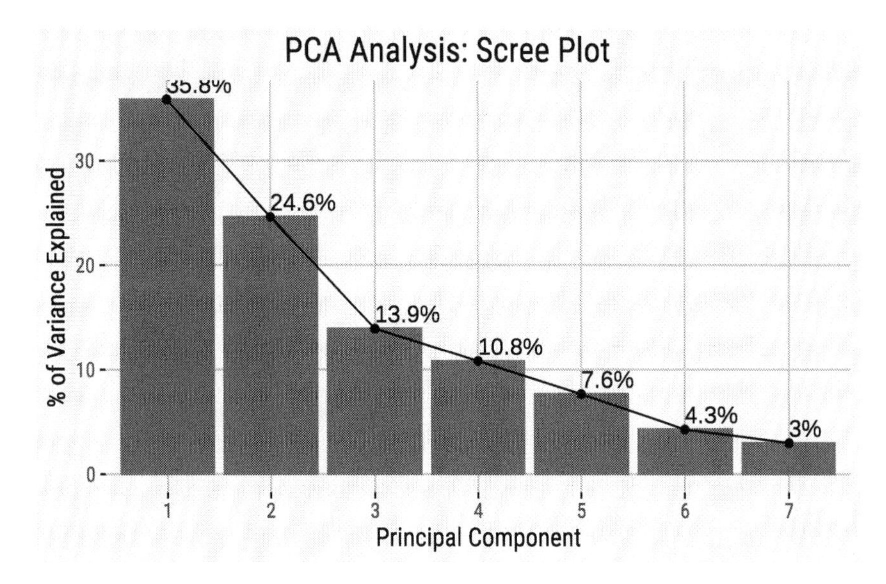

Figure 5.10: Using a Scree plot to help determine the correct number of components

In the plot, each bar represents the amount of variance explained for each principal component with the included numeric value at top. In this case, there is a fairly evident "elbow" in the plot at the third principal component. Compared to the rapid growth in the cumulative gain in the first three principal components, the final six provide diminishing returns.

We can view the contribution to the variables of each component in a combined manner using the `fviz_contrib` function and then arranging the results together using the `plot_grid` function from `cowplot`.

```
pc1 <- fviz_contrib(rushers_pca, choice = "var", axes = 1)
pc2 <- fviz_contrib(rushers_pca, choice = "var", axes = 2)
pc3 <- fviz_contrib(rushers_pca, choice = "var", axes = 3)

plot_grid(pc1, pc2, pc3)
```

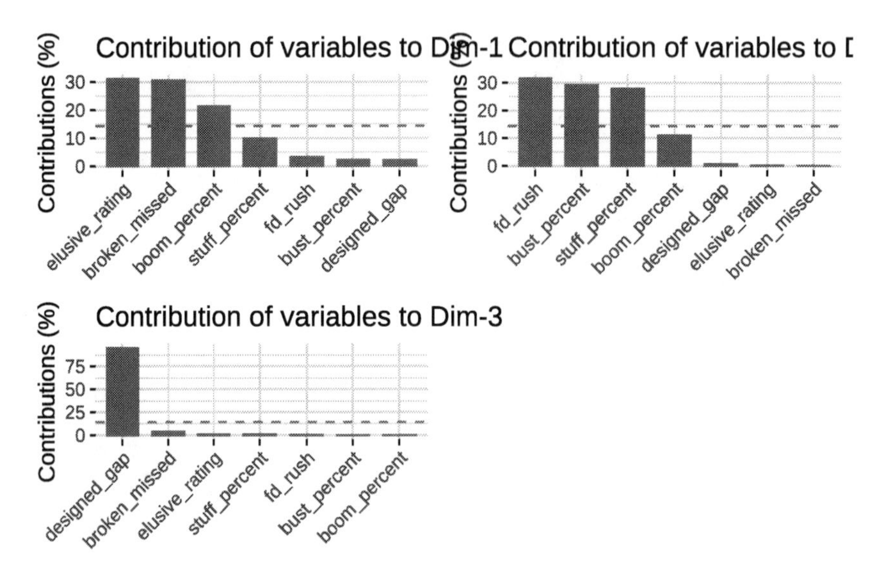

Figure 5.11: Viewing the contribution of each variable to each component

The results mirror the information we gathered from `Rotation` when we first explored the results of `rusher_pca`. However, the plots include the expected average contribution of each variable as represented by the dashed line. For each principal component, a variable that has a contribution higher than this average contribution is considered to be important in explaining the variance within that component.

With the PCA completed and knowing that four components account for an acceptable amount of variance in the data, the K-means process can be conducted.

5.3.1.3 Conducting the K-means Process

As a result of the PCA process, we know we will be grouping the running back into four distinctive clusters, so we will create a value in our environment called `k` and set it to `3`.

```
k <- 3
```

We then must extract the values created during the PCA process and write them into a new data frame in order to pass the information into the `kmeans()` function. This information is stored within `rushers_pca$x`.

```
pca_scores <- rushers_pca$x
```

We write the PCA results into a data frame called `pca_scores` and are now able to conduct the K-means process on the data, passing the correct number of `K` into the function's required arguments.

```
set.seed(1928)
rushing_kmeans <- kmeans(pca_scores, centers = k)
```

Compared to other models in this chapter, a K-means process completes almost instantly. After, the created `rushing_kmeans` provides access to eight components pertaining to the results, including `cluster`, `centers`, `totss`, `withinss`, `tot.withinss`, `betweenss`, `size`, `iter`, and `ifault`. For our purposes, we can view the cluster assigned to each running back. After, we write this information out into a data frame called `cluster_assignment` and then merge it back into the original data frame.

```
rushing_kmeans$cluster
```

```
     Derrick Henry         Josh Jacobs      Saquon Barkley
                 3                   1                   2
        Nick Chubb       Miles Sanders Christian McCaffrey
                 1                   2                   1
       Dalvin Cook        Najee Harris     Jamaal Williams
                 3                   3                   2
   Ezekiel Elliott      Travis Etienne           Joe Mixon
                 2                   1                   2
Kenneth Walker III        Alvin Kamara       Dameon Pierce
                 1                   3                   1
     Austin Ekeler        Tony Pollard         Aaron Jones
                 3                   1                   1
   Tyler Allgeier Rhamondre Stevenson       Isiah Pacheco
                 1                   1                   2
Brian Robinson Jr.      D'Onta Foreman    David Montgomery
                 2                   2                   3
Leonard Fournette     Devin Singletary     Jonathan Taylor
                 3                   2                   3
        Cam Akers      Jeff Wilson Jr.          A.J. Dillon
                 3                   2                   2
     James Conner       Raheem Mostert      Latavius Murray
                 2                   1                   2
```

```
cluster_assignment <- rushing_kmeans$cluster

rushing_kmeans_data$cluster <- cluster_assignment
```

The results of the K-means process placed, for example, Derrick Henry, Josh Jacobs, and Nick Chubb all into cluster 2 and Najee Harris, Tony Pollard, Aaron Jones into cluster 3. To better visualize the traits and attributes associated with each cluster, we can conduct some preparation and then produce a plot.

```
kmean_dataviz <- rushing_kmeans_data %>%
  rename(c("Elusiveness" = elusive_rating,
           "Broken/Missed" = broken_missed,
           "1st Downs" = fd_rush,
           "Stuffed" = stuff_percent,
           "Desi. Gap" = designed_gap,
           "Boom %" = boom_percent,
           "Bust %" = bust_percent))

kmean_dataviz <- kmean_dataviz %>%
  mutate(cluster = case_when(
    cluster == 1 ~ "Cluster 1",
    cluster == 2 ~ "Cluster 2",
    cluster == 3 ~ "Cluster 3"))

kmean_data_long <- kmean_dataviz %>%
  gather("Variable", "Value", -player, -player_id, -cluster)


ggplot(kmean_data_long, aes(x = Variable, y = Value,
       color = cluster)) +
  geom_point(size = 3) +
  facet_wrap(~ cluster) +
  scale_color_brewer(palette = "Set1") +
  gghighlight(use_direct_label = FALSE) +
  nfl_analytics_theme() +
  theme(axis.text = element_text(angle = 90, size = 8),
        strip.text = element_text(face = "bold"),
        legend.position = "none")
```

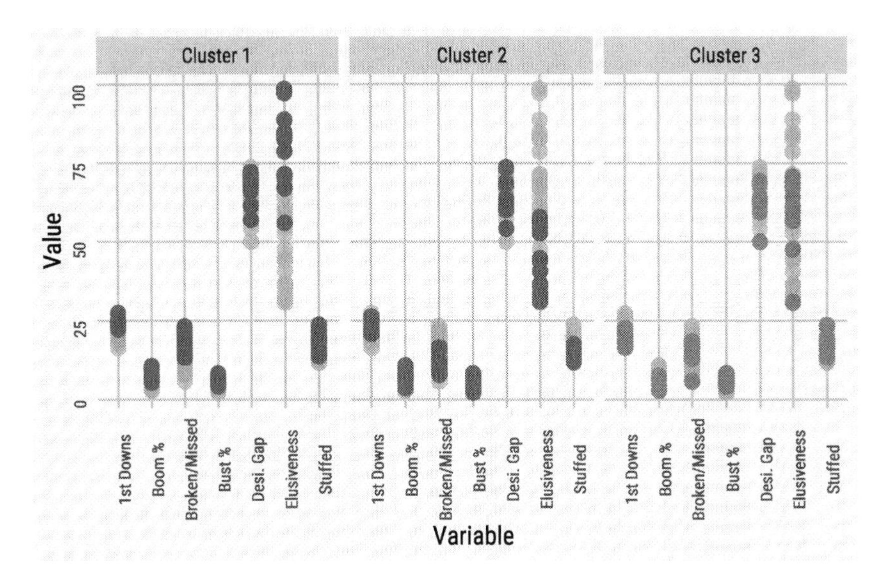

Figure 5.12: Visualizing the results of the k-means clustering process

When visualized, we are able to see and note the differences in the attributes and traits among the clustered running backs.

1. **Cluster 1 – "Feast or Famine"**
    1. Highest contribution in 1st down runs
    2. Average broken/missed tackles
    3. High use of designed gap
    4. Just as likely to boom as bust per run
    5. Most elusiveness among the clusters
    6. Gets stuffed at line more than others
2. **Cluster 2 – "Blue Collar Backs"**
    1. Also high contribution in 1st down runs
    2. Does not break or cause missed tackles
    3. Not very elusive running backs
    4. Rarely gets stuffed at the line of scrimmage
3. **Cluster 3 – "Cautious Carriers"**
    1. Low contribution in 1st down runs
    2. Low boom percentage
    3. High bust percentage
    4. Average elusiveness
    5. Occasionally gets stuffed at the line of scrimmage

To complete the process, we can take the completed model and transition the data into a presentable gt table. Before creating it, though, we must complete a bit of data preparation, such as merging in player headshots and team logos.

```
roster <- nflreadr::load_rosters(seasons = 2022) %>%
  select(pff_id, team, headshot_url) %>%
  mutate(pff_id = as.numeric(as.character(pff_id)))

teams <- nflreadr::load_teams(current = TRUE) %>%
  select(team_abbr, team_logo_wikipedia)

gt_table_data <- rushing_kmeans_data %>%
  left_join(roster, by = c("player_id" = "pff_id"))

gt_table_data <- na.omit(gt_table_data)

gt_table_data <- gt_table_data %>%
  left_join(teams, by = c("team" = "team_abbr"))

gt_table_data <- gt_table_data %>%
  select(player, cluster, headshot_url, team_logo_wikipedia)

gt_table_data <- gt_table_data %>%
  mutate(cluster = case_when(
    cluster == 1 ~ "Cluster 1 - Feast or Famine",
    cluster == 2 ~ "Cluster 2 - Blue Collar Backs",
    cluster == 3 ~ "Cluster 3 - Cautious Carriers"))
```

First, we use the `load_rosters` and `load_teams` function from `nflreadr` to bring in the necessary columns that contain the information for player headshots and team logos (matching on `pff_id` for roster information and the `team` column to bring in logos). Unfortunately, a small number of the running backs are missing headshot URLS[1]. Rather than including missing images in the table, we will use `na.omit()` to drop these players. Next, we create a data frame called `gt_table_data` that includes just the `player`, `cluster`, `headshot_url`, and `team_logo_wikipedia`. Last, we format the `cluster` information from numeric values to a character-based description based on the previous created titles for each.

With the data addition and preparation completed, we can use both the `gt` and `gtExtras` package to create the table.

[1]At the time of writing this book, this was an ongoing issue. Most missing headshots in the data are those belonging to rookies. The issue is a result of an issue with how the data is collected and a change to the API that provides access to the information.

```
gt_table_data %>%
  mutate(cluster = fct_relevel(cluster,
                     c("Cluster 1 - Feast or Famine",
                       "Cluster 2 - Blue Collar Backs",
                       "Cluster 3 - Cautious Carriers"))) %>%
  arrange(cluster, player) %>%
  gt(groupname_col = "cluster") %>%
  tab_spanner(label = "Clustered Running Backs",
              columns = c("player", "headshot_url",
                          "team_logo_wikipedia")) %>%
  cols_align(align = "center",
             columns = everything()) %>%
  cols_label(headshot_url = "",
           team_logo_wikipedia = "Team") %>%
  gt_img_rows(columns =
           team_logo_wikipedia, height = 25) %>%
  gt_img_rows(columns = headshot_url, height = 25) %>%
  cols_label(player = "Running Back") %>%
  tab_source_note(source_note =
                    md("**An Introduction to NFL Analytics
                    with R**<br>*Brad J. Congelio*")) %>%
  gtExtras::gt_theme_pff()
```

CLUSTERED RUNNING BACKS

RUNNING BACK		TEAM
CLUSTER 1 - FEAST OR FAMINE		
Aaron Jones		
Christian McCaffrey		
Josh Jacobs		
Nick Chubb		
Raheem Mostert		
Rhamondre Stevenson		
Tony Pollard		
Travis Etienne		

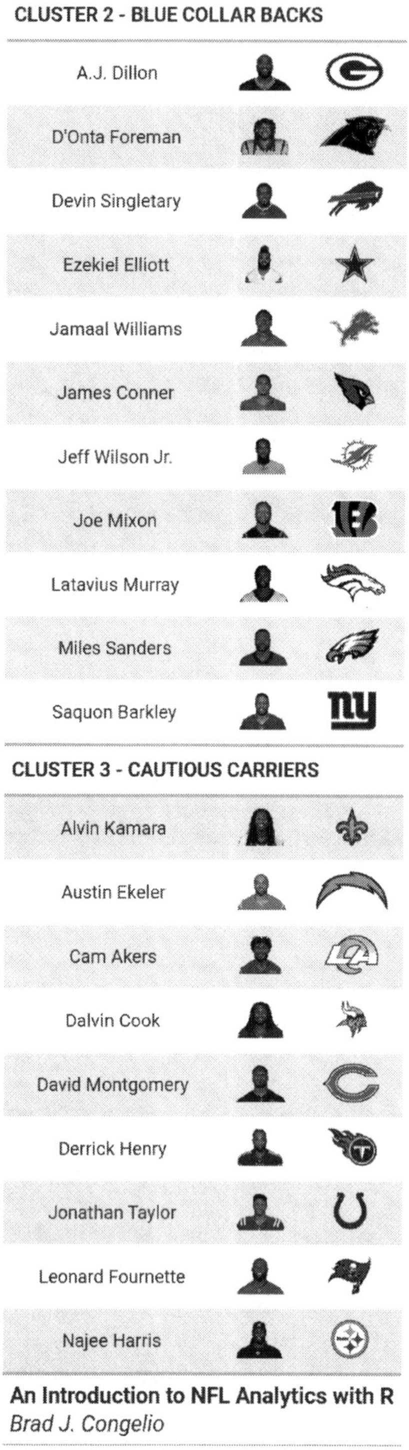

CLUSTER 2 - BLUE COLLAR BACKS

A.J. Dillon
D'Onta Foreman
Devin Singletary
Ezekiel Elliott
Jamaal Williams
James Conner
Jeff Wilson Jr.
Joe Mixon
Latavius Murray
Miles Sanders
Saquon Barkley

CLUSTER 3 - CAUTIOUS CARRIERS

Alvin Kamara
Austin Ekeler
Cam Akers
Dalvin Cook
David Montgomery
Derrick Henry
Jonathan Taylor
Leonard Fournette
Najee Harris

An Introduction to NFL Analytics with R
Brad J. Congelio

Result of the running back k-means process

5.3.2 Creating an XGBoost Model for Rushing Yards over Expected

Back in January of 2021, Tej Seth posted an article to the Michigan Football Analytics Society blog that outlined his vision for creating a "public expected rushing yards model." The structure of his model, as explained by Seth, was inspired by the prior work of Michael Egle, an honorable mention in both the 2021 and 2023 NFL Big Data Bowls, who previously used the college football equivalent of open-source data (`cfbfastR`) to create an RYOE model for the collegiate game.[2] In Tej's case, his approach to creating an NFL-centric RYOE model culminated with the creation of his RYOE Shiny App that allows anybody to explore his RYOE metric by season or team and even through three-way rusher comparisons.

Despite a slightly intimidating title, rushing yards over expected is a fantastic entry point into exploring model creation and analysis in NFL analytics – in fact, the growing number of "over expected" metrics in the NFL are all great ways to begin learning about and understanding advanced modeling. Robby Greere, the owner of nfeloapp.com – a website that provides "data-driven analytics, picks, and predictions for the NFL" – explains that over expected metrics are a increasingly popular avenue in which analysts can "paint a more accurate picture of performance by adjusting familiar statistics like 'completion percentage' or 'yards per rush' for conflating factors like degree of difficulty or game text" (Greere, 2022).

Some of these metrics, like completion percentage over expected (CPOE), are widely accepted. Specifically, CPOE calculates how likely any quarterback's pass is going to be complete or incomplete compared to other passing attempts. It is considered "widely accepted" because the metric itself is considered "stable" in that the R Squared retains a strong correlation for individual quarterbacks across several seasons. In fact, as Greer points out, the R Squared value for CPOE for just one season is 0.226 which is extremely strong based on NFL analytics standards.

On the other hand, RYOE – based on Greer's analysis – maintains an R Squared value below 0.15 until a running back's fourth season, wherein the average improves to 0.263 (an otherwise stable value). But that does not mean that RYOE is not a metric worth further exploration. The effectiveness of any one metric to account for factors such as degree of difficulty or game text largely relies on our ability to provide adequate feature engineering – specifically, how much relevant data the machine learning model can ingest to begin making predictions.

[2]Thanks to Tej Seth for briefly chatting with me over Twitter regarding the history of both Michael's RYOE model and his own.

Because of that, significant machine learning models have been built with information provided by the NFL's Big Data Bowl as it is the one chance that the public receives to feature engineer with the NFL's tracking data (which provides a player's position, speed, direction, etc. via tracking data that is recorded every 1/10th of a second). Unfortunately, only small windows of data exist from the Big Data Bowl releases and, as a result, we are often required to find creative ways to provide further context to each play/player for the machine learning model.

To showcase this idea, we are going to begin exploring ways to add additional feature engineering to Tej Seth's already fantastic Rushing Yard Over Expected model. While not the most stable metric, as mentioned, the idea of RYOE is generally easy to understand for even the most novice analyst. Broadly, given what we know about every other rushing play during a specific span of seasons, what is the most likely amount of yards a running back is going to gain on a specific rushing play as predicted by the model on other similar situations?

That difference is rushing yards over expected.

Using Tej's Shiny app, we can explore all seasons between 2015 and 2022 for those running backs that had a minimum of 755 rushing attempts.

RANK	PLAYER	TEAM	RUSHES	EPA/RUSH	YARDS PER CARRY	EXPECTED YARDS	RUSHING YARDS OVER EXPECTED
1	Nick Chubb		1244	-0.01	5.31	4.61	0.66
2	Aaron Jones		1088	0.03	5.12	4.49	0.61
3	Chris Carson		790	-0.07	4.50	4.14	0.33
4	Derrick Henry		1914	-0.01	4.78	4.47	0.28
5	Austin Ekeler		820	-0.04	4.51	4.23	0.26
6	Jonathan Taylor		781	0.01	5.08	4.80	0.25
7	Christian McCaffrey		1031	-0.02	4.66	4.39	0.24
8	Mark Ingram II		1267	-0.02	4.56	4.33	0.20
9	Dalvin Cook		1304	-0.07	4.62	4.43	0.16
10	Josh Jacobs		1088	-0.10	4.38	4.21	0.13

Figure 5.13: Example results from Tej Seth's RYOE model

According to Tej's model, since 2015, Nick Chubb of the Cleveland Browns earned – on average – 0.66 over expected. Aaron Jones is closely behind with 0.61 over expected and then a significant drop occurs for the third and fourth players.

To understand how Tej engineered his model and to begin exploring other possible features to feed into the model, we can dive into his publicly available code.

> **! Important**
>
> It is important to immediately point out that Tej built his RYOE model using the `xgboost` package, whereas we will construct ours using `tidymodels`.
>
> While the underlying eXtreme Gradient Boosting process is largely the same with both approaches, the necessary framework we will construct with `tidymodels` differs *greatly* from the coding used with the `xgboost`package.
>
> The `xgboost` package is a standalone package in R that provide an implementation of the eXtreme Gradient Boosting algorithm. To that end, it offers a highly efficient and flexible way to train gradient boosting models for various machine learning tasks, such as classification, regression, and ranking. The package provides its own set of functions for training, cross-validation, prediction, and feature importance evaluation.

Just like the model we will be building in this chapter, Tej constructed his model via eXtreme Gradient Boosting.

Which may lead to a very obvious question if you are new to machine learning: **what exactly *is* eXtreme Gradient Boosting?**

5.3.2.1 eXtreme Gradient Boosting Explained

eXtreme Gradient Boosting is a powerful machine learning technique that is particularly good at solving supervised machine learning problems, such as classification (categorizing data into classes, for example) and regression (predicting numerical values).

eXtreme Gradient Boosting can be thought of as an "expert team" that combines the knowledge and skills of multiple "individual experts" to make better decisions or predictions. Each of these "experts" in this context is what we call a decision tree, which is a flowchart structure used for making decisions based on a series of question about the data.

Once provided data, XGBoost seeks to iteratively build a collection of "bad" decision trees and then build an ensemble of these poor ones into a more accurate and robust model. The term "gradient" comes from the fact that the algorithm uses the gradient (or the slope) of the loss function (a measure of how well the model fits the data) to guide the learning process.

5.3.2.2 eXtreme Gradient Boosting with Tidymodels

As always, the first step in the modeling process is gathering the data and conducting the necessary feature engineering. In the case of our XGBoost model, we will gather play-by-play data from the 2016–2022 season and begin the work of creating additional metrics from the information contained in the data.

5.3.2.2.1 Data Preparation and Feature Engineering

```
pbp <- nflreadr::load_pbp(2016:2022)

rush_attempts <- pbp %>%
  filter(season_type == "REG") %>%
  filter(rush_attempt == 1, qb_scramble == 0,
         qb_dropback == 0, !is.na(yards_gained))

def_ypc <- rush_attempts %>%
  filter(!is.na(defteam)) %>%
  group_by(defteam, season) %>%
  summarize(def_ypc = mean(yards_gained))

rush_attempts <- rush_attempts %>%
  left_join(def_ypc, by = c("defteam", "season"))
```

After gathering the play-by-play data from 2016–2022 and doing a bit of preparation, we conduct our first bit of feature engineering by determining the average yards per carry allowed by each defense by season and then merge the results back into the main `rush_attempts` data frame.

Aside from the typical contextual variables such as down, yards to go, score, time remaining, etc., we can use the `load_participation` function from `nflreadr` to include information regarding what formation both the offense and defense were in, per play, as well as the personnel on the field for each and the total number of defenders in the box.

```
participation
<- nflreadr::load_participation(seasons = 2016:2022) %>%
  select(nflverse_game_id, play_id, possession_team,
         offense_formation,offense_personnel,
         defense_personnel,defenders_in_box)

rush_attempts <- rush_attempts %>%
  left_join(participation,
            by = c("game_id" = "nflverse_game_id",
                   "play_id" = "play_id",
                   "posteam" = "possession_team"))
```

After collecting the information for each `play_id` in the data from 2016–2022, we again use `left_join()` to bring it into the `rush_attempts` data frame, joining by the matching `game_id`, `play_id`, and `posteam`. Before continuing with the feature engineering process, we will create a secondary data frame to work with called `rushing_data_join` that will allow us to bring player names back into the data after the modeling process is complete.

```
rushing_data_join <- rush_attempts %>%
  group_by(game_id, rusher, fixed_drive) %>%
  mutate(drive_rush_count = cumsum(rush_attempt)) %>%
  ungroup() %>%
  group_by(game_id, rusher) %>%
  mutate(game_rush_count = cumsum(rush_attempt)) %>%
  mutate(rush_prob = (1 - xpass) * 100,
         strat_score = rush_prob / defenders_in_box,
         wp = wp * 100) %>%
  ungroup() %>%
  mutate(red_zone = if_else(yardline_100 <= 20, 1, 0),
         fg_range = if_else(yardline_100 <= 35, 1, 0),
         two_min_drill = if_else(half_seconds_remaining
                   <= 120, 1, 0)) %>%
  select(label = yards_gained, season, week, yardline_100,
         quarter_seconds_remaining, half_seconds_remaining,
         qtr, down, ydstogo, shotgun, no_huddle,
         ep, wp, drive_rush_count, game_rush_count,
         red_zone, fg_range, two_min_drill,
         offense_formation, offense_personnel,
         defense_personnel, defenders_in_box,
         rusher, rush_prob, def_ypc, strat_score,
         rusher_player_id, posteam, defteam) %>%
  na.omit()
```

There are multiple new features being created within our data in the above code:

1. We first group the data by `game_id`, `rusher`, and `fixed_drive` and then create a new column titled `drive_rush_count`. This column calculated, using the `cumsum()` function the cumulative total of rushes, per running back, on each the offensive drives in the game.
2. The pre-snap probability that the play is going to be a rush is calculated in the created `rush_prob` column that uses the `xpass` variable to determine the likelihood. The `strat_score` column attempts to quantify the team's decision to run based on the just calculated `rush_prob` against the total number of defenders in the box.
3. Finally, numeric `1` and `0` values are provided based on if the offense is in the `red_zone`, in `fg_range`, or if it is in a two-minute drill scenario.

We can continue to add to the data frame using information from the `next_gen_stats()` function, specifically information pertaining to the percent of rushing attempts that had eight defenders in the box and each running back's average time from handoff to the line of scrimmage.

```
next_gen_stats <- load_nextgen_stats(seasons = 2016:2022,
                                     stat_type = "rushing") %>%
  filter(week > 0 & season_type == "REG") %>%
  select(season, week, player_gsis_id,
         against_eight = percent_attempts_gte_eight_defenders,
         avg_time_to_los)

rushing_data_join <- rushing_data_join %>%
  left_join(next_gen_stats,
            by = c("season", "week",
                   "rusher_player_id" = "player_gsis_id")) %>%
  na.omit()
```

Last, we will conduct a bit of engineering on both the `offense_personnel` and `defense_personnel` we previously added into the data. Currently, the information for each is structured, for example, as `1 RB, 1 TE, 3 WR`. Instead, we can create a new column for each position with the numeric value indicating the number on the field for each play.

```
rushing_data_join <- rushing_data_join %>%
  mutate(
    ol = str_extract(offense_personnel,
           "(?<=\\s|^)\\d+(?=\\sOL)") %>% as.numeric(),
```

```
      rb = str_extract(offense_personnel,
             "(?<=\\s|^)\\d+(?=\\sRB)") %>% as.numeric(),
      te = str_extract(offense_personnel,
             "(?<=\\s|^)\\d+(?=\\sTE)") %>% as.numeric(),
      wr = str_extract(offense_personnel,
             "(?<=\\s|^)\\d+(?=\\sWR)") %>% as.numeric()) %>%
    replace_na(list(ol = 5)) %>%
    mutate(extra_ol = if_else(ol > 5, 1, 0)) %>%
    mutate(across(ol:wr, as.factor)) %>%
    select(-ol, -offense_personnel)

  rushing_data_join <- rushing_data_join %>%
    mutate(dl = str_extract(defense_personnel,
             "(?<=\\s|^)\\d+(?=\\sDL)") %>% as.numeric(),
           lb = str_extract(defense_personnel,
             "(?<=\\s|^)\\d+(?=\\sLB)") %>% as.numeric(),
           db = str_extract(defense_personnel,
             "(?<=\\s|^)\\d+(?=\\sLB)") %>% as.numeric()) %>%
    mutate(across(dl:db, as.factor)) %>%
    select(-defense_personnel)
```

We use the `str_extract` function to identify the abbreviation for each position, and then place the associated number with each with the specific `play_id`. The offensive personnel includes the additional output of including a column for offensive linemen, but only if there are more than six downed linemen on the line of scrimmage. In this case, we use the `ol` column to determine if there is an extra lineman on the field (indicated by the new `extra_ol` variable).

```
  rushing_data_join <- rushing_data_join %>%
    filter(qtr < 5) %>%
    mutate(qtr = as.factor(qtr),
           down = as.factor(down),
           shotgun = as.factor(shotgun),
           no_huddle = as.factor(no_huddle),
           red_zone = as.factor(red_zone),
           fg_range = as.factor(fg_range),
           two_min_drill = as.factor(two_min_drill),
           extra_ol = as.factor(extra_ol))

  rushes <- rushing_data_join %>%
    select(-season, -week, -rusher, -rusher_player_id,
           -posteam, -defteam) %>%
    mutate(across(where(is.character), as.factor))
```

In the last step of data preparation, we use `filter()` to remove any plays that took place in overtime, turn the created binary variables into factors using `as.factor()` and then create a data frame called `rushes` to use in our model that excludes identifying information and turns any remaining character variables into factors as well.

5.3.2.2.2 Model Creation in `tidymodels`

To begin creating the model, we will follow much the same steps as we did with our multinomial regression by following the `tidymodels` framework which needs split data, folds for cross-validation, a recipe, a model specification, a grid for hyperparameter tuning, and a workflow that combines all these elements. Using the `rushes` data frame, we will use the `rsample` package to create an initial split of the data and then produce the model's training and testing data sets from that split. Lastly, we utilize the `vfold_cv()` function to create folds of the training data to pass into the hyperparameter tuning grid.

```
set.seed(1988)

rushing_split <- initial_split(rushes)
rushing_train <- training(rushing_split)
rushing_test <- testing(rushing_split)
rushing_folds <- vfold_cv(rushing_train)
```

With the data split into the required components, we can create the recipe for our XGBoost model.

```
rushing_recipe <-
  recipe(formula = label ~ ., data = rushing_train) %>%
  step_dummy(all_nominal_predictors(), one_hot = TRUE)
```

The formula in our recipe is `label ~ .` (which is what the rushing yards gained column was renamed) and all other columns in the data frame are the predictive variables. We are also using `step_dummy` to one hot encode all the model's nominal predictors, which are those variables that contain two or more categories but do not contain any intrinsic order.

The next required component to push into the model's `workflow()` is the model specification.

```
rushing_specs <- boost_tree(
  trees = tune(),
  tree_depth = tune(),
  min_n = tune(),
  mtry = tune(),
  loss_reduction = tune(),
  sample_size = tune(),
  learn_rate = tune(),
  stop_iter = tune()) %>%
  set_engine("lightgbm", num_leaves = tune()) %>%
  set_mode("regression")
```

Given that we are building an XGBoost model, we will specify the use of the `boost_tree()` function from the `parsnip` package which will create a decision tree, wherein each tree depends on the results of the previous trees. Within the `tidymodels` framework, there are six engines that can drive a `boost_tree()` model: `xgboost`, `C5.0`, `h20`, `lightgbm`, `mboost`, and `spark`. As you will notice, rather than using the default `xgboost` engine, we are opting to run the model using `lightgbm`.

> **Note**
>
> Why are we using `lightgbm` instead of `xgboost` as the engine for the model?
>
> Using `lightgbm` and `xgboost` on the same data will provide nearly identical results, as they are both decision tree frameworks at their core. However, `lightgbm` performs significantly faster than `xgboost`, as it uses a histogram-based process for optimization that reduces the amount of data required to complete one tree in the ensemble.
>
> This speed increase is the result of `lightgbm` building the tree in a vertical growth pattern while `xgboost` does so in a horizontal fashion. While the vertical ensemble-building approach is unquestionably faster, it is prone to overfitting in the training process (but this can be adequately addressed through hyperparameter tuning). The caveat is that despite using `lightgbm`, the training process still required 4.5 hours when tuning for hyperparameters over 100 grids.

Regardless of using `lightgbm` or `xgboost` as the model's engine, we have the ability to tune the same items, including `tree`, `tree_depth`, `min_n`, `mtry`, `loss_reduction`, `sample_size`, `learn_rate`, and `stop_iter`. The `lightgbm` engine, specifically, can tune for the `num_leaves` in the vertical growth pattern.

After creating the model's specifications, we pass the required `tune()` information into `grid_latin_hypercube()` and then pass the created recipe and model specifications into our `workflow()`.

```
rushing_grid <- grid_latin_hypercube(
  trees(),
  tree_depth(),
  finalize(mtry(), rushes),
  min_n(),
  num_leaves(),
  loss_reduction(),
  sample_size = sample_prop(),
  learn_rate(),
  stop_iter(),
  size = 100)

rushing_workflow <-
  workflow() %>%
  add_recipe(rushing_recipe) %>%
  add_model(rushing_specs)
```

With a completed workflow that contains the recipe and the model's engine and tuning requirements, we can tune the hyperparameters using the `tune_grid()` function.

```
rushing_tune <- tune_grid(rushing_workflow,
                          resample = rushing_folds,
                          grid = rushing_grid,
                          control_grid(save_pred = TRUE))
```

Depending on your computing power, the process of tuning the hyperparameters can take a significant amount of time, even when using `DoParallel` to conduct the process under parallel processing. Manually setting for 100 trees over 30 grids took just 5 minutes, while increasing it to 1,000 trees over 60 grids took 96 minutes. Tuning for the number of trees over 100 grids took, as mentioned, 4.5 hours.

With the hyperparameter tuning process complete we can view the best performing ones based on RMSE (Root Mean Square Error).

```
best_params <- rushing_tune %>%
  select_best(metric = "rmse")

best_params
```

```
# A tibble: 1 x 10
   mtry trees min_n tree_depth learn_rate loss_reduction sample_size
  <int> <int> <int>      <int>      <dbl>          <dbl>       <dbl>
1     5   644    37          8     0.0257         0.0359       0.533
# i 3 more variables: stop_iter <int>, num_leaves <int>,
#   .config <chr>
```

The values in the output of `best_params` are the result of the hyperparameter tuning process and represent the values for each item that provided the best performance on the withheld validation set.

1. **mtry (5)** – this represented the number of variables that are randomly sampled as candidates at each split in the ensemble building process
2. **trees (644)** – the `trees` item is the total number of trees within the model, with each subsequent tree helping to correct the errors made in prior trained trees
3. **min_n (37)** – this is the minimum number of observations that need to reside in a node before the model permits a split to occur
4. **tree_depth (8)** – this is the maximum number of nodes allowed in any one tree
5. **learn_rate (0.0257)** – the `learn_rate` represents the "shrinkage" within the model and controls the contribution of each tree in the model, wherein lower rates require more trees but generally provide more robust results
6. **loss_reduction (0.0359)** – this is the minimum loss reduction the model requires before making a split in the trees. A larger value equates to a more conservative model
7. **sample_size (0.533)** – the `sample_size` value is the fraction of the training data used to sample the building of each tree within the tuning process.
8. **stop_iter (9)** – This is the number of iterations permitted if the validation scores does not make enough improvement
9. **num_leaves (71)** – the total number of leaves permitted in each node before a new tree must be created

In totality, the tuning of our model's hyperparameters suggest that it is fairly complex, as indicated by the number of `trees` and `tree_depth`. Specifically, the high number of `trees` suggests that the model is likely detecting and capturing very subtle patterns within the data. The low `learning_rate` and `stop_itr` will help the model avoid overfitting, while the moderate `sample_size` and `mtry` values indicate the potential for a robust final result.

With the tuning process complete and the best hyperparameters selected, we can take those parameters and pass them into a final workflow to verify the model on the testing data using only those metrics.

```
rushing_final_workflow <- rushing_workflow %>%
  finalize_workflow(best_params)

final_model <- rushing_final_workflow %>%
  fit(data = rushing_test)

rushing_predictions <- predict(final_model, rushing_data_join)

ryoe_projs
<- cbind(rushing_data_join, rushing_predictions) %>%
  rename(actual_yards = label,
         exp_yards = .pred)
```

In the above code, we are creating a second `workflow` titled `rushing_final_workflow` that will still contain the information from `rushing_recipe` and `rushing_specs` but will now only conduct the process using the tuned hyperparamters. We then take the final workflow and fit it to the testing data.

When the final fitting process is complete, which is typically much faster than the tuning process, we can take the results of `final_model` and use the `predict()` function to create a combined data frame and then, in the next step, use `cbind()` to create a file containing all of the projections created by the model called `ryoe_projs`.

At this point, the modeling process is complete and all there is left to do is explore the results To do so, we will use basic commands from `tidyverse` to make some necessary manipulations, merge in team color information from `nflreadr` and then visualize the data.

```
mean_ryoe <- ryoe_projs %>%
  dplyr::group_by(season) %>%
  summarize(nfl_mean_ryoe = mean(actual_yards) -
  mean(exp_yards))

ryoe_projs <- ryoe_projs %>%
  left_join(mean_ryoe, by = c("season" = "season"))

ryoe_projs <- ryoe_projs %>%
  mutate(ryoe = actual_yards - exp_yards + nfl_mean_ryoe)

for_plot <- ryoe_projs %>%
  group_by(rusher) %>%
  summarize(
```

```
    rushes = n(),
    team = last(posteam),
    yards = sum(actual_yards),
    exp_yards = sum(exp_yards),
    ypc = yards / rushes,
    exp_ypc = exp_yards / rushes,
    avg_ryoe = mean(ryoe)) %>%
  arrange(-avg_ryoe)

teams <- nflreadr::load_teams(current = TRUE)

for_plot <- for_plot %>%
  left_join(teams, by = c("team" = "team_abbr"))


ggplot(data = for_plot, aes(x = yards, y = exp_yards)) +
  stat_poly_line(method = "lm", se = FALSE,
                 linetype = "dashed", color = "black") +
  stat_poly_eq(mapping = use_label(c("R2", "P")),
               p.digits = 2, label.x = .35, label.y = 3) +
  geom_point(color = for_plot$team_color2,
             size = for_plot$rushes / 165) +
  geom_point(color = for_plot$team_color,
             size = for_plot$rushes / 200) +
  scale_x_continuous(breaks = scales::pretty_breaks(),
                     labels = scales::comma_format()) +
  scale_y_continuous(breaks = scales::pretty_breaks(),
                     labels = scales::comma_format()) +
  labs(title =
  "**Actual Rushing Yards vs. Expected Rushing Yards**",
       subtitle = "*2016 - 2022 | Model: LightGBM*",
       caption =
       "*An Introduction to NFL Analytics with R*<br>
       **Brad J. Congelio**") +
  xlab("Actual Rushing Yards") +
  ylab("Expected Rushing Yards") +
  nfl_analytics_theme() +
  geom_text_repel(data = filter(for_plot, yards >= 4600),
                  aes(label = rusher),
                  box.padding = 1.7,
                  segment.curvature = -0.1,
                  segment.ncp = 3, segment.angle = 20,
                  family = "Roboto", size = 4,
                  fontface = "bold")
```

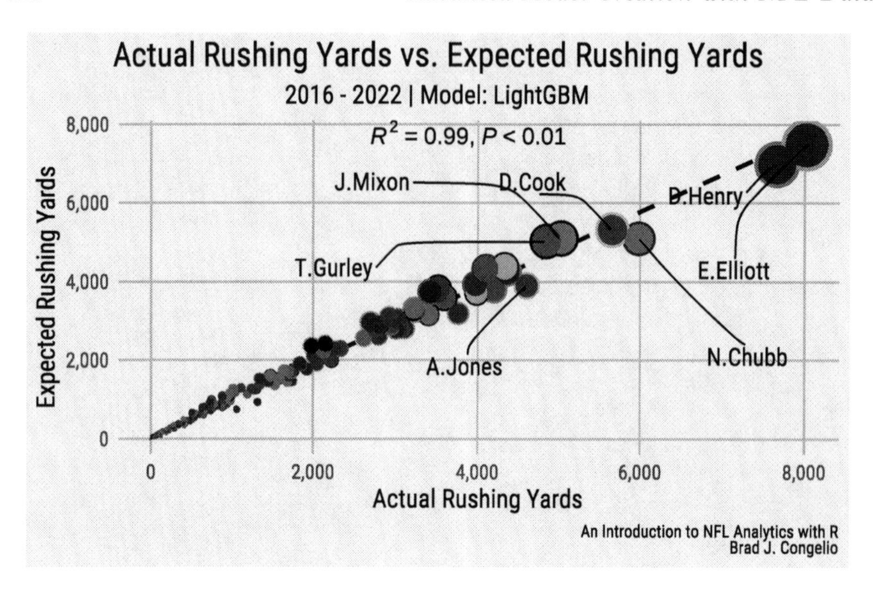

Figure 5.14: Actual yards vs. expected yards

The graph shows the relationship between `Actual Rushing Yards` and `Expected Rushing Yards`. Those running backs that outperformed the model's expectations (that is, more actual rushing yards than expected rushing yards) fall below the line, while those that underperformed the model are above the line of best fit. Nick Chubb, for example, accumulated a total of 5,982 rushing yards while the model expected him to gain just 5,097 (a net difference of 885 yards over expected). On the other side of the line of best fit is Joe Mixon, whose 5,025 actual rushing yards fell short of the model's expectation of 5,156 (a net difference of −131 rushing yards over expected). Those rushers very close to being on the line of best fit had an expected rushing yards prediction nearly identical to their actual performs (such as Dalvin Cook's 5,656 actual yards and 5,342 expected yards).

However, these results are over all of 2016–2022. We can use the results stored in our `ryoe_projs` data frame to produce more granular results to see which running backs performed the best against the model's expectation based on a number of different factors. For example, let's view the data by season and determine which rusher had the largest positive difference between their actual yards and expected yards.

```
diff_per_season <- ryoe_projs %>%
  group_by(season, rusher) %>%
  summarize(
    rushes = n(),
```

```
    team = last(posteam),
    yards = sum(actual_yards),
    exp_yards = sum(exp_yards),
    yards_diff = yards - exp_yards)

diff_per_season <- diff_per_season %>%
  left_join(teams, by = c("team" = "team_abbr"))

diff_per_season <- diff_per_season %>%
  group_by(season) %>%
  mutate(is_max_diff = ifelse(yards_diff ==
          max(yards_diff), 1, 0))
```

To plot the leaders per season required just two additions to the current `ryoe_projs` data frame. We first use `summarize()` to gather each running back's rushes, team, actual yards, and expected rushing yards per season and then created the `yards_diff` column which is simply the difference between the player's `yards` and `exp_yards`. In order to label just the leader's name in the plot for each season, we use `mutate()` to create the `is_max_diff` column, where a rusher is provided a numeric `1` if their `yards_diff` was the highest positive value for the season.

```
ggplot(diff_per_season, aes(x = yards, y = exp_yards)) +
  stat_poly_line(method = "lm", se = FALSE,
                 linetype = "dashed", color = "black") +
  stat_poly_eq(mapping = use_label(c("R2", "P")),
               p.digits = 2, label.x = .20, label.y = 3) +
  geom_point(color = diff_per_season$team_color2,
             size = diff_per_season$rushes / 165) +
  geom_point(color = diff_per_season$team_color,
             size = diff_per_season$rushes / 200) +
  scale_x_continuous(breaks = scales::pretty_breaks(),
                     labels = scales::comma_format()) +
  scale_y_continuous(breaks = scales::pretty_breaks(),
                     labels = scales::comma_format()) +
  geom_label_repel(data = subset(diff_per_season,
                                 is_max_diff == 1),
                   aes(label = rusher),
                   box.padding = 1.5, nudge_y = 1, nudge_x = 2,
                   segment.curvature = -0.1,
                   segment.ncp = 3, segment.angle = 20,
                   family = "Roboto", size = 3.5,
                   fontface = "bold") +
```

```
labs(title =
"**Rushing Yards over Expected Leader Per Season**",
     subtitle = "*2016 - 2022 | Model: LightGBM*",
     caption =
     "*An Introduction to NFL Analytics with R*<br>
     **Brad J. Congelio**") +
xlab("Actual Rushing Yards") +
ylab("Expected Rushing Yards") +
facet_wrap(~ season, scales = "free") +
nfl_analytics_theme() +
theme(strip.text = element_text(face = "bold",
family = "Roboto", size = 12),
      strip.background = element_rect(fill = "#F6F6F6"))
```

Figure 5.15: RYOE leaders per season

Last, given the success of Jonathan Taylor's 2021 season, it may be interesting to plot the trajectory of both his cumulative actual rushing yards and cumulative expected rushing yards against the his carries. Given the difference between his actual rushing yards and expected rushing yards was 414, there should be a increasingly growing gap between the lines on the plot that represent each.

```
j_taylor_2021 <- ryoe_projs %>%
  filter(rusher == "J.Taylor" & posteam == "IND" &
   season == 2021) %>%
  reframe(
    rusher = rusher,
    team = last(posteam),
    cumulative_yards = cumsum(actual_yards),
    cumulative_exyards = cumsum(exp_yards))

j_taylor_2021$cumulative_rushes =
as.numeric(rownames(j_taylor_2021))

j_taylor_image
<- png::readPNG("./images/j_taylor_background.png")
```

In order to create the `j_taylor_2021` data frame, we first create a new variable titled `cumulative_rushes` that is copying the numeric row names so that each subsequent rushing attempt increases this number, resulting in his total number of carries in the season. Similarly, the `cumulative_yards` and `cumulative_exyards` are calculated using the `cumsum` function which sums the results, play by play, in a rolling fashion.

Lastly, in order to provide a bit of "eye candy" to the plot, we use the `readPNG()` function from the `png` package to read in an image of Taylor rushing the football. We will place this image into the plot.

```
ggplot() +
  geom_line(aes(x = j_taylor_2021$cumulative_rushes,
                y = j_taylor_2021$cumulative_yards),
            color = "#002C5F", size = 1.75) +
  geom_line(aes(x = j_taylor_2021$cumulative_rushes,
                y = j_taylor_2021$cumulative_exyards),
            color = "#A2AAAD", size = 1.75) +
  scale_x_continuous(breaks = scales::pretty_breaks(n = 12),
                     labels = scales::comma_format()) +
  scale_y_continuous(breaks = scales::pretty_breaks(n = 10),
                     labels = scales::comma_format()) +
  annotate(geom = "text", label = "Cumulative Actual Yards",
  x = 200, y = 1050,
           angle = 30, family = "Roboto", size = 5) +
  annotate(geom = "text", label = "Cumulative Expected Yards",
  x = 200, y = 700,
           angle = 30, family = "Roboto", size = 5) +
```

```
annotation_custom(grid::rasterGrob(j_taylor_image,
                            width = unit(1,"npc"),
                            height = unit(1,"npc")),
                  175, 375, 0, 1000) +
labs(title = "**Jonathan Taylor: 2021 Cumulative Actual
    Yards vs. Expected Yards**",
     subtitle = "*Model: **LightGBM** Using
     ***boost_trees()*** in ***tidymodels****",
     caption =
     "*An Introduction to NFL Analytics with R*<br>
     **Brad J. Congelio**") +
xlab("Cumulative Rushes in 2021 Season") +
ylab("Cumulative Actual Yards and Expected Yards") +
nfl_analytics_theme()
```

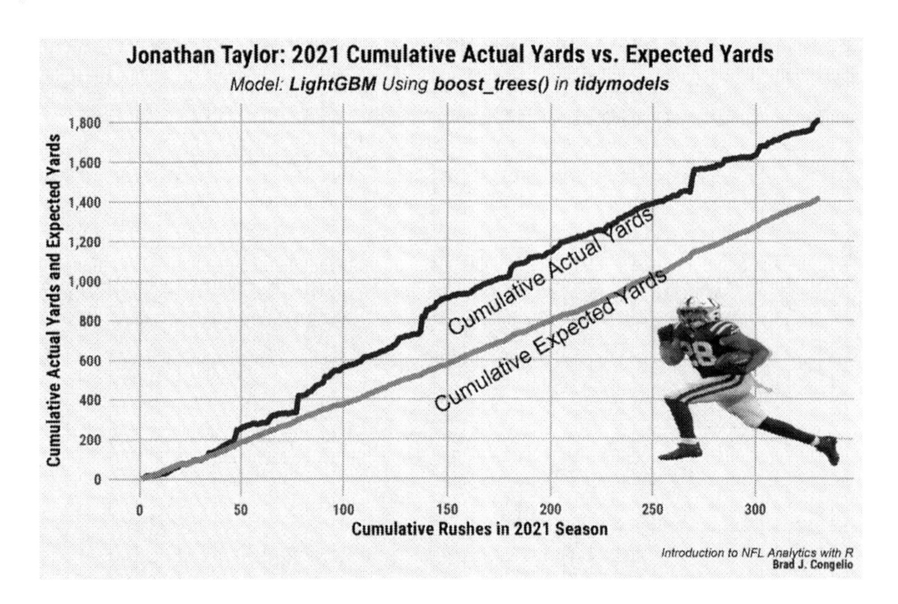

Figure 5.16: Jonathan Taylor's 2021 cumulative actual yards vs. expected yards

5.4 Exercises

The answers for the following answers can be found here: http://nfl-book.bradcongelio.com/ch5-answers.

5.4.1 Exercise 1

First, run the code below to create the data frame titled `fieldgoal_regression`

```
pbp <- nflreadr::load_pbp(2015:2022) %>%
  filter(season_type == "REG")

fieldgoal_regression <- pbp %>%
  filter(play_type == "field_goal" & field_goal_result !=
  "blocked") %>%
  select(play_type, field_goal_result, kick_distance) %>%
  mutate(field_goal_result = ifelse(
    field_goal_result == "made", 1, 0))
```

Using the data, construct a generalized linear model (`glm()`) with the `family` set to `binomial`. Use `field_goal_result` as the target variable and `kick_distance` as the predictor. View the results using `summary()`.

5.4.2 Exercise 2

Reproduce the above model into a data frame called `fieldgoal_regression_2` and include information pertaining to the field playing surface, the temperature, and the wind. Use `na.omit()` to exclude those plays with missing information. After, rerun the `glm()` model and view the results using `summary()`.

Appendices

A

NFL Analytics Quick Reference Guide

A.1 Air Yards

Air yards is the measure that the ball travels through the air, from the line of scrimmage, to the exact point where the wide receivers catches, or does not catch, the football. It does not take into consideration the amount of yardage gained after the catch by the wide receiver (which would be *yards after catch*).

For an example, please see the below illustration. In it, the line of scrimmage is at the 20-yardline. The QB completes a pass that is caught at midfield (the 50-yardline). After catching the football, the wide receiver is able to advance the ball down to the opposing 30-yardline before getting tackled. First and foremost, the quarterback is credited with a total of 50 passing yards on the play, while the wide receiver is credited with the same.

However, because air yards is a better metric to explore a QB's *true* impact on a play, he is credited with 30 air yards while the wide receiver is credited with 20 yards after catch.

In the end, quarterbacks with higher air yards per attempt are generally assumed to be throwing the ball deeper downfield than QBs with lower air yards per attempt.

There are multiple ways to collect data pertaining to air yards. However, the most straightforward way is to use `load_player_stats`:

```
data <- nflreadr::load_player_stats(2021)

air.yards <- data %>%
  filter(season_type == "REG") %>%
  group_by(player_id) %>%
```

DOI: 10.1201/9781003364320-A

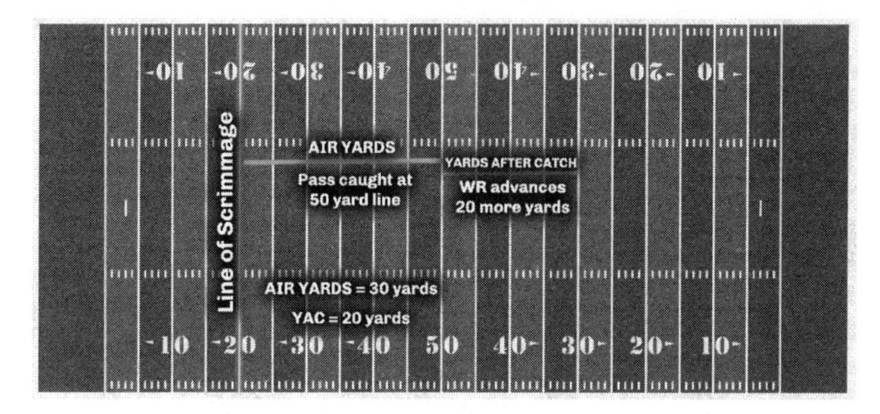

Figure A.1: Visual representation of air yards

```
  summarize(
    attempts = sum(attempts),
    name = first(player_name),
    air.yards = sum(passing_air_yards),
    avg.ay = mean(passing_air_yards)) %>%
  filter(attempts >= 100) %>%
  select(name, air.yards, avg.ay) %>%
  arrange(-air.yards)

air.yards
```

```
# A tibble: 42 x 3
   name        air.yards avg.ay
   <chr>           <dbl>  <dbl>
 1 T.Brady          5821   342.
 2 J.Allen          5295   311.
 3 M.Stafford       5094   300.
 4 D.Carr           5084   299.
 5 J.Herbert        5069   298.
 6 P.Mahomes        4825   284.
 7 T.Lawrence       4732   278.
 8 D.Prescott       4612   288.
 9 K.Cousins        4575   286.
10 J.Burrow         4225   264.
# i 32 more rows
```

In the above example, we can see that Tom Brady led the NFL during the 2021 regular season with a comined total of 5,821 air yards which works out to an average of 342 air yards per game.

A.2 Air Yards Share

A receiving statistics, air yards share is the measure of a player's share of the team's total air yards in a game/season. This metric can be found using `load_player_stats()`.

```
nfl_stats <- nflreadr::load_player_stats()

total_ay_share <- nfl_stats %>%
  filter(position == "WR") %>%
  group_by(player_name) %>%
  summarize(
    total_rec = sum(receptions, na.rm = TRUE),
    ay_share = sum(air_yards_share, na.rm = TRUE)) %>%
  filter(total_rec >= 100) %>%
  arrange(-ay_share) %>%
  slice(1:10)

total_ay_share
```

```
# A tibble: 10 x 3
   player_name total_rec ay_share
   <chr>           <int>    <dbl>
 1 A.Brown           101     8.17
 2 D.Adams           100     7.15
 3 J.Jefferson       135     7.07
 4 T.Hill            126     7.01
 5 C.Lamb            121     6.79
 6 D.Metcalf         100     6.69
 7 D.Smith           111     6.11
 8 S.Diggs           119     6.08
 9 J.Chase           107     5.78
10 A.St. Brown       106     4.66
```

A.3 Average Cushion

The average cushion measures the distance, in yards, between a WR/TE and the defender lined up against them at the line of scrimmage. This metric is included in the `load_nextgen_stats()` function.

```
nextgen_stats
<- nflreadr::load_nextgen_stats(stat_type = "receiving")

wr_cushion <- nextgen_stats %>%
  filter(week == 0 & season == 2022 & receptions >= 100) %>%
  select(player_display_name, avg_cushion) %>%
  arrange(-avg_cushion) %>%
  slice(1:10)

wr_cushion
```

```
# A tibble: 8 x 2
  player_display_name avg_cushion
  <chr>                     <dbl>
1 Chris Godwin               6.68
2 CeeDee Lamb                6.56
3 Amon-Ra St. Brown          6.52
4 Tyreek Hill                6.38
5 Travis Kelce               6.28
6 Davante Adams              5.55
7 Justin Jefferson           5.43
8 Stefon Diggs               5.36
```

A.4 Average Separation

Average separation measures the distance (in yards) between the receivers and the nearest defender at the time of catch/incompletion.

```
nextgen_stats
<- nflreadr::load_nextgen_stats(stat_type = "receiving")

wr_separation <- nextgen_stats %>%
  filter(week == 0 & season == 2022 & receptions >= 100) %>%
  select(player_display_name, avg_separation) %>%
  arrange(-avg_separation) %>%
  slice(1:10)

wr_separation
```

```
# A tibble: 8 x 2
  player_display_name avg_separation
  <chr>                        <dbl>
1 Tyreek Hill                   3.31
2 Amon-Ra St. Brown             3.10
3 Justin Jefferson              3.09
4 CeeDee Lamb                   3.07
5 Chris Godwin                  2.98
6 Davante Adams                 2.95
7 Travis Kelce                  2.88
8 Stefon Diggs                  2.83
```

A.5 Average Depth of Target

As mentioned above, a QB's air yards per attempt can highlight whether or not he is attempting to push the ball deeper down field than his counterparts. The official name of this is **Average Depth of Target** (or ADOT). We can easily generate this statistic using the `load_player_stats` function within `nflreader`:

```
data <- nflreadr::load_player_stats(2021)

adot <- data %>%
  filter(season_type == "REG") %>%
  group_by(player_id) %>%
  summarize(
    name = first(player_name),
    attempts = sum(attempts),
```

```
    air.yards = sum(passing_air_yards),
    adot = air.yards / attempts) %>%
  filter(attempts >= 100) %>%
  arrange(-adot)

adot
```

```
# A tibble: 42 x 5
   player_id  name        attempts air.yards  adot
   <chr>      <chr>          <int>     <dbl> <dbl>
 1 00-0035704 D.Lock           111      1117 10.1
 2 00-0029263 R.Wilson         400      3955  9.89
 3 00-0036945 J.Fields         270      2636  9.76
 4 00-0034796 L.Jackson        382      3531  9.24
 5 00-0036389 J.Hurts          432      3882  8.99
 6 00-0034855 B.Mayfield       418      3651  8.73
 7 00-0026498 M.Stafford       601      5094  8.48
 8 00-0031503 J.Winston        161      1340  8.32
 9 00-0034857 J.Allen          646      5295  8.20
10 00-0029604 K.Cousins        561      4575  8.16
# i 32 more rows
```

As seen in the results, if we ignore Drew Lock's 10.1 ADOT on just 111 attempts during the 2021 regular season, Russell Wilson attempted to push the ball, on average, furthest downfield among QBs with at least 100 attempts.

A.6 Completion Percentage Over Expected (CPOE)

At the conclusion fo the 2016 season, Sam Bradford, the quarterback of the Minnesota Vikings, recorded the highest completion percentage in NFL history, connecting on 71.6% of his attempts during the season. However, Bradford achieved this record by averaging just 6.4 yards per attempt. Bradford's record-breaking completion percentage is suddenly less impressive when one realizes that he was rarely attempting downfield passes.

Because of this example, we can conclude that a quarterback's completion percentage may not tell us the whole "story." To adjust a quarterback's completion percentage to include such contextual inputs such as air yards, we can turn to using completion percentage over expected (CPOE). A pre-calculated metric based on historical attempts in similar situations, CPOE take into account multiple variables, including: field position, down, yards to go, total air yards, etc.

The CPOE metric in the `nflverse` was developed by Ben Baldwin with a further explanation of it here: nflfastR EP, WP, CP, xYAC, and xPass Models. The data is included in the `load_pbp()` function of `nflreadR`.

```
pbp <- nflreadr::load_pbp(2022) %>%
  filter(season_type == "REG")

cpoe_2022 <- pbp %>%
  group_by(passer) %>%
  filter(complete_pass == 1 |
           incomplete_pass == 1 |
           interception == 1,
         !is.na(down)) %>%
  summarize(total_attempts = n(),
            mean_cpoe = mean(cpoe, na.rm = TRUE)) %>%
  filter(total_attempts >= 300) %>%
  arrange(-mean_cpoe) %>%
  slice(1:10)

cpoe_2022
```

```
# A tibble: 10 x 3
   passer       total_attempts mean_cpoe
   <chr>                 <int>     <dbl>
 1 G.Smith                 571      5.68
 2 P.Mahomes               648      3.59
 3 J.Brissett              366      2.86
 4 J.Burrow                605      2.74
 5 J.Hurts                 460      2.73
 6 D.Jones                 467      2.32
 7 T.Lawrence              582      1.45
 8 T.Tagovailoa            399      1.43
 9 J.Herbert               697      1.35
10 K.Cousins               640      1.26
```

A.7 DAKOTA

A QB's DAKOTA score is the adjusted EPA+CPOE composite that is based on the coefficients which best predicted the adjusted EPA/play in the prior year. The DAKOTA score is available in the `load_player_stats()` function within the `nflverse`.

```
nfl_stats <- nflreadr::load_player_stats()

mean_dakota <- nfl_stats %>%
  filter(position == "QB") %>%
  group_by(player_name) %>%
  summarize(
    total_cmp = sum(completions, na.rm = TRUE),
    mean_dakota = mean(dakota, na.rm = TRUE)) %>%
  filter(total_cmp >= 250) %>%
  arrange(-mean_dakota) %>%
  slice(1:10)

mean_dakota
```

```
# A tibble: 10 x 3
   player_name  total_cmp mean_dakota
   <chr>            <int>       <dbl>
 1 P.Mahomes          507       0.181
 2 D.Prescott         309       0.149
 3 J.Hurts            365       0.142
 4 T.Tagovailoa       259       0.135
 5 J.Burrow           486       0.133
 6 G.Smith            424       0.127
 7 J.Allen            407       0.126
 8 J.Goff             382       0.116
 9 K.Cousins          455       0.114
10 D.Jones            356       0.112
```

A.8 Running Back Efficiency

A running back's efficiency is measured by taking the total distance traveled, according to NextGen Stats, per the total of yards gained on the run. A lower number indicates a more North/South type runner, while a higher number indicates a running back that "dances" and runs laterally relevant to the line of scrimmage.

```
nextgen_stats
<- nflreadr::load_nextgen_stats(stat_type = "rushing")

rb_efficiency <- nextgen_stats %>%
  filter(week == 0 & rush_attempts >= 300) %>%
  select(player_display_name, efficiency) %>%
  arrange(efficiency)

rb_efficiency
```

```
# A tibble: 12 x 2
   player_display_name efficiency
   <chr>                    <dbl>
 1 Jonathan Taylor           3.17
 2 Josh Jacobs               3.33
 3 Derrick Henry             3.45
 4 Ezekiel Elliott           3.49
 5 Ezekiel Elliott           3.57
 6 Derrick Henry             3.57
 7 Ezekiel Elliott           3.65
 8 Dalvin Cook               3.74
 9 Nick Chubb                3.80
10 Derrick Henry             3.85
11 Najee Harris              3.97
12 Le'Veon Bell              4.17
```

A.9 Expected Points Added (EPA)

Expected Points Added is a measure of how well a team/player performed on a single play against relative expectations. At its core, EPA is the difference in *expected points* before and after each play. Because of this, on any given play, a team has the ability to either increase or decrease the *expected points*, with EPA being that specific difference. Importantly, EPA includes various contextual factors into its calculation such as down, distance, position on the field, etc. Then, based on historical data, an estimation of how many points, on average, a team is expected to score on a given situation is provided.

For instance, if the Chiefs are on their own 20-yard line with a 1st and 10, the *expected points* might be 0.5 based on historical data. This is the *expected points* before the play. If Mahomes completes a 10-yard pass, and now it is 1st and 10 on their own 30-yard line, the *expected points* for this new situation may increase to 0.8.

Therefore, the *expected points added* (EPA) of that 10-yard pass would be `0.8 - 0.5 = 0.3 points`. The resulting positive number indicates that the pass was beneficial and increased the team's *expected points*.

The `EP` and `EPA` values are provided for each play in the play-by-play data.

```
ep_and_epa <- nflreadr::load_pbp(2022) %>%
  filter(season_type == "REG" & posteam == "KC") %>%
  filter(!play_type %in% c("kickoff", "no_play")) %>%
  select(posteam, down, ydstogo, desc, play_type, ep, epa)

ep_and_epa
```

```
# A tibble: 1,252 x 7
   posteam  down ydstogo desc                       play_type    ep    epa
   <chr>   <dbl>   <dbl> <chr>                      <chr>     <dbl>  <dbl>
 1 KC          1      10 (15:00) (Shotgun) 25--~ run       0.931  0.841
 2 KC          2       1 (14:27) (Shotgun) 15--~ run       1.77  -0.263
 3 KC          1      10 (13:52) 25-C.Edwards--~ run       1.51   0.454
 4 KC          2       3 (13:15) (Shotgun) 25--~ run       1.96   1.37
 5 KC          1      10 (12:36) (Shotgun) 15--~ pass      3.34  -0.573
 6 KC          2      10 (12:30) (Shotgun) 15--~ pass      2.76   1.32
 7 KC          1      10 (11:54) 15-P.Mahomes ~ pass       4.09  -0.178
 8 KC          2       7 (11:11) (Shotgun) 25--~ run       3.91  -0.634
 9 KC          3       7 (10:28) (Shotgun) 15--~ pass      3.28   2.03
10 KC          1       9 (9:47) (Shotgun) 15-P~ pass       5.30  -0.565
# i 1,242 more rows
```

As well, the `load_player_stats()` function provides calculated `passing_epa`, `rushing_epa`, and `receiving_epa` per player on a weekly basis.

```
weekly_epa <- nflreadr::load_player_stats() %>%
  filter(player_display_name == "Tom Brady") %>%
  select(player_display_name, week, passing_epa)

weekly_epa
```

```
# A tibble: 18 x 3
   player_display_name  week passing_epa
   <chr>               <int>       <dbl>
 1 Tom Brady               1       1.05
 2 Tom Brady               2       2.43
 3 Tom Brady               3       1.08
 4 Tom Brady               4       9.51
 5 Tom Brady               5      13.0
 6 Tom Brady               6       8.72
 7 Tom Brady               7      -2.38
 8 Tom Brady               8       4.88
 9 Tom Brady               9      -4.88
10 Tom Brady              10      15.3
11 Tom Brady              12      -1.71
12 Tom Brady              13       0.435
13 Tom Brady              14      -9.47
14 Tom Brady              15       5.47
15 Tom Brady              16      -1.86
16 Tom Brady              17      17.3
17 Tom Brady              18       3.00
18 Tom Brady              19      -3.68
```

A.10 Expected Yards After Catch Success (xYAC Success)

xYAC Success is the probability that a play results in positive EPA (relative to where the play started) based on where the receiver caught the ball.

```
pbp <- nflreadr::load_pbp(2022) %>%
  filter(season_type == "REG")

xyac <- pbp %>%
  filter(!is.na(xyac_success)) %>%
  group_by(receiver) %>%
  summarize(
    completetions = sum(complete_pass == 1, na.rm = TRUE),
    mean_xyac = mean(xyac_success, na.rm = TRUE)) %>%
  filter(completetions >= 100) %>%
  arrange(-mean_xyac) %>%
  slice(1:10)

xyac
```

```
# A tibble: 8 x 3
  receiver    completetions mean_xyac
  <chr>               <int>     <dbl>
1 S.Diggs               100     0.872
2 T.Kelce               106     0.872
3 J.Jefferson           124     0.863
4 T.Hill                125     0.856
5 C.Lamb                103     0.832
6 A.St. Brown           103     0.810
7 C.Godwin              103     0.788
8 A.Ekeler              106     0.548
```

Considering only those receivers with 100 or more receptions during the 2022 regular season, Stefon Diggs and Travis Kelcoe had the highest expected yards after catch success rate, with the model predicting that just over 82% of their receptions would result with a positive EPA once factoring in yards after catch.

A.11 Expected Yards After Catch Mean Yardage (xYAC Mean Yardage)

Just as above XYAC Success is the probability that the reception results in a positive EPA, xYAC Mean Yardage is the expected yards after catch based on where the ball was caught. We can use this metric to determine how much impact the receiver had after the reception against what the xYAC Mean Yardage model predicted.

```
pbp <- nflreadr::load_pbp(2022) %>%
  filter(season_type == "REG")

xyac_meanyardage <- pbp %>%
  filter(!is.na(xyac_mean_yardage)) %>%
  group_by(receiver) %>%
  mutate(mean_yardage_result =
           ifelse(yards_after_catch >= xyac_mean_yardage,
                  1, 0)) %>%
  summarize(total_receptions = sum(complete_pass == 1,
                                   na.rm = TRUE),
            total_higher = sum(mean_yardage_result,
                               na.rm = TRUE),
            pct = total_higher / total_receptions) %>%
  filter(total_receptions >= 100) %>%
  arrange(-pct) %>%
  slice(1:10)

xyac_meanyardage
```

```
# A tibble: 8 x 4
  receiver     total_receptions total_higher   pct
  <chr>                   <int>        <dbl> <dbl>
1 T.Kelce                   106           51 0.481
2 A.Ekeler                  106           47 0.443
3 A.St. Brown               103           44 0.427
4 J.Jefferson               124           47 0.379
5 S.Diggs                   100           37 0.37
6 C.Lamb                    103           35 0.340
7 T.Hill                    125           41 0.328
8 C.Godwin                  103           29 0.282
```

A.12 Pass over Expected

Pass Over Expected is the probability that a play will be a pass scaled from 0 to 100 and is based on multiple factors, including yard line, score differential, to who is the home team. The numeric value indicates how much over (or under) expectation each offense called a pass play in a given situation.

For example, we can use the metric to determine the pass over expected value for Buffalo, Cincinnati, Kansas City, and Philadelphia on 1st down with between 1 and 10 yards to go.

```
pbp <- nflreadr::load_pbp(2022) %>%
  filter(season_type == "REG")

pass_over_expected <- pbp %>%
  filter(down == 1 & ydstogo <= 10) %>%
  filter(posteam %in% c("BUF", "CIN", "KC", "PHI")) %>%
  group_by(posteam, ydstogo) %>%
  summarize(mean_passoe = mean(pass_oe, na.rm = TRUE))

ggplot(pass_over_expected, aes(x = ydstogo, y = mean_passoe,
                               group = posteam)) +
  geom_smooth(se = FALSE, aes(color = posteam), size = 2) +
  nflplotR::scale_color_nfl(type = "primary") +
  scale_x_continuous(breaks = seq(1,10, 1)) +
  scale_y_continuous(breaks = scales::pretty_breaks(),
          labels = scales::percent_format(scale = 1)) +
  nfl_analytics_theme() +
  xlab("1st Down - Yards To Go") +
  ylab("Average Pass Over Expected") +
  labs(title = "**Average Pass Over Expected**",
     subtitle = "*1st Down: BUF, CIN, KC, PHI*",
     caption = "*An Introduction to NFL Analytics with R*<br>
     **Brad J. Congelio**")
```

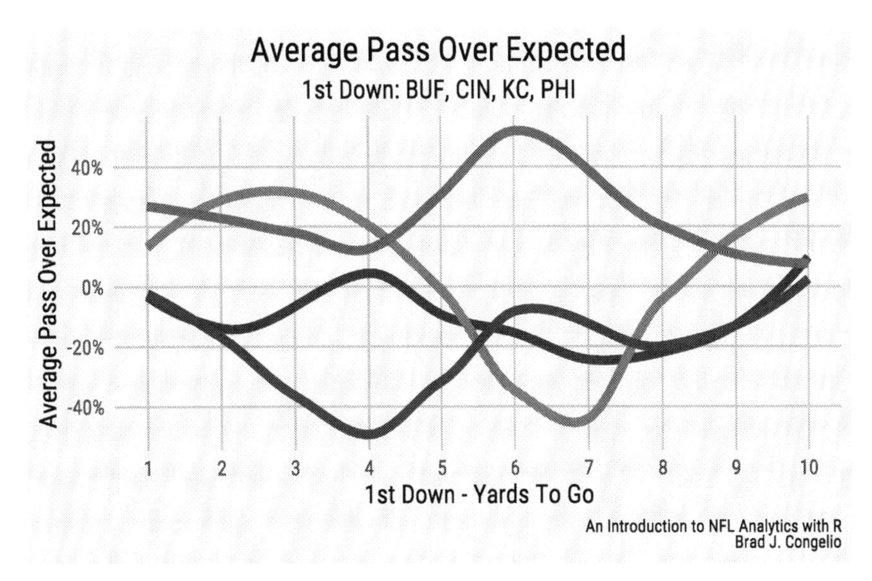

Figure A.2: Pass over expected values on 1st down for Buffalo, Cincinnati, Kansas City, and Philadelphia

Noticeably, the Chiefs pass well over expected (especially compared to the other three teams) when it is 1st down with six yards to go.

A.13 Success Rate

Prior to the formulation of EP and EPA, success rate was calculated based on the percentage of yardage gained on a play (40% of the necessary yards on 1st down, 60% on 2nd down, and 100% of the yards needed for a 1st down on both 3rd and 4th downs). However, modern success rate is determined simply by whether or not the specific play had an EPA greater than 0. Using success rate allows us to determine, for example, whether an offensive unit is stronger in rushing or passing attempts, as well as serving as a baseline indicator of a team's consistency.

```
pbp <- nflreadr::load_pbp(2022) %>%
  filter(season_type == "REG")

success_rate <- pbp %>%
  filter(play_type %in% c("pass", "run")) %>%
  group_by(posteam, play_type) %>%
  summarize(mean_success = mean(success, na.rm = TRUE)) %>%
  filter(posteam %in% c("BAL", "CIN", "CLE", "PIT"))

success_rate
```

```
# A tibble: 8 x 3
# Groups:   posteam [4]
  posteam play_type mean_success
  <chr>   <chr>            <dbl>
1 BAL     pass             0.423
2 BAL     run              0.484
3 CIN     pass             0.492
4 CIN     run              0.439
5 CLE     pass             0.433
6 CLE     run              0.449
7 PIT     pass             0.434
8 PIT     run              0.470
```

As seen in the output of `success_rate`, the teams in the AFC North were generally evenly matched. The Ravens had more success running the ball (.48 to .42) while the Bengals found more success in the air (.49 to .43). The Browns' success rate on passing and rushing attempts were nearly equal (.43 to .44).

A.14 Time to Line of Scrimmage

Measured by NextGen Stats, this is a calculation (to the 1/10th of a second), regarding how long it take the running back to cross the line of scrimmage.

```
nextgen_stats
<- nflreadr::load_nextgen_stats(stat_type = "rushing")

avg_los <- nextgen_stats %>%
  filter(week == 0 & season == 2022 &
```

```
             rush_attempts >= 200) %>%
    select(player_display_name, avg_time_to_los) %>%
    arrange(avg_time_to_los) %>%
    slice(1:10)

  avg_los
```

```
# A tibble: 10 x 2
   player_display_name avg_time_to_los
   <chr>                         <dbl>
 1 Jamaal Williams                2.64
 2 D'Onta Foreman                 2.64
 3 Joe Mixon                      2.66
 4 Saquon Barkley                 2.70
 5 Derrick Henry                  2.70
 6 Najee Harris                   2.72
 7 Ezekiel Elliott                2.72
 8 Rhamondre Stevenson            2.73
 9 Alvin Kamara                   2.75
10 Austin Ekeler                  2.76
```

B

Further Reading Suggestions

This book is, of course, not comprehensive when it comes to highlighting the R programming language. Because of this, the list below are suggested readings that will increase your knowledge/skill of R.

B.1 Introduction to R Programming Books

1. R for Data Science: Import, Tidy, Transform, Visualize, and Model Data
2. Hands-On Programming with R: Write Your Own Functions and Simulations
3. The Book of R: A First Course in Programming and Statistics
4. Learning R: A Step-by-Step Function Guide to Data Analysis
5. The Art of R Programming: A Tour of Statistical Software Design
6. Advanced R (Second Edition)

B.2 Data Visualization in R and Visualization Guides

1. R Graphics Cookbook: Practicl Recipes for Visualizing Data
2. Storytelling with Data: A Data Visualization Guides for Business Professionals

DOI: 10.1201/9781003364320-B

3. Better Data Visualizations: A Guide for Scholars, Researchers, and Wonks

B.3 Sport Analytics Guides/Books

1. The Midrange Theory: Basketball's Evolution in the Age of Analytics
2. Analyzing Baseball Data with R (2nd edition)
3. A Fan's Guide to Baseball Analytics: Why WAR, WHIP, wOBA, and other Advanced Sabermetrics Are Essential to Understanding Modern Baseball
4. The Book: Playing the Percentages in Baseball
5. The Hidden Game of Baseball: A Revolutionary Approach to Baseball and Its Statistics
6. The Hidden Game of Football: A Revealing and Lively Look at the Pro Game, With New Stats, Revolutionary Strategies, and Keys to Picking the Winners
7. Mathletics: How Gamblers, Managers, and Fans Use Mathematics in Sports
8. Basketball Data Science: With Applications in R
9. Data Analytics in Football (Soccer): Positional Data Collection, Modelling, and Analysis

References

Awbrey, J. (2020). *The future of NFL analytics.* Retrieved from https://www.samford.edu/sports-analytics/fans/2020/The-Future-of-NFL-Data-Analytics

Bechtold, T. (2021). *How the analytics movement has changed the NFL and where it has fallen short.* Retrieved from https://theanalyst.com/na/2021/04/evolution-of-the-analytics-movement-in-the-nfl/

Big data bowl: The annual analytics contest explores statistical innovations in football. (n.d.). Retrieved from https://operations.nfl.com/gameday/analytics/big-data-bowl/

Bushnell, H. (2021). *NFL teams are taking 4th-down risks more than ever - but still not often enough.* Retrieved from https://sports.yahoo.com/nfl-teams-are-taking-4th-down-risks-more-than-ever-but-still-not-often-enough-163650973.html

Carl, S. (2022). *nflplotR.* Retrieved from https://nflplotr.nflverse.com/

Fortier, S. (2020). *The NFL's analytics movement has finally reached the sport's mainstream.* Retrieved from https://www.washingtonpost.com/sports/2020/01/16/nfls-analytics-movement-has-finally-reached-sports-mainstream/

Greere, R. (2022). *Over expected metrics explained – what are CPOE, RYOE, and YACOE.* Retrieved from https://www.nfeloapp.com/analysis/over-expected-explained-what-are-cpoe-ryoe-and-yacoe/

Heifetz, D. (2019). *We salute you, founding father of the NFL's analytics movement.* Retrieved from https://www.theringer.com/nfl-preview/2019/8/15/20806241/nfl-analytics-pro-football-focus

Kirschner, A. (2022). *The rams' super bowl afterparty turned into a historic hangover.* Retrieved from https://fivethirtyeight.com/features/the-rams-super-bowl-afterparty-turned-into-a-historic-hangover/

Kozora, A. (2015). *Tomlin prefers 'feel over analytics'.* Retrieved from http://steelersdepot.com/2015/09/tomlin-prefers-feel-over-analytics/

Mock, T. (2020). *Intro to tidyodels with nflscrapR play-by-play.* Retrieved from https://jthomasmock.github.io/nfl_hanic/#1

Rosenthal, G. (2018). *Super bowl LII: How the 2017 philadelphia eagles were built.* Retrieved from https://www.nfl.com/news/super-bowl-lii-how-the-2017-philadelphia-eagles-were-built-0ap3000000912753

Silge, J. (n.d.). *Tidymodels.* Retrieved from https://tidymodels.org

Stikeleather, J. (2013). *The three elements of successful data visualizations.* Retrieved from https://hbr.org/2013/04/the-three-elements-of-successf

Wickham, H. (2022). *Tidyverse packages.* Retrieved from https://www.tidyverse.org/packages/

Wikipedia. (2023). *R package.* Retrieved from https://en.wikipedia.org/wiki/R_package

Index

PGIL2023USA